577 R982o   FV
RUTTEN
THE ORIGIN OF LIFE BY NATURAL
CAUSES
                            59.25

THE ORIGIN OF LIFE

The Origin of Life
by Natural Causes

M.G. RUTTEN[†]

Professor of Geology
State University of Utrecht, Utrecht, The Netherlands

ELSEVIER PUBLISHING COMPANY *Amsterdam London New York, 1971*

ELSEVIER PUBLISHING COMPANY
335 JAN VAN GALENSTRAAT
P.O. BOX 211, AMSTERDAM, THE NETHERLANDS

AMERICAN ELSEVIER PUBLISHING COMPANY, INC.
52 VANDERBILT AVENUE
NEW YORK, NEW YORK 10017

LIBRARY OF CONGRESS CARD NUMBER: 73-118255
ISBN: 0-444-40887-8
WITH 150 ILLUSTRATIONS AND 27 TABLES.

PRINTED IN BELGIUM

A portrait of Professor A. Oparin, made by A. Biro during the "Third International Conference on the Origin of Life", 1970, at Pont-a-Mousson, France.

Preface by A. Oparin

The problem of the origin of life has, of recent years, become of enormous interest to the whole of humanity. It attracks the attention not only of scientists in different countries and disciplines, but also of the layman. It is now broadly accepted the the process of the origin of life on earth is a natural phenomenon accessible to scientific research. The evolution of carbon compounds was basically the process which took place in the universe long before the appearance of our solar system and which then developed during the formation of the earth as a planet—during the formation of its crust, hydrosphere and atmosphere.

Man's ever-increasing penetration into space has rendered possible the discovery of carbon compounds of increasing complexity in various objects from outer space. This allows us to make judgements as to the pre-terrestrial stage of the evolution of organic substances and gives rise to the supposition that at the beginning of its formation, the earth obtained large amounts of organic substances from the solar system.

On the other hand, widespread model experiments on the abiogenic synthesis of organic compounds and on the formation from these compounds of different open multimolecular systems (precursors of organisms or probients), as well as the comparative biochemical study of the most primitive contemporary living beings, have allowed us theoretically to picture the evolution of these substances and systems along the pathway of the origin of life.

In parallel with model experiments, it also seems to be attractive, however, to follow the different stages of this evolution in consecutive order and to link up the increasingly complex carbon compounds from the bowels of the earth with those of the most primitive forms of life, fossils of which are to be found in Precambrian rocks. It is a very complicated and as yet unsolved problem, but the efforts of many geologists and paleontologists are directed towards its solution.

Professor Rutten was a pioneer in the field. Already in 1962 he published his book *The Geological Aspects of the Origin of Life* which was accepted by scientists of the whole world as a latest word on this problem.

It was indeed fortunate for the world of science that Professor Rutten was able to complete this new book before his untimely death. The book is now offered to the reader; it is an excellent generalisation of all the modern scientific data concerning the problem of the origin of life seen from the geological point of view.

We know that there are scientists who are "dead" during their lifetime—their creative work is outdated. On the other hand there are scientists who are "alive" even after their physical death. Professor Rutten belongs to this last category. His book *The Origin of Life* is a pledge of the fact that his ideas will live on in the minds and hearts of his adherents and disciples for a long time to come.

Preface by M. Schidlowski

It was with deep regret that members of the geological fraternity all over the world learnt of the sudden death of Professor M. G. Rutten on October 13, 1970. Having been in contact with the deceased immediately before his death and knowing of his work on the final draft of the present book, I felt obliged to help in editing the latter. It should be noted that—except for obvious typing errors and some minor changes—the original text has been left substantially unchanged, thus bearing exclusively the touch of the author. Respect to the integrity of the original manuscript was paid even in those cases where, in my opinion, a somewhat modernized version seemed to be desirable.

In providing us with a concise synopsis of the intriguing story of early life on earth, Professor Rutten has served an obvious need. Very few of those interested in this topic will have the time (or the opportunity) to acquaint themselves with the bulk of the original literature. In addition, the specialist confronted with single aspects of the problem will welcome a digest informing him briefly on the other parts of the story—in full realization of the shortcomings such a synopsis must necessarily entail. The references listed at the end of each chapter (covering the range from organic chemistry to geology) will give the reader an idea of the tremendous effort which went into the making of this compilation, and we must be grateful to the author for having undertaken this laborious task.

Being a geologist by profession, it was natural that Rutten stressed the importance of the contemporary environment for the development of early life, thereby reemphasizing the role of geology as an environmental science *par exellence*. Life being always intimately linked up with its environment, i.e. food, air, and water (which constitute a totality in dynamic equilibrium), knowledge of the paleo-environment is fundamental for the understanding of early organic evolution. For instance, geological evidence pertaining to atmospheric composition in Precambrian times will furnish important pieces in the puzzle of early life. It is one of the advantages of this book that it reviews the available geological (or paleo-environmental) evidence within the framework of the problem as a whole.

Reflecting as it does the present state of affairs in a rapidly expanding field of research, this review serves as an excellent introduction to the problems related to the origin of life. Those more familiar with the topic will gratefully acknowledge being provided by the author with a book which must not be *studied*, but which can be *read*.

Foreword

This book deals with the modern ideas regarding the possibility that life on earth, and elsewhere in the universe, may have originated by a gradual transition from a lifeless environment, and without the intervention of divine creation or other supernatural events.

Great will be the number of people who do not see at all how this could be a problem worthy of study. For most of those who believe in divine or supernatural powers, also believe that life has been created, not evolved. On the other hand, many scientists have over the last few decades been intensively studying the possibility of a natural origin of life. The results of these studies are fascinating and form a subject well worth knowing. So, independently of the fact whether one believes it to have happened one way or the other, all of us ought to have an inkling of what is going on in this field. The subject is not only of interest to theologians, philosophers and scientists, but to every human being. For even more than organic evolution, which has been so much in the public eye since Darwin, it broaches the question of how—or why, as some would, perhaps, like to put it—we are here.

The subject of this book is not only a very wide, but also a controversial one. It has caught the interest of many a philosopher and theologian during the past two thousand years. But it is only during this century that a more general interest in scientific circles has been aroused. The answer to our question remains at present still shrouded in our ignorance about precise events in the far distant early history of the earth. However, the meagre facts will be marshalled and upon this basis we can speculate about various possibilities. Although we are not yet able to prove the validity of one or the other of these theoretical possibilities, it is already possible to present a coherent theory of how life on earth could have evolved through natural causes. The speed of the development of our ideas on the origin of life has been tersely indicated by J. B. S. HALDANE (1965): "It is not so long ago that we thought that man had been specially created and that maggots rose from rotten cheese by spontaneous generation. Now we believe that human beings have evolved and that spontaneous generation, whenever it occurs, is a rare event."

It is understandable when, notwithstanding this rapidly accumulating insight into the possibility of an origin of life through natural causes, people who believe in an origin of life through creation remain unconvinced. For does not the belief in creation imply that such an origin can neither be proved, nor disproved by scientific methods? So one is quite justified, if one is so disposed, to state that "We still think that life was originally created by God", as did the *Harvester* (Vol. XLII, May, 1963) in its review of a forerunner of this book.

What we propose to do is to study whether an origin of life through natural causes is, let us say, absolutely impossible, or perhaps probable, or even highly probable. And, as we will find the latter case to be true, we may study when and where, and under what conditions it might have happened, and along what sort of pathways the gradual transition from non-living to living may have proceeded. This book will therefore not enter into the interpretative side of the question, whether life was created or has evolved. It will do no more than to present the scientific background relating to a possible development of life through natural causes.

This study will not only take us back into the very beginning of the geologic history of the earth, but it will also make contact with almost every aspect of the natural sciences. Astronomy, physics, chemistry, biology, geology, together with their multitudinous offspring such as biophysics and biochemistry, all have their say in this matter. If, as a geologist, I have nevertheless tried to review the problems by all these disciplines, I have endeavoured this only in the spirit of the British physicist J. D. Bernal, who once remarked: "True, there is not at present, and probably never will be, any single person who can himself command all these techniques, but this does not make the task an impossible one." The result nevertheless depends in a very large way on the great number of people who have helped me with discussions, critical reading of parts of the manuscript, suggestions for further exploration, references to the literature and permission to publish their results. In this vein I feel honoured to acknowledge the help of Mrs. P. van der Kruk and A. A. Manten, and Drs. Ph. Abelson, E. S. Barghoorn, L. V. Berkner, H. P. Berlage, J. D. Bernal, D. B. Boyland, M. Calvin, G. Claus, P. E. Cloud Jr., G. L. Davis, Miss S. Deutsch, G. Eglinton, H. Faul, A. Fischer, F. W. Fitsch, P. A. Florschütz, S. W. Fox, H. R. von Gaertner, R. M. Garrels, R. G. Gastil, M. F. Glaessner, S. S. Goldich, M. Gravelle, E. ten Haaf, P. E. Hare, M. Härme, W. Henderson, T. C. Hoering, H. J. Hofmann, A. Holmes, R. Hooykaas, J. L. Hough, M. L. Jensen, M. Lelubre, L. C. Marshall, A. Matisto, W. E. McGovern, W. G. Meinschein, G. Millot, M. G. J. Minnaert, R. C. Moore, W. W. Moorhouse, B. Nagy, C. B. van Niel, W. Nieuwenkamp, A. I. Oparin, J. Oró, J. Th. G. Overbeek, H. D. Pflug, C. Ponnamperuma, H. N. A. Priem, P. Ramdohr, S. I. Rasool,

Mrs. L. Sagan, M. Schidlowski, J. W. Schopf, R. Schuiling, L. G. Sillèn, A. Wilson.

REFERENCE

HALDANE, J.B.S., 1965. Introduction of Dr. Oparin. In: S. Fox (Editor), *The Origin of Pre-biological Systems*. Academic Press, New York, N.Y., p. 89.

M. G. RUTTEN

I am greatly indebted to Dr. M. Schidlowski, who did the proofreading for may late husband's book. Without his kind cooperation and great help it would have been extremely difficult to get this work ready for printing.

H. C. RUTTEN-VAN BERGHEM

Contents

Chapter 1 | Introduction

1. GEOLOGISTS AND THE ORIGIN OF LIFE

The origin of life ought to have interested me, as a naturalist specialized in geology, from my early days. This has not been so. My early interests were captured to such an extent by other aspects of the geologic history of the earth that the origin of life has remained outside my personal field for a long time and only through a happy coincidence did I become aware of the intriguing qualities of this subject. I believe my case to be typical for most geologists of my generation, and even for many other scientists. It seems therefore permissible to begin the introduction of this book with this personal note, as it can be used as a starting point for an understanding of the recent wave of scientific interest in our subject.

I eventually became interested through a number of searching questions put to me by the late Dutch microbiologist Prof. A. J. Kluyver of Delft. Only then did I realize how much thought had been given to this problem by certain groups of scientists, notably microbiologists and biochemists, astronomers and some physicists. It became clear that geology had to reconsider its attitude. Not only do we have to re-evaluate the existing facts and investigate how they can be incorporated into these modern viewpoints, but we also have to think about the directions in which geologic research can be oriented to help to unravel this subject. This will all be fully treated in this book and it will become clear that, as has been remarked in the Preface, the development of our knowledge and insight has been very rapid over the last decades. But before going any further I think it is wise to explore a little further into the backgrounds leading to the fact that the origin of life has remained for so long a closed book to geologists and other scientists.

As stated above, the interest of geologists and other scientists can easily be captured, to the point of saturation, by other fields of research. In the case of geology, it is well known that the history of the earth covers an enormous time span. Moreover, the data on which this history has to be founded are often very poor. Consequently geologists have tended to concentrate their studies on those periods of the earth's history which are relatively well known, and have often neglected the more obscure periods. In the case of the problems relating to the history of life this means that geology has supplied us with a fair amount of data in the way of fossilised representatives of former life which existed over the last half billion years. Although even here the paleontological record is still very incomplete, this has led to a concentrated effort in the study of the evolution of life on earth during its later days, that is during the last half billion years of its history, from the Cambrian onwards. In comparison, geological data about the origin of life, which reaches much farther back in time, to three billion years and more, are fragmentary in the extreme. The whys and wherefores will be set forward later in this book, but the bare fact remains that there are incomparably more geologic data available about the later evolution of life than about its earlier period and about a possible period of transition from non-living to living.

To this difference in the availability of data for the study of the evolution of life as compared with that of the origin of life must be added the resistance which such research, be it into the evolution or the origin of life, has so often encountered from church-people. Naturalistic views on these subjects have been, and in some cases still are, anathema to many of the stricter church members, clergy and laity alike. Geologists, in trying to overcome this dogmatic barrier against their legitimate scientific search, have of course concentrated on their strongest case, that is on the evolution of life during the last half billion[1] years.

They became even more strengthened in this choice, when they became aware of the fact that the natural evolution of life is often fairly easily reconciled with the more popular versions of church dogma, when the origin of life is pushed back to some nebulous beginnings, some *generatio spontanea*. If, however, evolution is coupled with an all-embracing theory, holding that natural processes govern all of life, not only its evolution, but also its origin, the resistance is found to increase markedly. In the first view indicated, creation can be thought of as separate from natural evolution. During discussions following lectures on the evolution of life, I have often gained the impression, as if in this way creation can be accepted on the faith of church dogma, natural evolution of life once created on the faith of the paleontological record.

---

[1] I have followed the American custom of calling 10^9, or 1,000,000,000, one billion. This is the equivalent of *un milliard* (French) and *eine Milliarde* (German). This is done because absolute dating in geology has since World War II mainly developed as an American science. So, in calling a thousand million a billion and not a milliard, I follow the prevailing custom in the relevant literature. As 10^9 is written as G in the international system of measures, one billion years can also be written as 1 G years. One billion years is also indicated as an *aeon* or *eon*.

10^6, or one million years is written as 1 m.y. (1 m.a.—*million d'années*—in the European literature).

So geologists have had on the one hand the possibility of studying the evolution of life on earth from the data handed down in the paleontological record, on the other the study of the origin of life, for which the data were very scarce. Moreover, research into the evolution of life is generally accepted[1] whereas research into the actual origin of life, implying a doubt towards creation, provokes criticism on immaterial grounds. A criticism which cannot be effectively answered, as it is in the study of evolution, owing to lack of data. A certain defeatism consequently has reigned for a long time in geology. Although perhaps never clearly expressed, the general attitude has been: "Let us concentrate on the study of the evolution of life from the paleontological record. We may then leave the origin of life either to creation or to some *generatio spontanea*. For want of data the subject is not ripe for scientific research. As long as this is the case, geologists had better leave it alone, as it might turn out to be too hot to handle."

2. THE BIOLOGICAL APPROACH

This situation has now completely changed. The impetus for this change has come, not from geologists, but from the interest biologists, together with some physicists and astronomers, have taken in the subject. Moreover, this interest has been perceptibly quickened since World War II. It is now no longer geology which is asked for an answer to the problem of the origin of life. Quite to the contrary, it is biology, mainly basing itself on biochemistry and microbiology, which has arrived at certain definite conceptions about this subject. On theoretical grounds a possible mode of origin of life through natural causes has been arrived at, and further research is eagerly pursuing the pathways and the environmental circumstances which must have been operative during this transition.

This does in itself not prove that life did really originate on earth in this way. But it provides us with an acceptable hypothesis. The question is now about the geologic setting of the origin of life, about the possibility whether the processes postulated by biological research have indeed had their place in the geologic history of the earth. As stated already in the Preface, and as will follow from this book, the new situation has in turn had a great influence on geologic research into the origin of life, which is largely of a recent date.

[1] How strongly uphill the grade still is, in trying to convince people of the reality of organic evolution, even in a country always thought to be progressive, like the United States of America, is illustrated by the emotional title of a paper by SIMPSON (cf. 1964): "One hundred years without Darwin are enough." Whilst only recently, more than twenty years after the famous "Monkey Trial" in Dayton, Tennessee, all laws against the teaching of natural evolution have been revoked in the U. S. A. (SPRAGUE DE CAMP, 1969). But American children are still taught not to listen to such perverted ideas. "Uncle Arthur", for instance, in his widely distributed bedtime stories, summarily dismisses the whole idea of organic evolution, and has the teacher berated who dares to tell of such things (MAXWELL, 1964, p. 168).

The whole problem is, as we saw, controversial because it touches the roots of religious and ethical conceptions of every educated person. In the preceding section I mentioned the resistance geologists have always encountered from church-people in their studies of the history of life. On the other hand, the reason why Russia under the leadership of Prof. A. I. Oparin was a pioneer in the new wave of interest in the origin of life undoubtedly was not purely scientific either. Such research was expressly furthered under the influence of Marxist doctrine. This becomes the more apparent when we realize that Oparin's staunchest supporters were the British biophysicist Prof. J. B. S. Haldane and the British physicist (or better perhaps, crystallographer) Prof. J. D. Bernal, who both, at least during a part of their scientific careers, supported the Marxist doctrine. Instead of offering a protection to popular church teachings, the Russian attitude was prompted by the wish to attack such thinking. The goal was a completely materialistic theory of life; not only of its evolution, but also of its origin. Or, to put it perhaps too succinctly, to do away with creation altogether.

I have thought it necessary to touch upon this religious, antireligious and ethical background of the problem of the origin of life, because it is so over-whelmingly alive in many minds and also because it is so difficult to abstract oneself from one's own background in trying to attain as high a measure of objectivity as possible. It has been a pleasant surprise to me, when reading the more recent literature on our subject, to note the objective and academic quality of most research. The literature seems to be completely free from either Marxist or religious dialectics. It reports on scientific experiments and on astronomical and on geological and on theoretical phenomena, based on these facts. If several of the latter may seem highly speculative, they are nevertheless deduced from scientific data only and purely along scientific lines of reasoning. In this book I will try to maintain a similar standard of objectivity.

3. ABOUT THIS BOOK

An earlier, and quickly outdated, version of this book (RUTTEN, 1962), although in general receiving a favourable reception in the scientific press, has failed to reach the general public. From private discussions I gained the impression that the more general sections were too much mixed up with detailed technical accounts to be palatable to the general reader. Since our subject must be of interest to that rather vague personality, the general reader, I have tried to make things easier for him in using two types of letterpress, bold face for the general accounts, small type for detailed matter. In this I have failed because it turned out that the organisation of the book would become too spotty so, the only thing I can do at this moment is to stress most strongly that an understanding of the detailed accounts is not necessary for an understanding of the general story. Even where it has not been expressly stated, the reader may without harm skip the technical material, such as formulas and the like, and resume his reading at the point where the text becomes again comprehensible.

A short summary of the organisation of this book leads us to the following main points. As the origin of life lies so far back in the geologic history, we must first understand how geology goes about in reconstructing this history from the data supplied by the rocks of the earth's crust. This is sketched in Chapters 2 and 3. In Chapter 2 we will see how the philosophy underlying this reconstruction is based on the principle of actualism. This states that the processes now operative in the atmosphere, the hydrosphere and the crust of the earth have also been operative in the past, whereas it is not necessary to postulate the influence of hypothetical processes, now no longer operative. Or, in short: "The present is the key to the past." Chapter 3 summarizes the modern methods of radiometric age determination, making use of the so-called "physical clocks" contained within some types of rocks. In this way we are able to deduce what has happened at a given time for a rock of certain characters to have been formed.

These two chapters are the foundation for our further studies. They have no direct relation to our special problem and the same material will be found, in more or less detail, in any textbook of physical geology. They are, however, necessary as a background to these further studies, because they give an answer to the question of how the information supplied by the rocks can be used to reconstruct the geologic history and how events thus deduced can be dated.

Chapter 4 relates how biologists have studied the possibility of an origin of life through natural causes, thereby starting the present wave of scientific interest. At the risk of becoming tedious through repetition, we might well state here already that this research has taught us that such a natural origin is impossible under the present oxygenic atmosphere. Moreover, we have learned that it is only possible under a primeval anoxygenic atmosphere of reducing character. All of the newer research into the origin of life has consequently been directed towards the question, whether the earth has had such a primeval anoxygenic atmosphere. And if so, when this was so, what were the environmental characters during this period.

In Chapter 5 we then learn that from the data of astronomy it is clear that reducing environments characterize the interstellar dust, the planetary atmospheres, etc. The earth is found to be exceptional amongst the planets in possessing an atmosphere with a major amount of free oxygen.

Chapter 6 describes how in chemical experiments with a simulated anoxygenic environment, compounds which are at present only synthesized by life can form inorganically. Such compounds, which here will be called *organic* are in the literature often designed as *abiogenic*. The main result of these experiments is that, once an anoxygenic environment is simulated, such "organic" compounds will form in an amazing variety and by the use of almost any kind of energy. Once those circumstances prevail, the inorganic synthesis of "organic" molecules will be the rule and not an exception. As an appendix to this chapter a short discussion of the modern analytical techniques used in organic chemistry is added for the benefit of the reader not well versed in chemistry. It is only by the amazing specificity and accuracy of these new analytical methods that the modern chemical research into the origin of life has become possible.

Chapters 7 and 8 treat of the various steps which must have been intermediate between the formation of the "organic" molecules under the primeval atmosphere and the development of life. Although it is not known what steps have exactly been taken, one can still form a general idea of the pathways which must have led from non-living to living. The recurrent theme here is that life will have been much less complicated in its metabolism during these early days, than it is at present after having been perfected over a period of time of the order of three billion years. This subject is then rounded off in Chapter 9, in which the further evolution of life is dealt with.

In Chapter 10 the environments of early life are studied in a schematic way. It is realised that the early compounds formed inorganically must have become concentrated and conserved, whereas both isolation and mixing must have taken place in alternating periods. Such processes must have been in operation to ensure chemical reactions between the compounds originally formed, thus in turn leading to further developments in the transition from non-living to living.

Chapter 11 describes how only in certain regions of the earth, the so-called *old shields*, non-metamorphosed rocks of Precambrian age can be found, whilst in Chapter 12 the main fossils found in these rocks are described. We will see how life may manifest itself either by the formation of macroscopic biogenic deposits, such as algal limestones, deposited by organisms during their metabolism, or by molecular biogenic deposits, that is by organic molecules preserved in part in contemporary sediments, or by real fossils. Real fossils have been recorded from rocks 2 billion years old, whereas biogenic deposits and molecular fossils are known from rocks over 2.7 billion years old, thus attesting to the presence of life on earth even at that early date. Moreover, probable fossils and molecular fossils have been reported from sediments over 3.2 billion years old, that is from the oldest sediments at present known on earth. This must, however, still be confirmed by further research.

Chapter 13 discusses the indications supplied by the nature of the sediments. Several types of sediments are found to occur only in the Early and Middle Precambrian, and this can be correlated with the existence of an anoxygenic primeval atmosphere. Other sediments, only occurring since the Late Precambrian, must, on the other hand, have been formed under an oxygenic atmosphere.

According to provisional dating the anoxygenic primeval atmosphere held sway until 1.8 billion years ago, whereas the oxygenic atmosphere has been present from 1.45 billion years onward. Between these dates the transition from the primeval anoxygenic to the present oxygenic atmosphere must have taken place.

After a discussion of various geological processes, related in one way or another to the origin of life, in Chapter 14, the difference between the two atmospheres is further discussed in Chapter 15. It is seen that the atmosphere of the earth can never have had more than one thousandth of its present amount of oxygen produced inorganically, by the dissociation of water. Everything above this level must have been produced by life, through the dissociation of carbon dioxide during organic photosynthesis. All, or virtually all, of the free oxygen of our atmosphere has therefore been produced by life. Moreover, a tentative definition of the primeval anoxygenic atmosphere can be given as an atmosphere containing not more than 1‰ of the amount of free oxygen our present atmosphere contains.

A tentative outline of the history of atmospheric oxygen and carbon dioxide, which summarizes the earlier discussions, is then presented in Chapter 16.

Chapter 17 treats of the possibility of extraterrestrial life, a subject much discussed since in 1961 organic compounds and "organized elements"—a cautious term for fossils—have been reported from a special group of meteorites. Despite the intensive research carried out since, this thesis has as yet not been proved. But it has not been disproved either and in my opinion an objective evaluation of the facts still points to the probability of an extraterrestrial biogenic origin of these remains.

Chapter 18, finally, contains the conclusions. As stated already in the Preface, the origin of life through natural causes, although not proved, has become highly probable. This once so chimaeric idea has now developed to a fullfledged theory, based on a surprising number and variety of facts, supplied by all the natural sciences. Moreover, a case can be made that pre-life and early life have been coexistent on earth for a period of about two billion years, during which time the transition from non-living to living may have taken place several, or even many, times.

REFERENCES

MAXWELL, A. S., 1964. *Uncle Arthur's Bedtime Stories, 1*. Review and Herald Publ., Washington, D. C., 192 pp.

RUTTEN, M. G., 1962. *The Geological Aspects of the Origin of Life*. Elsevier, Amsterdam, 146 pp.

SIMPSON, G. G., 1949. *The Meaning of Evolution*. Mentor Books, New York, N. Y., 192 pp.

SIMPSON, G. G., 1964. *This View of Life*. Harcourt, New York, N. Y., 308 pp.

SPRAGUE DE CAMP, L., 1969. The end of the monkey war. *Sci. Am.*, 220 (2):15–21.

Chapter 2 | The Principle of Actualism

1. BASIC PRINCIPLES IN THE RECONSTRUCTION OF GEOLOGIC HISTORY

The origin of life, it has already been stated in the Introduction, somehow or other took place in an early period of the history of the earth. So, before we proceed with a study of the remains of early life, we should have an inkling of how the events of geologic history are reconstructed from the data preserved in the rocks of the earth, and how these events, once reconstructed, can be dated. The first question is discussed in this chapter, whilst dating in geology is treated in Chapter 3.

As we will see, the main point in the question of establishing basic principles for the reconstruction of geologic history lies in the fact that it is difficult to apply strict definitions. Geology, as an environmental and a historical science, studies events each more complex and unique than those studied by the physical sciences. This difference between the environmental and the physical sciences is not often clearly understood. In our problem, it is, however, essential for the understanding of the reconstruction of the history of life and as such it deserves an explicit treatment.

As we will see, geologic thought is at present based on the principle of actualism (or uniformitarianism, as is the less apt and more cumbersome term in Anglo-Saxon languages). This tells us to look for the influences of processes now operative on and in the earth to explain the events of geologic history. Or, as it is often schematized, it tells us that "the present is the key to the past".

2. ACTUALISM AND UNIFORMITARIANISM

The two words heading this section have almost the same meaning. They express the assumption of a certain amount of continuity in geological history. Anglo-Saxons mainly use the word *uniformitarianism*, but continental Europeans denote the same principle in their own languages with words like *actualisme*

(French) or *Aktualismus* (German). These words lend themselves easily to a sort of pseudo-translation into *actualism* as the English version.

Apart from this variation in the usage of these two words in various languages, there is, however, also an intrinsic distinction. The two words highlight two sides of one and the same line of reasoning which is fundamental to all modern geology. As the laws of nature are unchanging with time, the physical processes which determined the events of geologic history have been invariant with time. These events must therefore have been continuous and uniform over the past, which is indicated by the word *uniformitarianism* (HARRIS and FAIRBRIDGE, 1970). On the other hand, once the assumption is accepted that the laws of nature are invariant with time, one may study the processes actually going on and extrapolate these back into the geological history to interpret the events of the past and to embody the actual findings of geology into a genetical and historical picture. Hence the word *actualism*.

If, in this book, I prefer the word "actualism", this is not because I am not an Anglo-Saxon myself, but because for one thing I think that "actualism" better expresses the meaning of the principle in question, since we are driving at an elucidation of the earth's history from the processes we actually see or infer to be in operation now. Whereas on the other hand the word "uniformitarianism" is so often thought to indicate "uniformity". The latter meaning has never been in the minds of those who assumed that uniformitarianism was the basic principle underlying geologic history, but it obscures the relevant literature to such an extent that it becomes difficult to understand for all but the specialist. I will cite here one example. Although the existence of Ice Ages was well known to those proposing the principle of uniformitarianism, even to our day we find that "uniformitarianism" is confounded with "uniformity" to the extent of the following quotation: "Hutton's theory of uniformitarianism has been widely applied. However, strict application of this theory to the study of paleoclimatology leads to considerable difficulty (because) climate is a fluctuating phenomenon." (OPDYKE, 1962.) Here the difference has been overlooked between the laws of nature, which are invariant or uniform, and the variable factors going into the equations based on these laws, which eventually determine the outcome of the physical processes which have formed the fabric of geologic history.

The blatant confusion which may arise from the use of the term uniformitarianism follows, for instance, also from the title *Uniformity and Simplicity* (ALBRITTON, 1967) and from several papers contained in a symposium volume on the principle of the uniformity of nature. In this volume it is repeatedly stated that the uniformity of physical laws is but a special case of the wider principle of simplicity, which means that the uniformity of said laws with time, being the simplest explanation, therefore is the true one. Or, in the words of the philosopher Prof. GOODMAN (1967): "In conclusion, then, the Principle of Uniformity dissolves into a Principle of Simplicity that is not peculiar to geology and pervades all science and even daily life."

Prof. Goodman then proceeds to state "The geologist works under no uncommon handicaps and enjoys no special privileges merely because his work may be historical or descriptive." We find here the common mistake, made not only by philosophers of science, of an extrapolation of the background of the exact sciences to that of the environmental ones. For biology the dangers inherent in such an unwarranted extrapolation have been set forth by the systematic zoologist Prof. MAYR (1968) as follows: "Theory formation in taxonomy, as elsewhere in biology, has been greatly handicapped by endeavours to squeeze biological concepts into the strait-jacket of concepts and theories developed by logicians and philosophers of physics. Many of the generalizations derived from the facts of the physical sciences are irrelevant when applied to biology. More importantly, many phenomena and findings of the biological sciences have no equivalent in the physical sciences and are therefore omitted from the philosophies of science that are based on physics. It is not only the enormous complexity of biological systems which requires concepts that have no analogue in the physical sciences, but, in particular, the fact that organisms contain a historically evolved genetic programme in which the results of three billion years of natural selection are incorporated. The uniqueness of almost all biological situations above the level of molecules (for example, individuals, populations, species and the like) is another phenomenon requiring a rather different emphasis in biology from that customary and appropriate in the physical sciences... Many of the terms used in the formation of theories in biology (for example, relation, species, classification, population and the like) have also been used (but often with an entirely different meaning) in the formation of theories in the non-biological sciences. As long as biologists attempted to use the definitions of these terms that had been customary in physical science, and as long as they stayed strictly within the framework of theory that was appropriate for physical science, it was impossible to accomodate the special demands of the biological situation."

Mutatis mutandis these remarks of Prof. Mayr hold good for all environmental sciences, and thus also for geology. I do therefore flatly disagree with Prof. Goodman, when, in the publication cited above, he states that, because "the geologist works under no uncommon handicaps and enjoys no special privileges, he can now forget the obsolete controversy over uniformitarianism... (and) can more profitably turn his philosophical efforts to fundamental and far-reaching problems in the theory of classification, simplicity, measurements and mapping". On the contrary, there is no way to classify, to measure and even to do all but the most elementary mapping, when we do not have a guide line to the fundamental principle(s) underlying the reconstruction of the earth's history from the data actually found.

In compliance with the uniformity of the laws of nature we must try to elucidate what are acceptable guide lines in reconstructing this history. If these guide lines can only be rather vaguely expressed—a criticism often voiced—and if they seem to be a far cry from the laws of physics known in so much detail and strictness, this results from the meagreness of geologic data on the one hand, from the complex nature and the uniqueness of all geologic phenomena on the other. It is, for instance, possible in physics to study over and over again the freezing and boiling of water under varying environmental circumstances and so to arrive at the physical law underlying these phenomena. In geology, on the other hand, we seldom know under what circumstances an experiment carried out by nature has taken place, and we have normally no possibility of duplicating it under the same or other environmental circumstances. It is sometimes possible to recognize an ancient sediment, such as a ground moraine, as having formed under below-freezing conditions. But even here the distinction with a mudflow formed at above-freezing temperatures remains difficult. With regard to the boiling point of water, it is mostly difficult to decide whether, say, ancient, volcanic processes, or the formation of ore veins, is due to vapours or to liquids.

So it follows that for the reconstructions of the geologic history we are dependent on data of a vagueness not generally met with in the physical sciences. This may help to explain the vagueness of the main principles underlying geology. Of the many principles proposed, I will only cite that of actualism and its major opponent, catastrophism. But before going into a more detailed account, I should like to interpolate a section on philosophizing in geology. This might lead to a better understanding of the normal geologist's attitude in these matters.

3. PHILOSOPHIZING IN GEOLOGY

To speak at all about philosophy in geology will perhaps seem unfortunate, to use an understatement, to some, because only rarely are geologists good philosophers. For a large part this stems from the fact that in the pursuit of geology one has to do with so many scattered data, with so many theoretical possibilities that it is hardly possible to build up any valid logical structure. But in our quest toward the origin of life we will have to penetrate far into the geological history of the earth. We will have to use geological data and interpretations and thus it is advisable to know how a geologist's mind works.

The phenomena our planet offers for study are so immensely varied that no mere description is sufficient to convey their full meaning. No word picture, even of a single outcrop of rocks, conveys all the information to the man who has not himself seen that spot. This is apparent from every travelogue of a geologist who, visiting for the first time regions about which he may have read extensively, finds his mental picture still nebulous. Take a European geologist to the Appalachians or the Rocky Mountains, an American to the Alps, the result will be the same. You will hear your man enthuse about his luck to be able to see these regions himself, and so to check by personal inspection his incomplete and unbalanced impressions from the literature.

In geological descriptions one normally finds only part of ths tory. Either some main line is stressed, which results in oversimplification, or exceptional details are diligently pointed out, whilst the author forgets that only a minor part of his readers can really take the general description for granted. Moreover, the impossibility to convey the full meaning by a description alone fosters the tendency to supply as many illustrations—maps, sections, photographs—as possible. And even then the tendency expressed above, to try to see things for oneself, forcefully expressed by a Swiss geologist as "hingehen und gucken" (go there and look), persists.

If then it is wellnigh impossible in the practice of geology, to express oneself in words, it will be clear that such a background does not lead easily to an aptitude for philosophical essays in which ultimate meanings are conveyed verbally. I hope this digression will have clarified why basic principles such as actualism are rarely discussed by geologists. In their early days, when they were postulated, they roused a flutter of protest and defense, but then the interest died down quickly. The normal geologist "lives" vaguely with such a fundamental principle, without knowing its exact meaning, and without having to defend or attack it during his day-to-day work. It forms the basis of his thinking, but it is rather far removed from his daily preoccupations. This may help to explain why, in the case of the principle of actualism (and this includes uniformitarianism) we had only one modern and thorough study in Prof. HOOYKAAS' *Natural Law and Divine Miracle* (1959)[1], which even then was not written by a geologist, as Hooykaas is a chemist by training. In recent years, however, the interest in such basic principles has grown. So we now have a book by HÖLDER (1960) and one edited by ALBRITTON (1963) in which actualism is discussed along with a number of other basic notions of geology, and finally there is the report on the symposium on uniformity and simplicity already mentioned (ALBRITTON, 1967).

It has already been indicated in the preceding section that it is difficult to give an exact definition of the basic principles in geology, such as that of actualism. One has to content oneself with a general outline, which leaves room for borderline cases. In practice, however, it is found that such borderline cases are limited both in number and in importance, so even the general outline is generally enough to work with. I will in this chapter stress this general outline and indicate possible

---

[1] The 1963 edition, in which the subtitle *The principle of uniformity in geology, biology and theology* has been promoted to title, is, except for an expanded introduction and a short homily at the end, a reprint of the first edition.

variations only as far as they are of importance for our study of the origin of life. For a more general discussion of actualism the reader is referred to the literature cited above.

4. CATASTROPHISM

As remarked already, actualism has replaced the earlier principle of catastrophism, the theory of successive catastrophes, which was the popular concept at the beginning of the last century, and which is commonly, although not quite justly (POTONIÉ, 1957) attributed to the French naturalist G. Cuvier (1769–1832).

The theory of catastrophism postulated catastrophes of various sorts, such as floods, major volcanic eruptions, and/or strong and sudden crustal movements, recurring at specific times in geological history, to be responsible for major breaks within this history. Catastrophes led to faunal crises, mass extinctions, orogenetic revolutions, sudden transgressions of the sea, and the like. Such catastrophes were presumed to be due to causes not at present operative on the rather stable earth. Even if these causes would ultimately be comparable to the present ones, their intensity would have been such that these intermittently occurring catastrophes could not be compared to the environment of the present day. It was consequently postulated that there has existed a marked difference between the present—and by implication much of the normal history of the earth—and the events occurring during such catastrophic episodes in the past.

The successive states of the surface of the earth, as reconstructed from the contemporary sediments, were thought to be so vastly different *inter se* and from the present, that only very large catastrophes, suddenly occurring at recurrent points in geologic history could explain these differences. The emphasis on worldwide sudden inundations and volcanic upheavals is of course based on the biblical version of the deluge and on the description of the volcanic catastrophe of the Vesuvius, as described by Pliny the Younger, both books being required reading for every 18th and 19th century European intellectual. So what is more sensible than to assume similar catastrophes to explain geologic phenomena in the past? The erection of a mountain chain then is nothing but a sudden catastrophic upheaval of the earth, a catastrophe not known to have occurred on our stable earth during all of our own history. Sudden inundations of the continents were due to earlier deluges, and the sudden extinction of a fossil fauna was the result of another, as yet not clearly defined major calamity.

Catastrophism, with its postulated recurrence of major disasters, each one, however, more or less comparable to the deluge and to other, historically known catastrophic events, was acceptable to the church, who viewed with a benevolent eye a theory which proposed a synthesis between the teachings of the bible and the new findings of the natural sciences.

5. ACTUALISM

Actualism does not deny that sudden catastrophes occur, and will have occurred over all geologic history. But they are never world-wide, nor is their effect comparable to the results of actualistic processes of small immediate effect but operating over large time spans. Fundamental to the philosophy of actualism is the conception of the extremely long time-spans available in geologic history. Actualism postulates that the processes now known or inferred to be operative in the atmosphere, the hydrosphere and the lithosphere have been operative all through geologic history. Moreover, it states that every phenomenon inferred to have occurred in the geological past must be the result of one or more of these processes. In other words, the evolution of the earth during its geological history is due to present causes. There is no need to postulate ancient causes.

Actualism explicitly refers only to the geological history of the earth. That is to the history which is based on the rocks of the earth. An earlier history of the earth, during the time when it was still molten, or when it formed a dense nebula of interstellar matter, has left no record in the rocks of the earth, and therefore by definition is pre-geological. A famous discussion of the ideas of the Scottish geologist James Hutton (1726–1797) by the physicist William Thomson, the later Lord Kelvin (1824–1907), is therefore irrelevant. Hutton discussed the results from his studies of the rocks, from which, for the geological history of the earth, he could see "no vestige of a beginning". Thomson, on the other hand, argued from general physics and astronomy that there *must* have been a beginning and that Hutton consequently had been wrong. Thomson, however, talked about the pre-geological history of the earth and therefore was in no position to discuss the results reported by Hutton.

Although the basis for actualism was laid by the work of Hutton, it was only about half a century later that this principle became generally accepted and replaced the principle of catastrophes. This was mainly the result of another Scottish geologist Charles Lyell (1797–1875). Hutton's ideas were not only contrary to the leading geologic theory of Neptunism, developed by the German geologist Abraham Gottlob Werner (1749–1817), but also to theology, which taught that one of the prerogatives of Divine Providence was the ability to set aside the laws of nature and to interfere with terrestrial activities in an arbitrary manner (HUBBERT, 1967). It is only through the influence of Lyell's great work on the *Principles of Geology: Being an attempt to explain the former changes of the Earth's surface by reference to causes now in operation* (1830–1833), that the principle of "uniformitarianism" became generally accepted.

As already stated in the second section of this chapter, actualism postulates uniformity of causes, not overall uniformity of conditions, nor uniformity of results. We also observed that this misinterpretation is more easily arrived at, when the term uniformitarianism is used instead of actualism. It has never been questioned that in the processes operative on earth larger and smaller variations in a quantitative sense do occur. To cite only a few of the major ones, for instance: variations with time in climate and in weathering, variations with time in intensity of crustal movements, in erosion, transportation and sedimentation, variations in volcanic activity, and so on. However, such variations, let it be stated once more, always remained of a quantitative nature.

The distinction between catastrophism and actualism is complicated by the fact that there is a big difference, however, difficult to define exactly, between what is a catastrophe according to human standards and the world-wide major catastrophes postulated by the theory of catastrophism. The main influence of Charles Lyell upon contemporary geologic thought was that he showed how small changes, operative over sufficiently long time spans, produced results equal to those which had been interpreted as due to some major catastrophic event, because they appear condensed in the geological record of the past. A volcanic eruption, an earthquake or a tsunami often have the most direct effect upon humanity, especially so if they happen to strike a densely populated area. The most striking example being perhaps the story of the local flood of a couple of major rivers —the Tigris and Euphrates—which perpetuated itself for thousands of years and as the Deluge became incorporated in the gospel.

Major catastrophes on the human scale cannot satisfy the requirements of catastrophism. In the example of floods we have the rain floods such as in Mesopotamia, tsunamis due to earthquakes around the Pacific, and storm floods in low-lying coastal areas such as Holland. Each and everyone of such floods may have been so important that it has entered the saga or the history of that particular area as a major calamity, extinguishing all previous life. The effects of even the largest of such floods will, however, only have been felt in a restricted area and none of them will have been capable, to alter, at a single time, the entire surface of the earth. Only when such floods occur repeatedly, will they gradually acquire geological significance. The same can be said for volcanic eruptions. Even the Plinian eruption of Vesuvius destroyed no more than three cities, Herculaneum, Pompei and Stabiae, all situated on its southwestern flank. And had not Pliny the Younger lost his uncle in the event, it is doubtful if we would have such a detailed account, even of this major eruption.

So a flood here, a quake there, even with a volcanic eruption thrown in, this is all compatible with actualism. Such major disasters on the human scale are small-sized when compared to the dimensions of the earth. They result from processes actually at work, and do not reach global size or importance.

To drive home this local character, both in area and in time, of even the major human catastrophes we can best return to Lyell's *Principles*. In the closing remarks of his first volume, following a description of the ravages resulting from volcanic activity in the Naples district, we read: "The signs of changes imprinted on it during this period may appear in after-ages to indicate a series of unparalleled disasters... If they who study these phenomena consider the numerous proofs of reiterated catastrophes to which the region was subject, they might, perhaps, commiserate the unhappy state of beings condemmed to inhabit a planet during its nascent and chaotic state, and feel grateful that their favoured race has escaped such scenes of anarchy and misrule." However, pursuing Lyell's narrative, we read: "What was the real condition of Campania (Napolitana) during those years of dire conclusion? A 'climate', says Forsyth, 'where Heaven's breath smells sweet and wooingly—a vigorous and luxuriant nature unparalleled in its productions— a coast which was once the fairy-land of poets, and the favourite retreat of great men. Even the tyrants of the creation loved this alluring region, spared it, adorned it, lived in it, died in it'. The inhabitants, indeed have enjoyed no immunity from the calamities which are the lot of mankind; but the principal evils which they have suffered must be attributed to moral, not physical causes—to disastrous events over which man might have exercised control, rather than to inevitable catastrophes which result from subterranean agency. When Spartacus encamped his army of ten thousand gladiators in the old extinct crater of Vesuv, the volcano was more justly a terror to Campania then it has ever been since the rekindling of its fires." (LYELL, 1875, Vol. I, pp. 654–655).

This emphasizes the difficulty of giving a strict definition of actualism. There is no cut and dried boundary between catastrophism and actualism. Presumably this would be the case if geology could supply figures on the precise rates of its processes. It would then be feasible to make distinctions such as: "Crustal movements of up to 1 cm/year are considered actualistic, anything above that rate belongs to catastrophism." As long as this is not the case every geologist must more or less rely on his own judgment, or "feeling" as to these rates, a thing which might be as variable as the psychology of the geologists concerned.

To take the floods: a flood of 1 m high is quite a normal thing. A flood 10 m high is already something quite exceptional, when measured against the experiences of our short life spans. But we would be at a loss to say if a flood 100 m high was still due to actual causes, if we ever found evidence for such a catastrophic deluge in geologic history. Luckily for the geologist, this difficulty does not occur in the practical application of actualism. Floods are normally between 1 m and 10 m high and only very rarely a little higher. Volcanoes normally destroy only part of their surroundings during one single eruption. And even the worst of quakes are very limited in the areal extent of their damage. It is only by continuous repetition, let us stress this basic fact again, not over short time-spans such as centuries, but over thousands and millions of years, that such catastrophes in the human sense, but quite minor events on the geological scale, aquire global significance.

To the layman, the physicist and the chemist this insistence on the concepts of actualism might well seem to be making a mountain out of a mole hill. What is more natural, one might ask, what more obvious, than using present-day causes to explain the past. Is all this insistence not just another example of the application of Occam's razor? The reason this is not so lies in the extreme compression of the geologic time scale in part in all geology; in geologic literature, and perforce, in most of the geologist's subconscious. The common saying that one year in human history may be compared with a million years in geology may indicate the importance of this condensation of geologic time. Every amateur film maker knows what a speed-up factor of one million would do to his movies. Even the slowest of crustal movements of, say, 1 mm/year will become catastrophic when viewed at the rate of 1 km/m.y. (one million millimeters per million years). In the following sections we will therefore explore the influence of this condensation of time common to all geologic writing and thinking. This condensation is, of course, a necessity. Nobody is mentually able to visualize even "only" one million years, whereas no time scale in years could find a place in a textbook. But even though it is a necessity, it is well to realise the pitfalls one can be led into by a condensation of time of that magnitude.

6. ACTUALISM AND TIME

More than anything else, I think it is Charles Lyell's description of great changes effectuated by slow-moving processes over a long time, which has won over geologists from catastrophism to actualism. Time is the ally of actualism. Given enough time, earth movements of 1 mm/year can create mountains or oceans. Given enough time, high mountain chains will be worn down by erosion and parts of the oceans filled by the sediments derived from these mountains. Given enough time, a volcano will in due time bury all of its surroundings under lavas and ashes, even if none of its single eruptions attains the violence of the Plinian eruption of Vesuvius. And given enough time, a new fauna may develop gradually out of an older one whilst the earlier forms die out almost imperceptibly.

If we condense time, a mountain chain may seemingly arise suddenly, an ocean may seem to "fall" into the surface of the earth in a single tremendous catastrophe. A volcano may be thought to change all of its surroundings in one stroke, whilst an older fauna may seem to be extinguished by a single catastrophe, to be replaced by a younger fauna newly created.

Amongst the examples found in Lyell's descriptions that of Pozzuoli near Naples is, perhaps, the most telling. In a couple of pictures and a small section, reproduced here as Fig.1–3, it is seen how marine inundations have left their mark by the boreholes of molluscs on the pillars of a ruin of Roman times, at Lyell's date regarded to represent an ancient temple of Serapis. Moreover, several

Fig.1. The columns in the so-called temple of Serapis at Pozzuoli near Naples, which, owing to historic changes of sea level, play a prominent part in the narrative of Charles LYELL (1875).

A marine transgression is clearly indicated by the dark boreholes of molluscs, which end abruptly at a couple of meters from the base of the columns. Scattered boreholes indicate a later, much higher, level of the sea, which did, however, last shorter, so that the number of boreholes is less than those made during the earlier but lower inundation. The water surrounding the base of the pillars is not the sea, which at present stands many meters below this level, but an artificial fresh-water pond which the town fathers of Pozzuoli for some reason maintain at this site.

Fig.2. View of the Bay of Balae near Naples, the region in which volcanic eurptions and slow crustal movements induced Charles Lyell to the now classical remarks on actualism. The town of Pozzuoli is situated at left (*1*), the columns of the so-called temple of Serapis in the left foreground (*2*). (From the original drawing by LYELL, 1875.)

ash falls and lacustrine deposits attest to intermediate stages, in which the sea had retreated from the land. For one thing all these variations occurred long after the time of the supposed Deluge, on the other hand they have taken place during historical time, when it is known that although volcanic activity occurred in its surroundings, the town of Pozzuoli itself was not affected. These changes in relative sea level of the Mediterranean, which have taken place during a period from which no catastrophes are known, was definite proof of the big effect small causes may have, if only consistently at work over long enough periods.

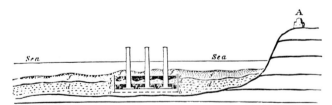

Fig.3. Section through the so-called temple of Serapis at Pozzuoli near Naples, showing the various sediments with which it has been covered, and which have since been excavated, together with the highest level the sea ever reached, as witnessed by the boreholes of marine molluscs on its pillars. See Fig.1.

   $a–b$ = ancient mosaic pavement of the floor of the building; $c–c$ = dark marine incrustations; $d–d$ = first filling up by a shower of volcanic ash; $e–e$ = fresh-water calcareous deposit; $f–f$ = second filling up by volcanic ashes. (From LYELL, 1875.)

7. TIME AND THE GEOLOGIST

It follows that when we study the processes governing the development of
the earth we are easily led astray by their slow tempo. In our human short-lived
and egocentric mind we think of the earth as stable and strong. Nothing is less
true.

Whole continents float in a heavier substratum, or, as we say in our scientific
jargon, are in isostatic equilibrium. In fact they seldom attain this equilibrium;
in the geologist's eye they are continually bobbing up and down. Volcanoes bring
a part of that substratum to the surface, their work being comparable to that
of global-scale moles. Meanwhile, all higher hills and mountains are attacked
by erosion, whilst the detritus which is chafed off these mountains is laid down in
lowland flats and in the seas. If we wish to study the history of the earth, we must
completely discard this egocentric way of thinking and accept the vast amounts
of time of the history of the earth as a common, everyday affair.

A geologist consequently sees no stable earth, but a constantly moving
picture. He sees a jumble of mountains, say, the Alps or the Rocky Mountains
forming, only to be immediately attacked by erosion. He may see wide seas expanding
over large parts of the continents, whilst elsewhere shallow seas dry up completely
and make place for a rich continental fauna and flora, perhaps accompanied by
spectacular volcanic events. Or, to take another example, geologists may see the
continents wandering over the earth's surface, or breaking up and partly founder-
ing to oceanic depths. But what we have to remember is that this is the earth's
history, as seen through a geologist's eye, used to a speed-up factor of one
million. All these "catastrophic" events on our old, our so very old earth, in reality
come about at a snail's pace.

Of course, one cannot, as a geologist, stress all the time the immense time-
spans in geological history, and the result is a certain schematization in geologic
writing. Many geologists express perhaps too freely what their imagination sees
when looking at the slow history of the earth through eyes expressly trained to
speed up this development one million times. To illustrate this, let us take just
one example, *orogeny* or the process of mountain building.

The creation of our present major mountain chains, such as the Alps, the
Himalaya and the Andes, has taken some 50 million years. During that period
earth movements were of the order of 1 mm/year. Before that time the earth's
crust was much more quiet and earth movements of 1 mm/century were normal.
Moreover, this earlier, more stable, period was of much longer duration. It lasted
between 100 million and 200 million years.

Now geologists are apt to speak, in their speeded-up version of the history
of the earth, of a "catastrophic period of mountain building", or of "revolutions
in the earth's crust" when describing that shorter, relatively unstable period of

mountain building, which lasted "only" 50 million years. We like to stress the difference between this period of mountain building and the earlier, longer, more stable period. In doing so, perhaps for want of words, perhaps owing to a sort of laziness, we definitely tend to overstress this point. We often forget to state explicitly that our catastrophes and revolutions, as described in the geologic literature, are a far cry from human catastrophes. No geologist will therefore deny the mobility of the earth, nor the existence of variations of speed and intensity of the different process in operation on and in the earth. This does, nevertheless, form no objection to adhere to the principles of actualism, and to use present causes to explain the past.

The common textbook treatment to reduce all geologic history of the Phanerozoic, including Paleo-, Meso- and Cenozoic, that is a time span of more than 500 million years, to a diagram of 6×8 inches on a single page, is another case in point. The same is true when a paleontologist depicts a so-called faunal crisis with a single line, although the faunal extinction he speaks of may have taken 50 million years (cf. NEWEL, 1967). Especially coming from a species like ours, not yet 50,000 years old, this is of course somewhat presumptive. But all the time such schematizations and condensations of geological time are common in geological literature.

8. VARIATIONS IN INTENSITY OF PROCESSES: "THE PULSE OF THE EARTH"

It is, consequently, in keeping with the principle of actualism, when we accept variations in intensity of the various processes active in the earth's history. We met already with the differences in crustal mobility during an orogenetic period and the normal quieter periods. Another well known example is that of the occurrence of Ice Ages. Although an ice-cap extending over most of northern Europe and northern North America is quite another proposition than the small vestiges of such an ice-cap as now found on Greenland, Iceland and Spitzbergen, the processes of ice-cap formation are not intrinsically different. There is a difference in size, in extension, only. The local climatic conditions, the surplus of snow-fall, the flowing-out of the ice towards the periphery, the formation of moraines, in short, everything one can think of, are the same in both cases. With the actual processes now at work in Greenland, one can interpret the much larger ice-caps which covered northern Europe and northern America in the recent past. Although of varying intensity, the processes are similar in kind.

Such variations with time of the intensity of actual processes have already been sung by Victor von Scheffel in 1868 in several of his geological stanzas in *Gaudeamus*. In "Der Basalt" we find a non-volcanic period described as follows:

Vulkanische Kraft war damals gehasst
Ob ihrer zerstörender Schläge.
Dem Ruhebedürfniss der Erde entsprach
Entwicklung auf feuchtem Wege.

Eintönig wogte die Flut und litt
Nichts Hartes mit scharfer Kante.
Die Felsen zerrieb sie zu Kieselstein.
Die Kiesel zerrieb sie zu Sande.

During the following period of volcanic activity a basalt volcano is formed:

... Was weiter geschah, man erfuhr es nie,
Doch plötzlich fasst ihn ein Wüten,
In feuriger Lohe schnob er heraus,
Seine Adern glühten und sprühten.

Lautrasend drang er nach oben vor
Und sprengte mit sengenden Gluten
Die Decke der Schichten, die wie ein Alp
Schwerlastend über ihm ruhten.

Und Schlag auf Schlag—dumpfkrachend Getös
Von tausend und tausend Gewittern
Die Erde barst, es durchzückte sie tief
Ein Schüttern und Zittern und Splittern.

Bis steil majestätisch der feurige Kern
Den klaffenden Spalten entsteiget,
Und trümmerbesäet sich Land und Flut
Dem Säulengewaltigen neiget.

In "Der Granit" a similar story is told about an earlier quiet period followed by a granite intrusion (mistakenly described as an eruption):

In unterirdischer Kammer
Sprach grollend der alte Granit:
"Da droben den wäss'rigen Jammer
Den mach' ich jetzt länger nicht mit.
Langweilig wälzt das Gewässer
Seine salzige Flut über's Land
Statt stolzer und schöner und besser,
Wird Alles voll Schlamm und voll Sand.

Das gäb' eine mitleidwerthe
Geologische Leimsiederei,
Wenn die ganze Kruste der Erde
Nur ein sedimentäres Gebräu.
Am End würd' noch Fabel und Dichtung
Was ein Berg—und was hoch und was tief;
Zum Teufel die Flötzung und Schichtung,
Hurrah! ich werd' eruptiv.''

These variations in intensity with time have been concisely expressed by the late Prof. UMBGROVE (1947) in the title of his textbook: *The Pulse of the Earth*. Based on the succession of orogenetic and more quieter periods in the history of the earth, that is on the *orogenetic cycle* (see Chapter 10), Umbgrove postulated that many other variations in intensity in the processes of the earth together formed a giant sort of pulse. Not every geologist adheres to Umbgrove's idea of a strict synchronization of many different processes on earth to a pulse beat with a period of around 250 million years. But we will see how the basic idea of a cyclic variation of the intensities of crustal movements in the orogenetic cycle has been a very fruitful one. Anyone who might maintain that such a pulse beat has something of a catastrophic nature and cannot be aligned with the principle of actualism, might be reminded of the fact that the duration of one single pulse of the earth is estimated to be of the order of magnitude of 250 million years, a speed hardly catastrophic according to human standards.

A major difficulty in the implementation of the ideas of actualism, we must finally note, is the fact that we do not now live in a normal period of the earth's pulse beat. We live at present in a period which is both post-orogenetic and inter-glacial, whereas the normal periods of the earth's history are non-orogenetic and non-glacial (RUTTEN, 1949, 1953). This distinction will be discussed in more detail in Chapter 11. Our present time is characterized on the one hand by a recent strong uplift of major parts of the continents, presumably as an aftermath of the Alpine orogeny, and on the other by the drowning of all continental shelves by the eustatic rise of the sea-level since the last glaciation.

Geology textbooks consequently place considerable emphasis on topics such as "Upland and mountain geomorphology", on "Glacial versus fluviatile erosion" and on "Coastal scenery", "Mechanical weathering" and "Clastic sedimentation". During normal periods of geologic history, the continents will have been base levelled by erosion, whilst the continental shelves were built up to wave base. There was hardly any mechanical weathering, nor clastic sedimentation and no glaciation at all. Conditions must have been much more similar to those now existing at very special points of the earth's surface, such as the north-western coast of Australia, the Bahamas, or along the coasts of the Red Sea.

Comparable chapters in geology textbooks for these more normal periods of the earth's history would read: "Peneplain geomorphology", "Fluviatile erosion", "Coastal swamps", "Chemical weathering" and "Limestone and evaporite sedimentation".

It follows that the difference in intensities of the earth's processes during its pulse beat has great influence on the morphology of our surroundings. It is, however, due to variations in intensity of actual causes only and does not offer a repudiation of the principle of actualism (RUTTEN, 1970).

9. THE PRINCIPLE OF ACTUALISM AND THE ORIGIN OF LIFE

The implications of actualism for our study of the origin of life are clear. We will have to look for natural causes of the same character as are in operation at present. We do not envisage some sudden event by which life appeared all at once as a full-fledged phenomenon on every corner of the earth. The origin of life will have covered an enormous time span, when measured against human standards. During this period development must have been slow, almost beyond imagination.

In its slowness, the origin of life may well have been infinitely varied. For all we know there may have been different parallel series of development in the transition from non-living to living. Possibly only a small number of these, or even only a single line, led to our present life. In its slowness the origin of life will have been subject to the same physical and chemical laws as life is today. As we shall see later in this book, it is probable that our atmosphere, the rivers and the oceans were quite different, in those early years, when the real *struggle for life*, the struggle for its origin, took place, from what they are now. But even if the environment was quite different, so different as to be unrecognizable by present-day standards, the laws of nature were the same. This permits us to extrapolate the findings of present-day biochemistry and microbiology into that distant past.

In exploring the fundamental principles which underlie the reconstruction of geologic history, we have made an important stride because we now know how geology arrives at its conclusions about events long past. We have found that the principle of actualism underlies all such conjectures, and what its implications are. In doing so we became impressed by the enormous amount of time elapsed since life began in the early days of the history of the earth. It is now time to gain some idea how these time spans are measured, and what is the reliability of such measurements. The next chapter will consequently deal with this aspect of the preliminary questions to be solved before we can turn our attention to our real problem, the origin of life.

REFERENCES

ALBRITTON, C. C. (Editor), 1963. *The Fabric of Geology*. Addison-Wesley, Reading, Mass., 372 pp.
ALBRITTON, C. C. (Editor), 1967. *Uniformity and Simplicity. A Symposium on the University of Nature—Geol. Soc. Am., Spec. Papers*, 89: 99 pp.
GOODMAN, N., 1967. Uniformity and simplicity. In: C. C. ALBRITTON (Editor), *Uniformity and Simplicity. A Symposium on the University of Nature—Geol. Soc. Am., Spec. Papers*, 89: 93–99.
HARRIS, S. A. and FAIRBRIDGE, R. W., 1970. Uniformitarianism. In: R. W. FAIRBRIDGE (Editor), *Encyclopedia of Applied Geology and Sedimentology*. Reinhold, New York, N. Y., in press.
HÖLDER, H., 1960. *Geologie und Paläontologie*. Orbis Academicus II, Freiburg, München, 566 pp.
HOOYKAAS, R., 1959. *Natural Law and Divine Miracle. A Historical–Critical Study of the Principle of Uniformity in Geology, Biology and Theology*. Brill, Leiden, 237 pp.
HUBBERT, M. K., 1967. Critique of the principle of uniformity. *Geol. Soc. Am., Spec. Papers*, 89:1–33.
LYELL, C., 1875. *Principles of Geology: Being an Enquiry how far the Former Changes of the Earth's Surface are Referable to Causes now in Operation*, 12th ed. John Murray, London, 655 pp. + 652 pp.
MAYR, E., 1968. Theory of biological classification. *Nature*, 220:545–548.
NEWELL, N., 1967. Revolutions in the history of life. In: C. C. ALBRITTON (Editor), *Uniformity and Simplicity—Geol. Soc. Am., Spec. Papers*, 89:63–92.
OPDYKE, N. D., 1962. Paleoclimatology and continental drift. In: E. K. RUNCORN (Editor), *Continental Drift*. Academic Press, New York, N. Y., London, pp. 41–66.
POTONIÉ, R., 1957. Vom Wesen der Geschichte der Geologie. *Geol. Jahrb.*, 74:17–30.
RUTTEN, M. G., 1949. Actualism in epeirogenetic oceans. *Geol. Mijnbouw*, 11:222–228.
RUTTEN, M. G., 1953. Shallow shelf sea sedimentation during non-glacial and a-tectonic times in geological history. *Compt. Rend. Congr. Géol. Intern., 19e, Alger*, 4:119–125.
RUTTEN, M. G., 1970. Principle of actualism. In: R. FAIRBRIDGE (Editor), *Encyclopedia of Earth Sciences*. Reinhold, New York, N. Y., 3: in preparation.
UMBGROVE, J. H. F., 1947. *The Pulse of the Earth*, 2nd ed. Nijhoff, The Hague, 358 pp.

Chapter 3 | Measuring Time in Geology

1. THE YEAR IN HUMAN AND IN GEOLOGIC HISTORY

Human history is based almost exclusively on the time elements of the day and the year, corresponding with one revolution of the earth itself, and one revolution of the earth around the sun respectively. These are the only two natural units used in measuring time. For a day might just as well be divided into 20 as into 24 hours, an hour into 50 or 100 minutes, instead of 60. The artificial character of this division of our natural time units is indicated by the fact that we divide the day by a duo-decimal and a senary system, but add the years according to the decimal system. To be consistent, we ought to use centuries of 144 years instead of 100 years.

Except for two methods which can be applied only to the very youngest geological history, neither the day nor the year are ever used as such in geology. For all geological events older than 20,000 years, time is measured on a different basis, because the total amount of time elapsed since a certain event is assessed directly. For convenience's sake such an amount of time is expressed in a certain number of years, but the years are not counted as such. As we will see, time is measured in geology by the so-called *physical clocks* based on the natural radio-activity of certain elements. The amount of decay of such elements is a direct indication of the total time elapsed since those elements were incorporated in a given rock. Using the physical constants of the decay of a particular element, one can calculate how many years are represented by a given amount of decay.

The two methods which measure time in geology by actually counting the years passed by are *dendrochronology* and the analysis of *varved clays*. They have no value for our problem because life appeared so far back in the history of the earth that we have to use dating methods that can measure amounts of time of millions and billions of years, not just some thousands and tens of thousands of years. Both methods have, however, been widely used in archeology and as such have become well known outside the circle of geologists. They are therefore mentioned here for those who wonder why they will not be used any further.

Dendrochronology is no more than the counting of tree rings, in which, moreover, the nature of these rings may give indications on variations in the contemporary climate. Although living trees over 3,000 years old have actually been found, this still does not bring us back into a significant part of the geological record.

Varved clays are finely layered clays which were deposited in lakes in front of the retreating ice-cap of the last Ice Age. The layering is yearly and results from the fact that during the summer meltwater from the ice-cap will bring more and coarser material than in winter. Moreover, when in winter the lakes froze over and the lake water became entirely tranquil underneath the ice, even the finest clay particles could settle down. By diligent counting, all over the area which once had been covered by the ice-cap, of the yearly pairs of coarser and thicker summer layers and thin and finer winter layers, a history of the last 20,000 years could be established. Although formerly widely used for studying the retreat of the ice-cap of the last Ice Age, even this method has now largely been superseded by a radiometric age determination, which uses the decay of "carbon fourteen", that is of ^{14}C. This leaves us only the radiometric dates to measure time in geology, methods which are further studied in this chapter.

2. RELATIVE AND ABSOLUTE DATING

Returning now to the measuring of time for the geological record we must stress from the outset that the art of actually measuring the time which has elapsed since a certain event is a very young one. It has been developed into a reliable method only after World War II, although the first tentative measurements were already performed in the latter part of the thirties.

Before that time all "dating" in geology was done by methods in which only the relative age of a given event, in relation to that of others, could be elucidated. One could arrive at conclusions such as "older than", or even such as "only slightly younger than", or "very much older than". This is now called *relative dating*, whereas the modern method by which the actual amount of time elapsed can be measured is called *absolute dating*.

This is of course common knowledge for a geologist. But just because it is so common, everyday usage, the fact that most geologic dating is but relative dating, is not always properly stated. As we are not interested in facts alone, but also in how they have been arrived at, I think it is well to thrash out properly this distinction between the normal everyday geologic method of relative dating and the means we now possess of measuring geological time in absolute terms, in the amount of years, of millions, or billions of years.

How recent this development of absolute dating is, can be attested by the fact that the first comprehensive treatise on absolute dating, Prof. Zeuner's highly original *Dating the Past* (cf. ZEUNER, 1958), appeared as late as 1946. And even so, the radiometric age determinations were at that time still in their infancy, and much stress was still laid on varve analysis.

3. RELATIVE DATING

Accordingly, in normal, everyday geology, only a relative or comparative age is used in dating the past. The methods used today essentially derive from the English surveyor and engineer William Smith (1769–1831). They are based on two main tenets, viz., the *principle of superposition* and the *fauna evolution with geologic time.*

Principle of superposition

The principle of superposition quite simply states that in any pile of sedimentary rocks, each bed was laid down on top of the next underlying bed. The younger bed consequently is always superposed on the older, and hence the name of this principle.

In this way a relative age can be assigned to a succession of layers of rock found, for example, in a hill scarp or in a bore hole. If it is possible to recognize the same succession in other hill sides or bore holes, a correlation can be established. And if in the latter localities other beds are exposed, either older beds below the known series, or younger beds on their top, then the local relative time scale can even be extended.

In some areas—for instance, in the London and Paris Basins, which formed a starting point for this type of age determinations—it is possible to follow such successions over quite long distances along the hill sides. But any interruption of such a series of exposures, for instance, a broad alluvial valley, or a lake or an ocean, limits the applicability of this method. It is often difficult, if not impossible, to tell which layer on one side of the gap is the exact counterpart of a layer on the other side. This may hold true even for narrow gaps such as the English Channel. The white chalk cliffs of Dover and of Cap Blanc Nez, which look so exactly similar at first sight, already show many details in their lithological succession which cannot be correlated across the intervening water body.

Organic evolution

It is here that organic evolution, both floral and faunal, comes in. Sediments laid down at widely different points on earth may contain fossilized remnants of the contemporary flora and fauna. Since most higher organisms have shown evolution over the geologic past, there have been at any time in the history of the earth organisms which existed for a short period only. If these are sufficiently distinct from both their ancestors and their descendants so as to enable us to recognize them specifically from their fossilized parts, organic evolution has kindly supplied geologists with what are called *index fossils*. By comparing such fossils comparative ages can be established for widely separated rock sections. The fact that normally the term "fauna evolution" is used in this context, when indeed "organic evolution", of both flora and fauna is meant, stems, incidentally, from the fact that a far greater number of such index fossils come from extinct faunae than from extinct florae. This is mainly the result of the fossilization process, which strongly favours the preservation of recognizable parts of animals over plants.

The basic idea of using organic evolution for dating the geologic past, be it in a relative way only, is thus delightfully simple. Why then, anyone familiar with geologists or geology will ask, why then is there divergence of opinion among geologists about almost any long-distance age correlation? The reason is that, although the basic idea is simple, its implementation is crowded with difficulties. Not only have parallel lines of development evolved in many groups of organisms throughout geologic time, so that some groups of organisms may show remarkable resemblances to forms which were not directly related, and which may have lived much earlier or later. But also most fossils do not occur in all sediments formed in one given time interval, because living things tend to concentrate in areas with suitable environmental conditions. Moreover, many groups with a relatively rapid organic evolution, which supply the best material for later index fossils, have never occurred in large numbers and their fossilized remains are rare.

4. ERAS OF RELATIVE AGE IN GEOLOGY

The details needed in establishing a relative time scale for the history of the earth do not necessarily affect our understanding of how it has become possible to date the past. We may safely leave the details of the implementation of the principle of organic evolution to the efforts of the stratigrapher and the paleontologist. For us it may suffice that by painstaking application of the two principles, that of superposition and that of organic evolution, an impressive body of facts has been assembled.

This permits dating of the geologic past in the relative time scale. All the commonly used divisions of geologic time, and even the main eras, such as Paleozoic, Mesozoic, and Neozoic, are based on evolutionary changes of life. Or, to be more exact, on the evolution of the fauna. For the names just mentioned, which divide geologic time into the eras of the Early, the Middle and the New Animal World, indicate that these periods and their limits were defined by the contemporary fauna, whereas the contemporary flora plays only a minor part in geologic dating.

Paleozoic, Mesozoic and Neozoic are in a way comparable to the Early History, the Middle Ages and the Newer History epochs of our human history. They are nowadays often combined into one single super era, that of the *Phanerozoic*, indicating the later part of the history of the earth, in which the animal world has come to full development.

Just as in human history, where we have prehistoric and protohistoric periods in which the normal methods of history are of no avail, we find in the geologic history of the earth a long early history in which the methods of relative dating cannot be applied for want of fossils[1]. This earlier history of the earth is called the *Precambrian*, because it precedes the Cambrian System, which, as follows from Table II (p. 48), is the earliest system of the Paleozoic and consequently also of the Phanerozoic.

The Precambrian has formerly been subdivided into a Proterozoic, in which period extremely scarce remains of life have been found, and an Archeic, comprising the period when there was no life on earth at all. The scarcity of the fossils, and their problematical nature, makes such a subdivision meaningless. None of the earlier divisions of the Precambrian as found in the literature which use the terms Proterozoic and Archeic can be trusted. For the Precambrian absolute dating is the only meaningful method. An example of such a newer subdivision of the Precambrian, in this case for Minnesota, will be given in Chapter 13.

As we shall see later our quest into the origin of life will eventually lead us into this earlier geological history. Before continuing, let me, however, stress firmly once more that all normal dating in geology is relative dating. It tells us, for instance, that beds of the Mesozoic are younger than beds of the Paleozoic. But any statement like "the Mesozoic began umpteen million years ago" is based on absolute dating.

5. RELATIVE AGE OF SEDIMENTS AND OF IGNEOUS ROCKS

It will have become apparent that relative dating can only be applied directly to sedimentary rocks. Igneous rocks, formed within the crust through the gradual cooling and crystallization of molten magma, do not follow the principle of superposition. Molten magma tends to break through the crust and thus defies the superposition law. And even when poured out as lava from a volcano over

---

[1] Perhaps with the exception of the Late Precambrian in which it might be possible to use the deposits formed by algae called stromatolites, as described in section 5 of Chapter 12.

surrounding sediments, thus following the principle of superposition, magma does not contain living organisms. Without fossils the second guiding principle, that of organic evolution, cannot be applied. So in normal geologic dating the age of igneous rocks is doubly relative. It can only be applied relative to relatively dated sediments. The age of igneous rocks is bracketed by the surrounding sediments in the following way. Igneous rocks are always younger than the sediments they penetrate. And conversely, igneous rocks are older than the sediments by which they are covered. We will see how in absolute dating this situation is reversed (Fig.8, p.45).

6. ABSOLUTE DATING

In absolute dating the ages of rocks are measured by determining the amount of time which has elapsed since their formation. This is normally expressed in millions (10^6) or in billions (10^9) of years. Apart from the two methods mentioned in the first section of this chapter, which go back for some 3,000 years and 20,000 years respectively, so for all older rocks, the one applicable method is that based on physical clocks. These employ the decay of naturally radioactive elements contained in the rocks.

Absolute dating forms part of the science that uses the occurrence in nature of stable and unstable components, called isotopes, of a number of elements (HOLMES, 1963; FAUL, 1966). Many other problems besides absolute dating can be attacked by *isotope geology*. One of the most spectacular of these is the Urey paleothermometer, which uses small variations of the stable isotopes of oxygen in carbonate shells. A full review of the possibilities of isotope geology, and of the various elements which possess natural isotopes can be found in RANKAMA (1954, 1963).

Isotope geology uses both stable and unstable isotopes to solve its various problems. Absolute dating, on the other hand, being based on the amount of decay of a certain element or isotope during its geological history, uses unstable isotopes exclusively.

7. PHYSICAL CLOCKS: RADIOACTIVE DECAY SERIES

All heavier natural elements are unstable, and so are several isotopes of the lighter elements. Their presence today is due to their exceedingly slow decay. The age of the rocks in which they are found, and even the age of the earth, is small, compared to their time of decay. They represent a left-over from the amount originally present at the formation of the earth.

A single atom of a radioactive isotope decays by a spontaneous process in its nucleus. The exact moment when this will happen is not predictable. But when large numbers of atoms of the same radioactive isotope are present, statistical laws will govern to overal decay process. The probability that a certain fraction of a given quantity of atoms of a radioactive isotope will decay in a certain time is constant for each isotope, and not dependent upon the quantity of atoms present. This probability is, moreover, different for every radioactive isotope. With the help of these statistical laws it is possible to predict how a given quantity of radioactive isotope will decrease with time.

As the disintregation of the individual atoms is a random process, the rate of decay of a certain amount of a radioactive isotope is proportional to the number of parent atoms. This constant of proportionality is called the decay constant (RUSSEL and FARQUAR, 1960).

The decay rate of a radioactive isotope can also be expressed conveniently in terms of its *half-life*. This is the time for any number of atoms of that isotope to decay to one half of its original value. The radioactive decay constant λ is related to the half-life T by the formula:

$$\lambda = \ln2 \cdot T, \text{ or: } T = \frac{0.693}{\lambda}$$

Incidentally, we cannot use the "full-life" time of a given radioactive isotope, because this is, in theory at least, the same for all radioactive isotopes, that is infinite. The decay of a radioactive isotope is a so-called exponential process. When from an original amount a of a given isotope, one half, $\frac{1}{2}a$, has been decayed in the half-life, only half of the remaining quantity, that is $\frac{1}{4}a$, will decay during the next half-life period, and so on.

The original, unstable, isotope is called the *mother* or *parent isotope* (P), whereas the ultimate stable isotope is called the *daughter isotope* (D). If we assign the symbol P_0 to the number of parent atoms present in a rock at the moment of its formation, and the symbols P_t and D_t to the number of parent and daughter atoms respectively now present in this rock, we can calculate the number of parent atoms present in the rock at the time of its formation by the simple formula:

$$P_0 = P_t + D_t$$

If we assume that there has been neither addition or loss of parent or daughter atoms from the rock sample during its geologic history, than the age t of this rock, that is the time elapsed since its formation, can be calculated by the formula:

$$t = \frac{1}{\lambda} \ln \left(\frac{D_t}{P_t} + 1 \right)$$

Apart from the analysis of rocks or of separate minerals for the included quantities of parent and daughter isotopes of a given decay series, several other methods, all based on radioactive processes, have been studied with a view to establish the age of rocks. Basically, these methods are, however, similar to those using the relationship between parent and daughter isotopes. Since these other methods are not very important for absolute dating at present, they will not be reviewed here. Any reader interested in them will find a first introduction in FAUL, (1966).

8. CONSTANCY OF RADIOACTIVE DECAY

In all dating methods using radioactive decay in tracing the history of the earth there is one basic assumption. This is that the decay rate did not vary over these long periods.

Theoretical physics tells us this is so. According to theoretical physics radioactive decay is a nuclear process which is innate to a given nucleus. Unlike the configuration of the peripheral electrons, which determines the chemical properties of an atom, the nucleus cannot be changed through external influences. Neither heat, cold or pressure, nor changes in the electrical and magnetical fields, all of which have changed within certain limits during the history of the earth, have any influence on a nuclear process such as radioactive decay. Experimental physics corroborates this doctrine of theoretical physics within the limits of the experiments devised so far. The limits in geology, however, are not the extremes of heat or cold, of gravity or of earth magnetics, or even of extreme pressures. Geology knows of one factor which can never be duplicated in the laboratory, i.e., the extreme length of time connate to geological processes. Is it justified, the geologist will ask, to extrapolate the findings of nuclear physics, a science two scores of years old, to geologic eras some billions of years in duration?

Pleochroic rings

Geology provides an answer to this question in the curious phenomenon of the pleochroic rings. Many minerals are found to certain small, darkly coloured spots, which under the microscope appear as a number of concentric rings (Fig.4). In polarized light these rings show colour schemes that are different from that of the most minerals, and which moreover change their colour, when rotated under the polarizing microscope. Hence their name of pleochroic rings. These rings are due to the destruction caused in the crystal lattice of the host mineral by decay products of included small amounts of radioactive elements.

Now in a decay process the emanation accompanying the decay of a certain

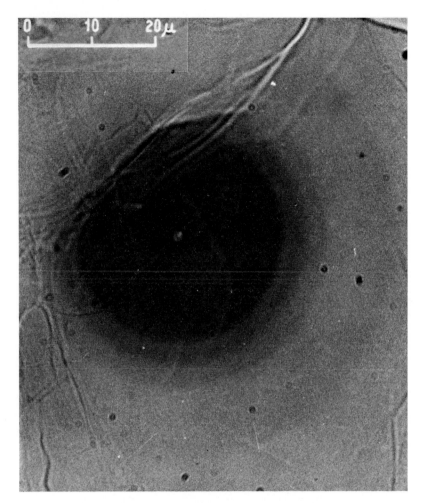

Fig.4. A system of pleochroït rings. (Courtesy Miss S. Deutsch.)

radioactive element always has the same energy, because this is a property of the nuclear decay process innate to the atom constituting that element or that isotope. If there is a small grain of a radioactive element included in a host mineral, the emanation of its decay will consequently always penetrate the same distance into the crystal lattice of the host mineral. It will form a spherical zone of destruction around the radioactive grain in the host mineral. In a thin section of the rock, as studied under the microscope, this sphere will appear as a ring or perfect circle (DEUTSCH et al., 1954).

In the uranium- and thorium–lead decay series, in which many intermediate steps occur between the parent mineral and the ultimate daughter mineral lead

(Fig.4), a series of emanations will occur. These successive emanations have quite different energies, and thus a series of concentric spheres of destruction will be formed around a grain of uranium or thorium imbedded in a host mineral. Such a series of spheres, or rings in a thin section, is schematically indicated in Fig.5.

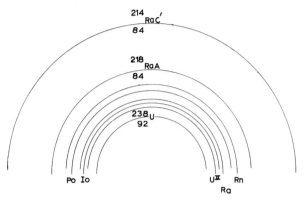

Fig.5. Schematic drawing of the pleochroic rings around an inclusion of uranium. In normal rock-forming minerals the radius of the outer sphere formed by the decay of RaC' is about 30 μ.

If, during the course of geologic history, the energy of the radioactive processes would have changed, the spheres of destruction of the pleochroic rings would have presented gradually changing radii with time. A slow change of the energy of the radioactive decay over the millions of years would have resulted in a blurred spot, not in a system of discrete concentric rings. Since the decay rate is directly proportional to the energy released during the decay process, we may rest assured that the decay rates of radioactive processes have not changed, even over the length of geologic time.

Of course, in nature, complications almost always occur. The radioactive grain included in a host mineral has a certain size, and the narrower of the rings will consequently overlap. Moreover, in some cases the radioactive element will tend to be scattered throughout the host mineral, in which case pleochroic haloes instead of pleochroic rings will develop. But the fact remains that beautiful pleochroic rings occur in many different minerals. So we can feel quite safe as to the reality of this phenomenon.

Consequently we may rightly assert that the physical clocks which operate on the decay of natural radioactive isotopes really measure absolute time. It is, however, somewhat presumptive to say that they date in years. The clocks may well have been working at the same rate for billions of years, but our methods of reading them are still far from the reliability achieved in laboratory physics. In rocks we are never quite sure that we do really measure exactly the remaining amount of the parent mineral, nor all the atoms of the ultimate stable isotope actually produced by radioactive decay. An error of, say, 5% may be expected in quite good rock age measurements. This margin, in a rock one billion years old, would already give us an uncertainty of fifty million years. So, at least for the Phanerozoic, the normal stratigraphic methods which ascertain the stratigraphic sequence of relatively older and younger rocks cannot altogether be replaced by absolute dating techniques. Instead, the latter only supply us with a number of fixes in which our relative stratigraphical data can be arranged. In modern geological dating both methods, absolute and relative, must go hand in hand.

9. RADIOACTIVE DECAY SERIES USED IN ABSOLUTE DATING

We have at present one radioactive decay process which can be used to date younger rocks, and five that are suitable for the dating of older rocks.

Younger rocks can be dated by the so-called radio-carbon or "carbon-fourteen" method, in which the decay from ^{14}C to ^{14}N is used. Radio-carbon has a half-life of about 5,570 years only, and even with the highest instrumental sophistification rocks older than about 50,000 years cannot be dated by this method.

The five decay processes used to date older rocks have been summarized in Table I and Fig.6. In order of ascending number of the parent element, they are the *potassium–argon*, the *rubidium–strontium* and the *thorium- and uranium–lead* methods. The last mentioned, although starting from three different radioactive parent elements, e.g., thorium and two different uranium isotopes, and although their decay processes are quite different, are always used together, because they have the same stable daughter element, lead. There is a difference, however, in that all three parents produce a different isotope of lead, as seen in Table I. On this difference the modern refinement of this method called the *lead–lead* method is based, which is described in section 15 of this chapter.

Of these methods the thorium- and uranium–lead methods are the oldest. Only after World War II has better instrumentation, mainly the development of the mass-spectrometers, led to techniques suitable for using the other decay series. Moreover, as stated above, the lead methods themselves have also been very much improved. It is therefore appropriate to intercalate a short description of the mass-spectrometer, without which absolute dating would still be in its infancy. But knowledge of how a mass-spectrometer works is not essential for the understanding of our main problem. So any reader wo finds the matter too technical may do well to skip the next section.

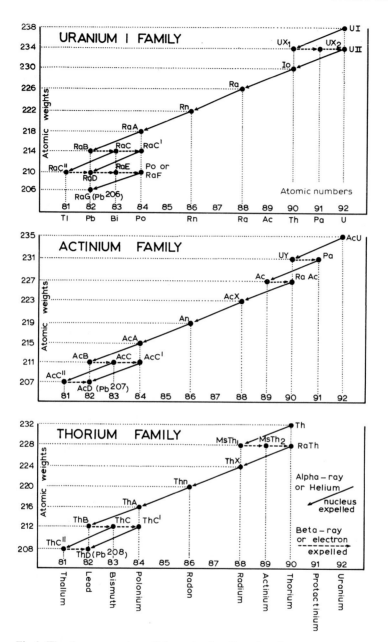

Fig.6. The decay processes of the natural radioactive elements of the uranium, actinium and thorium series. (From HOLMES, 1947.)

TABLE I

DECAY PROCESSES OF NATURAL RADIOACTIVE ELEMENTS USED IN ABSOLUTE DATING OF OLDER ROCKS

Parent element*	Decay process	Half-life ($\times 10^9$ years)	Final stable isotope
$^{40}_{19}$K potassium	electron capture	1.3	$^{40}_{18}$A argon
$^{87}_{37}$Rb rubidium	electron or β-ray emission	47	$^{87}_{38}$Sr strontium
$^{232}_{92}$Th thorium		14.1	$^{208}_{82}$Pb lead
$^{235}_{92}$U uranium	many intermediate steps	0.7	$^{207}_{82}$Pb lead
$^{238}_{92}$U uranium		4.5	$^{206}_{82}$Pb lead

* $^{40}_{19}$K denotes: 40 = atomic weight
                          K = atomic symbol
                          19 = atomic number

10. ISOTOPIC AGES AND MASS-SPECTROMETRY

Before World War II the only way of analyzing rocks was by the methods of classical analytical chemistry. A sample had to contain relatively large amounts of uranium, thorium and lead before being suitable for absolute age determinations, whereas the methods using the other decay series were not yet developed. Uranium ores, which contained the uranium mineral pitchblende, were favourite rock types, and an age determination of, say, a granite was out of the question. And even then the results were not really trustworthy, because a chemical analysis cannot differentiate between the various isotopes of one element. Such samples were analysed for their uranium, thorium and lead content. From the respective decay rates of the uranium and the thorium series it was then calculated what percentage of the lead present ought to be attributed to the decay of uranium and thorium respectively. Assuming that all lead found was formed only by these two decay processes, the age of the sample could be "established" by applying the formula for the half-life of the parent elements.

This obviously made little sense, because lead not only contains the isotopes ^{206}Pb, ^{207}Pb and ^{208}Pb, the stable endproducts of respectively ^{238}U, uranium, ^{235}U, actinium uranium and ^{232}Th, thorium, but also the isotope ^{204}Pb, which is not radiogenic at all. At that time one was of course aware of the fallibility of this assumption. But a little sense is better than no sense at all. And although later work has shown that many of the earlier measurements were far off the mark, at that time they served their purpose. They gave the first direct indication of the stupendous length of geologic time. In the early days of absolute dating this was in itself already of paramount importance. Geologists and astronomers had stressed their conclusions that the age of the earth and the universe must be inconceivably long, only to be met with a certain disbelief from a group of physicists. Here then were the direct measurements of the age of rocks, of matter that one could touch, which actually were that old.

With the mass-spectrograph and the mass-spectrometer, only developed after the war, it is now possible to separate the various isotopes of one single element. Moreover, it has become possible to analyse the relative abundance of isotopes in very small quantities, down to the order of one microgram (one millionth of a gram). All isotopes of a given element are identical in chemical properties. This means that they are identical in atomic number and in the number of their peripheral atoms, which are responsible for their chemical properties. Their difference lies exclusively in the structure of the nucleus, which is expressed in a difference of their atomic weight. It follows that isotopes of one element only differ in physical properties, a difference which can only be detected by physical methods of analyses. In the early days of isotope studies isotopes of a single element have been separated by using physical properties such as differences in the temperature and pressure of their boiling points. These differences are, however, too small to form a reliable basis for a quantitative separation of the minute amounts encountered in the analyses of rocks.

The instrument now in almost exclusive use in absolute dating is the *mass-spectrometer*. The technique used in this instrument is based on the deflection electrically charged particles undergo when passing through a magnetic field. Such particles, in this case the electrically charged—or ionized—individual atoms of the isotopes under study, will undergo a force perpendicular to both the magnetic field and their own direction, whilst the strength of this force is proportional to their mass. The result being that in a mass-spectrometer individual atoms of various isotopes of a single element, or of different elements, are separated according to their mass or atomic weight. The individual atoms of different mass pass through collimator slits at the end of the instrument, and can be counted separately by an electronic device (RUSSEL and FARQUAR, 1960). From a given sample containing atoms of different weight, the spectrum of the masses of the atoms can thus be detected. It is in this way not only possible to ascertain, in a quantitative way, which atoms are present in the sample, but also the relative abundance of these different species of atoms.

In practice, a beam of ions is sent through a high-vacuum tube, at right angles to a magnetic field. Upon entering the tube the ions are accelerated by an electrical high-potential field V, from which they acquire the energy $\frac{1}{2}mV^2$. Their trajectories through the tube will show slight differences in radius which are related to their mass by the formula:

$$m = e \cdot \frac{R^2 B^2}{2V}$$

where m = atomic mass, e = ion electric charge, R = radius of ion trajectory in the high-vacuum tube and B = magnetic field strength (cf. Fig.7). The number of atoms of different mass which follow these slightly different paths through the high-vacuum tube can be counted separately either by moving a collimator slit along the end of the instrument, or by letting these different atoms one after the other pass through the same slit, by slightly altering the strength of the magnetic field[1].

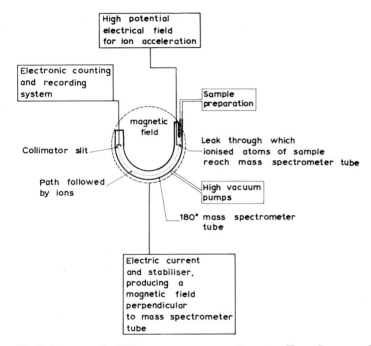

Fig.7. Diagram of a 180° gas-source mass-spectrometer. (From RUSSEL and FARQUAR, 1960.)

Mass-spectrometers are easiest to work with vapour mixtures. The samples of lead or any other element—with the exception of K, potassium—of which the relative isotopic composition must be ascertained, are introduced as a suitable

[1] Mass-spectrometers are also used in many other analytical techniques. We will meet with one of such applications, the analysis of carbon compounds from their break-up products in a mass-spectrometer, in Chapter 6, section 9, and in Chapter 12, section 11.

gaseous ionized chemical compound. These leak through a small orifice into the high-vacuum tube, at which time they are accelerated by the high-potential electric field. It follows that a mass-spectrometer basically consists of three units. First comes the system for handling and preparing the samples. This is a purely chemical apparatus, in which the samples are purified, enriched and converted into a gaseous compound suitable for mass spectrometry. Then follows the actual instrument, the high-vacuum tube to which the magnetic field is applied, and which is often small compared with the accessory parts. At the end of the instrument is the ponderous electronic equipment which counts the individual ions passing through the collimator slit at the end of the high-vacuum tube. These counts are amplified and recorded on charts in such a way that the mass-spectrometer gives a direct reading which does not require further processing.

The introduction of mass-spectrometry in absolute dating has led to great refinements in the relevant techniques. In one method, called *isotope dilution*, the sample is mixed with artificially prepared mixtures of known composition of the same isotopes, which makes it possible to analyse very small samples. This is of great importance in the application of the techniques of absolute dating, because nowadays a great variety of igneous rocks, and even some sediments, have become available for dating. We will, however, in this text not pursue this subject any further as this would lead us too far astray from our main topic. The interested reader is again referred to FAUL (1966), RANKAMA (1954, 1963) and especially to the Holmes memorial volume (HARLAND et al., 1964).

11. RELIABILITY OF ABSOLUTE DATING

As a result of this development in the techniques of absolute age determination of rocks, it is often possible to date the same rocks by different techniques. Even rocks containing minor quantities of the radioactive elements and their daughter products can often be dated by both the rubidium–strontium and the potassium–argon methods, while it is often also possible to use the thorium- and uranium–lead decay series. Moreover, there are frequent interlaboratory checks made on the same rocks to ensure reproducibility of the measurements. There is a tendency to consider as reliable only those ages that have been arrived at by such multiple measurements. I will follow this custom, and if I have not always quoted the very oldest ages published in the literature, my reason is that I have tried to limit myself to those dates that, as far as we know, can be fully trusted.

If we now look at the results, it is quite natural for anyone not acquainted with absolute dating to become bewildered by the figures produced. Ages of millions and billions of years are discussed as if the people had been present themselves, and it is quite natural to distrust such pronouncements. This reserve may, moreover, have been fostered by the fact that many of the earlier datings have proved erroneous. Moreover, many of these dates which have suffered quite considerable corrections, have tended to become corrected towards older and still older ages. So one might well have gained the impression that for every year isotope geologists studied absolute ages, the age of the rocks did rise a couple of million years. This is, however, but to be expected, because in these early years the isotope geologists tended to be very cautious, and the ages given were always on the conservative side.

My aim has been to show that absolute dating in geology is not just a kind of bluff in producing older and still older ages. It is not even guesswork any more. It is based not only on a number of sound physical laws, but on several parallel methods, by which the results can be checked independently. On the other hand, it must be admitted that these methods are still difficult to apply. They ask the utmost of present-day analysis and electronic instrumentation. Moreover, the rocks of the earth that contain the physical clocks have been subjected to all kinds of vicissitudes during the long years of their history. The main difficulty nowadays consequently is not to measure an "age" of a rock, but to determine what the meaning of that "age" is. It might really represent the date of its first formation, but it might also relate to some later event in the history of the rock, some recrystallization, or such influence, by which the original data of the clock have been destroyed so that it now only gives the date of that later event, and not the real age of the rock. It is on such questions of interpretation that most of the serious discussions in absolute dating nowadays center.

But let us skip details. In broad outline we may conclude that absolute dates are to be trusted as near as an approximation of the real age of the rocks in question, as we may possibly obtain at this moment.

12. ABSOLUTE AGE OF IGNEOUS ROCKS AND OF SEDIMENTS

As a counterpart to section 5 of this chapter, in which the relative age of sediments and of igneous rocks was discussed, we must now study how these two major rock classes are situated in absolute dating. Absolute dating is in most cases performed on igneous rocks and on ore minerals, whilst it can only be applied to sediments in certain, rather rare, circumstances, which will be indicated below. Absolute dating, it follows, is also opposite to relative dating in this aspect, because relative dating, which, as we have seen, is based on fossils and on the superposition principle, is always performed on sediments.

With the exception of the complications, such as later recrystallization, which were mentioned in the preceding section, the "age of a rock" or an ore mineral in absolute dating is the time which has elapsed since that rock crystallized, or since that mineral was deposited in an ore vein. At that date small amounts of radioactive elements may have been trapped and built into the crystal lattice of a growing mineral. From that date onwards the decay products of such radioactive elements also remain entrapped—at least in ideal cases—in the host mineral. Measurement of the relative amounts of parent and daughter element, either in separate mineral crystals, or in the whole rock, consequently will give us the age of that rock.

Sediments, on the other hand, may contain any amounts of decay products from earlier radioactive cycles, which have been weathered out of older rocks and have been washed into the sediment at the time of its formation. So a "whole rock" analysis of a sediment will not give the time elapsed since its sedimentation, that is the age of that particular sediment, but will normally yield an entirely different and non-related figure. Only in exceptional cases, for instance in very fine-grained sediments of the Swedish Ordovician, will such an analysis with the thorium- and uranium–lead method give a reliable figure, because the difference in weight between grains of thorium and uranium on the one hand and lead on the other, will ensure that no lead was deposited at the same site as the other elements. Such a sediment is in a way a virgin sediment which contains no blemish from earlier radioactive cycles.

Separate minerals in sediments normally are also unsuitable for a determination of the age of the sediment. Most mineral grains have been washed into the sediment and still carry both the radioactive parent element(s) and the radiogenic daughter element(s) from the time they were formed during the crystallization of an older igneous rock. They might, however, be used in a sophisticated approach to determine from what older rocks the material of the sediment has been derived.

The only sediments suitable for absolute dating are those rocks containing minerals that did form during the process of sedimentation. Such newly formed, or *authigenic minerals*, when containing radioactive elements, are comparable to the minerals growing by crystallization at the time of formation of an igneous rock or an ore vein. For the Late Precambrian and the Phanerozoic there is one mineral, that has been used in this way, viz. the greenish *glauconite*. This is a complex silicate, containing potassium and consequently suitable for absolute dating by the potassium-argon method. It forms on the bottom of warm, shallow seas during limestone sedimentation, presumably under some as yet unknown biochemical control. Glauconite is, however, only found with certainty in sediments less than a billion years old. So this method cannot be applied to those earlier periods of the history of the earth when the origin of life is thought to have taken place. For these earlier periods absolute dating has been sometimes performed on sedimentary uranium ores under the assumption that at least some of the uranium minerals found in these sediments are authigenic. We will meet with an example of this kind of absolute dating in section 10 of Chapter 13, but we might already note here that this method is not altogether a safe one, because many of the minerals in the uranium ores are definitely older and have been washed as pre-existing grains into the sediments at the time of their formation.

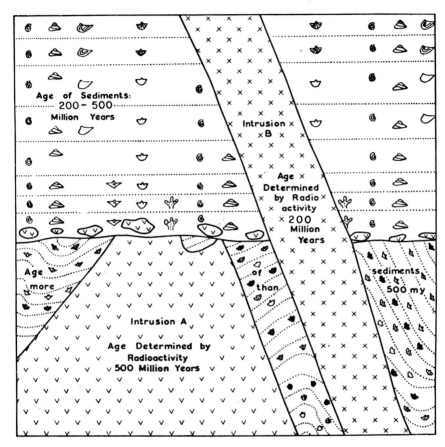

Fig.8. Schematic representation of relative and of absolute dating. A pile of older sedimentary rocks, which can be relatively dated by the enclosed fossils clearly showing organic evolution, has been folded during an orogeny. At the end of the orogenetic period magma intruded into the sediments. This now forms the intrusion A. The folded rocks, together with those of the intrusion, were later denudated and eroded. Upon this erosion surface a newer pile of sediments has been laid down nonconformably. It contains pebbles of the igneous rocks of intrusion A at its base, so in relative dating those rocks can be dated as older than the folded sediments and younger than the overlying horizontal sediments. The series of younger sediments can again be dated relatively by the contained fossils which show organic evolution. Since none of the fossils of the older folded series are found in the horizontal strata of the overlying series it can be concluded that the time elapsed between the deposition of the two series has been a long one. It can, however, not be stated how long this period was in years.

Absolute dating gives the age of the two intrusions as 500 million years and 200 million years respectively. The absolute age of the sediments can, on the other hand, be only given as "older than 500 million years" and "between 200 million and 500 million years" respectively.

To conclude, with the exception of those sediments just mentioned, absolute dating is performed on igneous rocks and on ore minerals. It follows that even in absolute dating the age of sediments is only determined in a relative way. Their age can only be bracketed as being "younger than intrusion A and older than intrusion B", just as in relative dating it was the igneous rocks, that were younger than sediment x and older than sediment y. This basic difference between absolute and relative dating is illustrated in Fig.8.

13. THE LONG EARLY HISTORY OF THE EARTH

As its main result absolute dating has made us aware of the extremely long early history of the earth, that part of its history about which we have only meagre data on contemporary life. Only when life on earth had developed organisms with hard parts, such as skeletons, did abundant fossilization become possible. This has begun, in a number of different zoological phylae, during the lower and middle parts of the Cambrian system. This is the first system of the Paleozoic era, and every geologist working with fossils feels there is a clean break in the history of the earth at that time. There is the later history of the earth, from the Cambrian onwards, during which one can nicely date rocks with fossils, and then there is that vague, earlier period, almost without fossils, the Precambrian. This break has in recent years even been more strongly indicated, because that later, fossiliferous, period, when life was "fully developed" on earth, is now taken together as the *Phanerozoic*. This consequently embraces the three separate eras of this later history, the Paleozoic, the Mesozoic and the Cenozoic (see Tables II and III). So the first division of the time of the history of the earth nowadays is into the earlier Precambrian, starting at the beginning of the geologic history (cf. pp.30–31), and the Phanerozoic, starting at the base of the Cambrian and continuing up to the present time.

This break, occurring at the base of the Cambrian system, has now been dated at 570 million years ago (ANONYMOUS, 1964). But the oldest crustal rock reliably dated is between 3,300 million and 3,400 million years old, whereas the oldest rocks on earth are supposed to be around 4,500 million years old (see next section). So the three eras of the Paleozoic, the Mesozoic and the Cenozoic together, with their 600 million year time-span, take up less than one fifth of the time during which our earth had already developed an outer crust more or less similar to the crust we live on now. That is, more than four fifths of the geologic history of the earth belong to that but vaguely documented earlier history of the Precambrian. It is during this earlier history that life appeared.

This difference is indicated in Table II and III. Table II lists the ages and the duration of the eras and systems of the Phanerozoic, since the beginning of

TABLE II

ABSOLUTE TIME-SCALES OF THE PHANEROZOIC BY KULP (1961) AND HOLMES (1959)[1]
(From HARLAND et al., 1964.)

Era	Period	Epoch	Kulp 1961	Holmes 1959
Cenozoic	Quaternary	Pleistocene	1	1
	Tertiary	Pliocene	13	11
		Miocene	25	25
		Oligocene	36	40
		Eocene	58	60
		Paleocene	63	70
Mesozoic	Cretaceous	Upper	110	
		Lower	135	135
	Jurassic	Upper	166	
		Middle/Lower	181	180
	Triassic	Upper	200	
		Middle/Lower	(230)	225
Palaeozoic	Permian	Upper/Middle	260	250
		Lower	280	270
	Carboniferous — Pennsylvanian			300
	Carboniferous — Mississippian		320	
	Devonian	Upper	345	350
			(365)	
		Middle	390	
		Lower	405	400
	Silurian		(425)	440
	Ordovician	Upper	445	
		Middle/Lower	500	500
	Cambrian	Upper	530	
		Middle/Lower		
?	?	?	?	?

Left and right axes: Millions of years (0, 50, 100, 150, 200, 250, 300, 350, 400, 450, 500, 550, 600).

[1] "Period" is used here instead of the word "system" used in the text. A later, and more detailed, time-scale can be found in the Arthur Holmes memorial volume (HARLAND et al., 1964)—the Geological Society Phanerozoic time-scale, which at present is the authoritative one. But as it is too detailed for our subject, and as the differences are minor—of the order of but a couple of millions—the more schematic table is presented here.

TABLE III

TIME SCALE OF THE GEOLOGICAL HISTORY[1]
(Dates of the main events since the origin of the earth in million years. From KULP, 1960.)

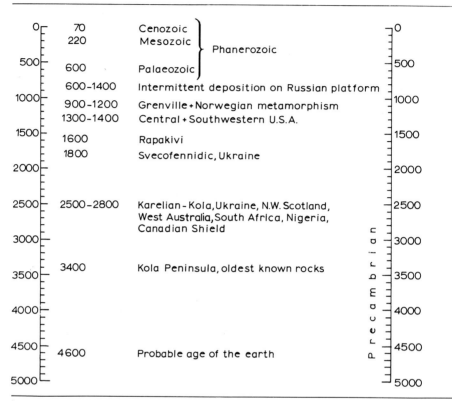

[1] As will follow from the text, several of the events are now dated slightly differently. This general table is mainly given to supply a handy review for comparison with Table II, and to stress the relatively short duration of the Phanerozoic. Differences as, for instance, between an age of 3,400 m.y. for the Kola Peninsula given here, and 3,000 m.y. given elsewhere, stem from the fact that a more detailed date of 3,340 m.y. will be given as 3,400 m.y. by one author as 3,300 m.y. by another, depending upon the degree of conservativeness of these authors.

the Paleozoic and its lowest system the Cambrian, that is of the time during which fossils were more or less abundant. Table III includes the earlier history, and the attention is immediately focussed on the long duration of the earlier history of the earth, when compared with the Phanerozoic. Fig.134 and 135 also indicate this difference. The first figure shows the development of life during the Phanerozoic, the second an inclusive, but of course much more schematic, view of the total history of the origin of life.

14. THE "OLDEST ROCKS"

It seems necessary to enquire at this point what we mean by "the oldest rocks". Of course, the "oldest" rock is no more than that rock sample in which the accumulation of radioactive decay products indicates the longest time elapsed since its formation, yet found in a rock. As such it gives a minimum value for the age of the earth. We have no direct way of measuring the age of the earth, and we will see how this can as yet be only assessed from circumstantial evidence.

It turns out that the ages of the oldest fossils found (Chapter 12), and the age of the oldest rocks and the age of the earth, as arrived at by different lines of reasoning, tend to approach each other. To avoid confusion we must therefore state what the meaning of these terms, often too loosely used, exactly implies.

For the origin of life on earth we are not so much concerned with the age of the earth itself, but with the age of the oldest rocks that could have supported life. That means: with the age of the rocks at the surface of the earth, the so-called *crustal rocks*.

One question leading to another, we must at this point introduce a slight digression on the overall structure and composition of the earth. Geophysical measurements have established that we may distinguish between three concentric layers within the earth, the core, the mantle and the crust, respectively.

The heavy core is supposed to be molten, and mainly formed by alloys of nickel and iron. The mantle is supposed to be solid and formed by heavy, ultrabasic rocks, that is rocks with little silica, SiO_2. They are thought to resemble the greenish serpentine often used as a marble finish in monumental buildings, and to be mainly formed by the mineral olivine. The crust, which has an average composition close to that of a granite, is by far the thinnest of the three concentric layers. Underneath the continents it is about 30 km thick, but in the deep ocean basins only several km of crust are present (Fig.9).

During its early development, the earth has probably had a more or less uniform composition, as it has been assembled in one way or another from stellar or intrastellar material of much the same composition. But the heavier elements would be most strongly attracted to the centre of gravity, that is to the centre of the earth, and tend to migrate inward. Consequently slow centripetal diffusion of the heaviest elements is thought to have led to the gradual separation of core and mantle. The outer surface of the earth would, on the other hand, suffer from erosion and related processes. It is thought that these have led, in time, and in some as yet not properly understood way, to the formation of the superficial crust. So, only after the earth had formed in one way or another, the separation into core, mantle and crust could start. Forcibly the age of the earth must be greater than that of the mantle and the crust.

Now if we look at crustal rocks first, we find that the 3.3 or 3.4 billion-year-

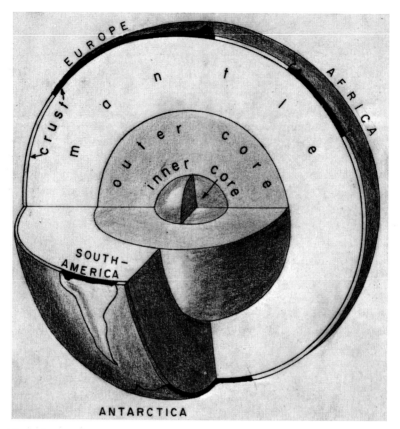

Fig.9. Schematic cross-section through the earth, showing the crust of the continents, the mantle and the core. The thickness of the crust and the altitude of the mountain belts is exaggerated.

old granites from the Kola Peninsula (Table III)[1] are the oldest crustal rocks known as yet. Geologists do agree, however, that these granites are not primary rocks. Instead they are thought to represent still older sediments which have been thoroughly metamorphosed—migmatized as the technical term is—during the orogeny which took place around 3.3 billion years ago.

In section 18 of Chapter 12 we will meet with the sedimentary Swaziland Series of South Africa, for which only a minimum age as "older than 3.2 billion years" can be given, because igneous rocks of that age are found to be intrusive into these sediments. The art of absolute dating has not advanced enough to decide if perhaps the South-African intrusives determined at age of 3.2 billion

[1] Still older datings from the African continent seem, as yet, not entirely reliable. To be on the safe side the figure of 3.3 billion years is accepted here. It is, however, probable, that it will prove to be too low.

years are more or less contemporaneous with the Kola Peninsula granites dated at 3.3 to 3.4 billion years. If this interpretation would be justified, the sediments of the Swaziland Series would in fact be the oldest crustal rocks known. If one takes a conservative attitude however, it is only possible to label these rocks as "the oldest sediments known so far".

There is, however, a possibility that tectonic movements have been so active that they have locally brought up parts of the mantle towards the earth's surface. One such place seems to be St. Pauls rocks, a group of shoals surrounded by a deep ocean trench in the middle of the southern Atlantic. St. Pauls rocks are formed by heavy, ultrabasic rocks, of the type as thought to belong in the mantle. Moreover, their age has been estimated at 4.5 billion years (STUBBS, 1965; MELSON et al., 1967)[1].

We may therefore provisionally set the age of the mantle at 4.5 billion years. Whereas the age of the crust, or at least of the oldest part of the crust, can be set at anything younger than 4.5 billion years and older than 3.3 billion years. So, even if it will be possible to find the vestiges of still older crustal rocks in the rocks which were metamorphosed around 3.3 billion years ago, it seems we have here reached an upper limit for the "oldest crustal rocks". Paraphrasing the famous remark by James Hutton (1726–1797), one might say, that, at last, "we see the vestige of a beginning".

Even if these still older sediments have been affected everywhere on earth by younger metamorphosis, it might still be possible to find evidence for the existence of these older sediments by studying the minerals which make up the Kola Peninsula granites. It is quite possible that some minerals have survived the strong metamorphism accompanying either the orogeny at 3.3 billion years ago, or a later orogenetic period. As we will see in Chapter 10, such a situation is commonly found in the younger orogenies.

It follows that the 3.3 billion years age of the Kola Peninsula granites is a minimum age for the oldest crustal rocks. We know crustal rocks must have been present earlier in the history of the earth, but we have as yet no indication on how much earlier they might have been there already.

Turning now to the mantle, we have as yet no tangible physical knowledge of the mantle. None of our deepest drill holes has so far penetrated the crust of the earth. Both in the U.S.A. and in the U.S.S.R. there have been projects to drill several holes to the mantle. The boundary between crust and mantle, as

---

[1] This is, however, not a direct age determination. Instead it is based on the relative amounts of the two stable isotopes of strontium, $^{87}Sr/^{86}Sr$ (the first one radiogenic and derived from ^{87}Rb, the other non-radiogenic) contained in the St. Pauls rocks. Using an approach similar to that described in the next paragraph for the lead–lead method of dating the age of the earth, it is concluded that these ultrabasic rocks have remained chemically intact over the last 4.5 billion years. There is not enough radioactive material in these rocks to base a direct age determination on (HART, 1964).

registered by seismographs, has been termed the "Mohorovičić discontinuity", after the Serbian geophysicist who first drew attention to the existence of this boundary in the interior of the earth. In the U.S.A. this has been shortened to the "Moho", and the project to drill a hole to the mantle has consequently been christened the "Mohole". But nothing has been heard from the Russian attempts to drill through the crust, and in the U.S.A. the Mohole Project has been abandoned after suffering severe political vicissitudes.

At the the same time, this value sets a lower limit for the age of the earth. The surprising fact is that this lower limit comes very close to the figure actually given for the age of the earth itself (cf. Table III) which is also very generally put at 4.5 billion years. It will therefore be necessary to study the circumstantial evidence on which all figures for "the age of the earth" are based. All newer estimates use a certain refinement of the uranium- and thorium–lead method of absolute dating, which comes under the name of *lead–lead method*, and which is explained in the next sections.

15. REFINEMENTS OF THE URANIUM- AND THORIUM–LEAD METHOD IN ABSOLUTE DATING

The refinements of the uranium- and thorium–lead method could only be arrived at by the introduction of the mass-spectrometer. As follows from Table I and from Fig.4, both uranium and its isotope actinium, together with thorium, have lead as their stable end product. In the days of classical chemical analysis there was no way to distinguish between the isotopes of lead, and so it was not known what part of the lead found in an uranium ore vein was due to the uranium isotopes, or to the thorium, or even if perhaps part of this lead was not formed at all as an end product of a radioactive decay series. In other words, how much of this lead was non-radiogenic, normal lead.

As stated before, with the help of the mass-spectrometer we can now measure how much of this lead is ^{208}lead derived from ^{232}thorium, how much is ^{207}lead derived from ^{235}actinium, and how much is ^{206}lead, derived by decay from ^{238}uranium. Moreover, we now know from every lead sample analyzed by a mass spectrometer, what is its relative amount of the normal, non-radiogenic ^{204}lead isotope. In absolute dating, this has led to the development of a so-called lead–lead method, in which only the relative amount of two lead isotopes has to be measured in a mass spectrometer. It is based on the fact, established by numerous chemical analyses, that in nature the two isotopes of uranium, i.e., uranium proper and actinium uranium, always occur in the ratio 139:1. So instead of the actual amounts of uranium and actinium found as parent elements in a given sample, we can use 1/139 uranium for actinium. In dividing the amounts of the daughter isotopes, the actual amount of parent isotopes is then no longer material, and the age t can be solved directly from the formula:

$$\frac{^{206}\text{Pb}}{^{207}\text{Pb}} = \frac{(e^{\lambda t}-1) \cdot {}^{238}\text{U}}{(e^{\lambda_1 t}-1) \cdot 1/139 \, {}^{238}\text{U}}$$

where λ and λ_1 are the respective decay constants of ^{238}uranium and ^{235} actinium.

16. THE AGE OF THE EARTH

For an estimate of the age of the earth another lead–lead method has been employed. The main fact used in this method is that by far the largest amount of heavy elements, including both lead and uranium and thorium, occur in the earth's crust as widely disseminated trace elements in the various rocks. Concentrations of these minerals in ore veins form only an insignificant portion of their total volume.

The British geologist A. Holmes and the German nuclear physicist F. G. Houtermans have independently proposed a model to use this fact of the geo-chemical distribution of the heavy elements in the earth's crust to calculate the age of the earth. This method has since become known as the "*Holmes-Houtermans model*". It uses the isotopic composition of so-called *common lead*, the galena of lead ore veins.

Based on the disseminated distribution of lead and of uranium and thorium in the earth's crust, the assumption is made that concentration of lead into ore veins is a very rare event. Consequently, one may postulate that the lead deposited in an ore vein has been concentrated only once during its life time.

Such concentration of lead into ore veins separates the lead from its earlier environment, from the rocks in which it was originally disseminated as a trace element. Since the temperature and pressure under which lead is mobilized differ from, and are in fact much lower than, those under which uranium and thorium are mobilized, the lead will also become separated from the uranium and thorium with which it was formerly associated in the rocks before concentration. As a result, once lead has been concentrated into ore veins there will be no more addition of radiogenic lead. Its isotopic composition will remain constant from now on. Galenas therefore are fossilized in their composition at the time when their concentration started. They indicate the composition of the lead disseminated as a trace element in the contemporary parent rock.

In the parent rock the isotopic composition of the lead had slowly changed with time, because of the addition of the radiogenic lead isotopes 206, 207 and 208, resulting from the decay of uranium and thorium present in the same rock. Dependent upon the original amounts of lead and of uranium and thorium present in the parent rock, and on the time elapsed since the formation of the rock the

isotopic composition of common lead derived from various parent rocks will
therefore show considerable differences.

In order to correlate those of the differences which are due to original
variations in the relative amounts of lead and of uranium and thorium, these
differences can best be expressed in the index μ. This represents the quotient of
the amount of ^{238}U and of ^{204}Pb, as extrapolated to present time:

$$\mu = \left(\frac{^{238}U}{^{204}Pb}\right)_{present}$$

For each value of the index μ the isotopic composition of common lead will
follow a definite path with time.

In order to calculate the age of the earth, it now becomes necessary to know
the isotopic composition of primeval lead. One could, of course, do some wishful
thinking, and assume that all of the primeval lead had to be the non-radiogenic
isotope ^{204}Pb. This is, however, a false assumption, because interstellar matter
presumably contains natural radioactive elements. In confirmation of this, it is
found that meteorites contain lead in which the radiogenic isotopes are present.
It follows that the primeval lead of the earth must already have had a mixed
composition.

HOUTERMANS (1960) now assumed that lead of iron meteorites which are
practically free of uranium and thorium give the best approximation of the
composition of the primeval lead of the earth. None of the 206, 207 and 208 lead
isotopes present in such meteorites could have derived from radiogenic decay
during the life time of that particular meteorite. The best analysis was supplied
by the Canyon Diablo meteorite, which has consequently been chosen to represent
the isotopic composition of the primeval lead of the earth.

In Fig.10 the growth lines of common lead of various values of the index μ
are drawn. The values actually found for μ vary from 9 to 11, with a sharp maximum at 10. Taking the Canyon Diablo meteorite as a starting point, isochrones
can be constructed. Their bisection with the growth lines for the index μ indicates
the composition of common lead separated from its parent rock, at that time.
The growth lines are limited by the isochron w, which is the oldest isochron possible, which as such indicates the age of the earth. The slope of this initial isochron
corresponds with a time interval of:

$$w = 4.49 \cdot 10^9 \text{ years}$$

and this is the value that leads to the generally accepted "age" of the earth of
4.5 billion years.

The basic assumption underlying the Holmes-Houtermans model, e.g. that
lead concentration has been a singular event, has not always found to be true.
In detail rather curious results have been obtained by this method, when the

geological history of a particular galena was not well known. But the rather critical attitude sometimes expressed toward the whole Holmes-Houtermans concept seems to me to go too far. Statistically there seems to be ample ground for the thesis of the separate and parallel evolution of common lead in various places on earth.

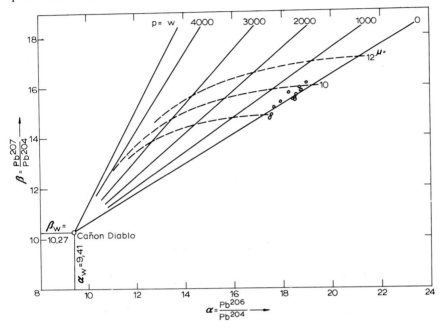

Fig.10. The history of common lead, according to the Holmes-Houtermans model. Fully drawn lines are isochrones in millions of years; dashed lines indicate the evolution with time of the isotopic composition of common leads of various indexes, dependent upon the original ratio of uranium and non-radiogenic lead in the parent rock. The w-isochron corresponds to 4.5 ± 0.3 billion years. (From HOUTERMANS, 1960.)

However, although the fundamental value of the Holmes-Houtermans model for estimating the age of the earth seems well established, slight changes in the numerical values employed in the calculations, may well have their effect at the ultimate value for the age of the earth arrived at. TILTON and STEIGER (1965), using lead of granites and galenas 2,700 million years old, in which later contaminations should be smaller, have, in fact, already calculated a value of 4,750 million years for the age of the earth. ULRYCH (1967), using oceanic basalts, even arrived at an age of $5,430\pm40$ million years, a figure which is, however, still in need of corroboration. So it seems well possible that the age of the earth could be somewhat higher than the 4.5 billion years commonly accepted. This would leave us some time for mantle rocks to crystallize and relieve us of the apparent paradox that the earth should be just as old as the rocks of the mantle. But on the strength of our present data, it seems not warranted to assume an age of the earth of more than 5 billion years.

This figure leads us to the realization of the amazingly quick pace of the early evolution of the earth. With the age of the earth set at not exceeding 5 billion years, with the age of the mantle at 4.5 billion years and the oldest part of the crust at over 3.3 billion years, we find that the earth has had a relatively rapid evolution during its early days (or better: during its early millions of years). This was followed by a sort of steady state of successive orogenetic cycles, without any major change, at least over the last 3.3 billion years of the history of the earth.

REFERENCES

ANONYMOUS, 1964. The Geological Society Phanerozoic time-scale. *Quart. J. Geol. Soc. London*, 120 s:260–262.

DEUTSCH, S., HIRSCHBERG, D. and PICCIOTTO, E., 1956. Étude quantitative des halos pléchroïques. *Bull. Soc. Belge Géol., Paléontol. Hydrol.*, 65:267–281.

FAUL, H., 1966. *Ages of Rocks, Planets and Stars*. McGraw-Hill, New York, N. Y., 109 pp.

HARLAND, W. B., SMITH, A. G. and WILCOCK, B., 1964. A symposium dedicated to Professor Arthur Holmes. *Quart. J. Geol. Soc. London*, 120 s:458 pp.

HART, R. S., 1964. Ultramafic rocks of St. Paul's islands. *Ann. Rept. Terrestrial Magnetism—* Carnegie Inst., Washington, D. C., 63:330–331.

HOLMES, A., 1947. The construction of a geological time-scale. *Trans. Geol. Soc. Glasgow*, 21:117–152.

HOLMES, A., 1959. A revised geological time-scale. *Trans. Edinburgh Geol. Soc.*, 17, III:183–216.

HOLMES, A., 1963. Introduction. In: K. RANKAMA (Editor), *The Precambrian*. Interscience, New York, N. Y., 1:XI–XXIV.

HOUTERMANS, F. G., 1960. Die Blei-Methoden der geologischen Altersbestimmung. *Geol. Rundschau*, 49:168–196.

KULP, J. L., 1960. The geological time-scale. *Intern. Geol. Congr., 21st, Copenhagen, Rept. Session, Norden*, 3:18–27.

KULP, J. L., 1961. Geologic time scale. *Science*, 133:1105–1114.

MELSON, W. G., JAROSEWICH, E., BOWEN, V. T. and THOMPSON, G., 1967. St. Peter and St. Paul Rocks: A high-temperature mantle-derived intrusion. *Science*, 155:1532–1534.

RANKAMA, K., 1954. *Isotope Geology*. McGraw-Hill, New York, N. Y., 535 pp.

RANKAMA, K., 1963. *Progress in Isotope Geology*. Interscience, New York, N. Y., 705 pp.

RUSSEL, R. D. and FARQUAR, R. M., 1960. *Lead Isotopes in Geology*. Interscience, New York, N. Y., 243 pp.

STUBBS, P., 1965. The oldest rocks in the world. *New Scientist*, 25:82.

TILTON, G. R. and STEIGER, R. H., 1965. Lead isotopes and the age of the earth. *Science*, 150:1805–1808.

ULRYCH, T. J., 1967. Oceanic basalt leads; a new interpretation and an independent age for the earth. *Science*, 158:252–256.

ZEUNER, F. E., 1958. *Dating the Past. An Introduction to Geochronology*, 4th ed. Methuen, London, 516 pp.

Chapter 4 | The Biological Approach

1. INTRODUCTION

We have seen in Chapter 2 how the principle of actualism is the basic philosophy underlying the reconstruction of the history of the earth from the data preserved in the rocks. Within certain limitations, actualism teaches us to accept extrapolation of actual processes into the geological past, even over time spans of millions and billions of years. So we must now first look into the results of biological studies of life at present. We must learn what characterizes life, what we may call living and what we do consider to be non-living, what actual processes are at work in living things, or, in other words, what is the metabolism of present life.

As stated already in the first chapter, it was biologists who have been most active in kindling the renewed interest in a possible origin of life through natural causes, without the intervention of divine creation. The pioneer in this movement was the Russian biochemist A. I. Oparin, who in 1924, in the first edition of the book which later became more widely known through its English translations under the title of *The Origin of Life on Earth* (OPARIN, 1938; see also OPARIN, 1961, 1964) already advanced the theoretical framework of our present ideas. Quite independently the British biophysicist J. B. S. Haldane became convinced that the anaerobic way of living was more primitive than the aerobic, which led him to conclusions which were quite similar to those of Oparin. These views found their integration in a pithy little book by the British physicist BERNAL (1951), who in his *The Physical Basis of Life* firmly anchored the biological views unto a solid basis of physics and chemistry.

However, the views formulated by the investigators mentioned above remained more or less in a rarefied atmosphere. Other investigators were interested, but no real participation was forthcoming. For one thing, the second world war interrupted the development of these ideas because during this period people were more bent on the efficient destruction of life than on thinking about its

possible origin. But in part this singular reticence may well have been brought
about by the general disinclination to tinker with the problem of the origin of
life, a disinclination referred to already in the first chapter.

The turning point in this attitude was, I believe, brought about by the first
symposium of the newly formed International Union of Biochemistry, which
was held in Moscow in 1957 (OPARIN, 1959). This symposium was devoted to the
problems relating to a possible origin of life through natural causes, as seen against
a broad background of the various biological sciences, physics, chemistry, geology
and astronomy. Not only were these problems at that date disseminated through
a much wider group of scientists, but it became clear that it must be possible to
develop experimental checks on the theories proposed.

The next step was the first realization of such an experimental check. In 1959
MILLER was able to synthesize "organic" molecules in an inorganic way, using
an environment specially designed to simulate the primeval atmosphere such as
postulated by the theories mentioned above, and about which we will come to
speak in Chapter 6.

It is true that earlier research had already led to similar results, but this
implies reactions studied during normal chemistry research, whilst the implications
in regard to the origin of life had never been realised by these earlier scientists.
The experiments of Miller, expressly set up to test these problems, therefore justly
received publicity.

This first experimental break-through aroused wide-spread interest, and
many other scientists began experimenting on similar lines. A second conference
in Moscow, on the occasion of the 5th Congress of the International Union for
Biochemistry (OPARIN, 1963) already showed a tremendous development, which
has proceeded at full speed.

Every detailed description of the state of the art will be outmoded, before
it comes into print. In Chapter 6 I will therefore confine myself to sketching the
general lines of this development of the experimental checks, which is mainly in
the hands of biochemists, chemists and "molecular" biologists. In 1965–1966 the
field was well reviewed by the proceedings of three conferences, reported respectively by BRYSON and VOGEL (1965); by FOX (1965), and by MAMIKUNIAN and
BRIGGS (1966). The last mentioned volume may be specially commended for its
full and detailed bibliography. Still newer reviews are those by PONNAMPERUMA
and GABEL (1969) and CALVIN (1969). J. M. Buchanan ("Chairman's remarks"
in FOX, 1965, p.101) summarized the results in this field as follows:

"The efforts of several groups over the past decade have shown that a large
number of compounds of biological interest may be synthesized from the simple
precursors, methane, ammonia, hydrogen and water, under prebiological conditions. This list of compounds includes organic acids, purines, pyrimidines, and
many of the amino acids...

We now have many, if not all of the compounds of small molecular weight that are the components of macromolecules... We must conclude that these compounds were formed because they are plausible reaction products thermodynamically and kinetically...

Investigators... are now confronted with the difficult problem of demonstrating how functional macromolecules could have arisen. The aggregation of low molecular weight units into biologically active macromolecules has in fact been demonstrated.[1] But the formation of biologically active units introduces a complexity of considerable magnitude. Yet short of supernatural intervention, we know that this must have happened—not by the chance appearance of a highly specialized enzyme, but by the evolution of macromolecules with reactivities that are both chemically advantageous and biologically reinforcing."

These experiments have become both very varied and sophisticated. It would interrupt the story too much to relate them in some detail in this chapter. They are therefore treated in Chapter 6, together with some basic formula of biochemistry. Chapter 6 consequently acquires a more technical flavour, whereas in this chapter the more general narrative is given.

2. NON-LIVING AND LIVING IN BIOLOGY

One of the main difficulties biologists have to cope with is the ultimate distinction between non-living and living. This difficulty does not so strongly arise in the higher organized life forms like man, or the animals and higher plants. No doubts are felt about living or dead in relation, for example, to the obituary column of our daily papers, although even here the exact definition of the time of death plays a major role in the fuss about transplantations. But in the so-called lower reaches of living matter difficulties crop up when one tries to distinguish non-living from living. Here it becomes difficult to draw an exact boundary between the more lowly, unicellular or non-cellular organisms on the one hand, and big, non-living molecules on the other. Or between extremely simple systems of metabolism and reproduction, which are very similar to chemical reactions, but still are performed by living things, and complicated chemical reactions between very large molecules, which, for a variety of reasons, still have to be considered non-living.

In these borderline cases it is impossible to separate by a strict definition the non-living from the living. For instance, even the fact that all living organisms contain protein, whilst part of their metabolism is based on a protein cycle, does not give us a watertight definition, although it forms an easy schematization. We can, however, think of possible life forms not based on protein. Or, more to

---

[1] This alludes to the results of S. W. Fox and collaborators, reviewed in Chapter 6.

the point, protein might be synthesized inorganically, as follows from the experiments of S. W. Fox and collaborators, reviewed in section 6 of Chapter 6.

We must, however, realize that this biological distinction between non-living and living requires a knowledge of the functions of life which will never be solved from the geological record. As will be indicated more fully in section 2 of Chapter 12, the geologist never studies the life he describes. He only finds its remnants, not only dead but fossilized. Of course, the biologist, and the biochemist in particular, normally also use dead material, such as sections from dead organisms, or dead cells, or even extracts from cells expressly broken up by laboratory techniques. In a way, using dead remains to study life, he is, as WINKLER (1960) once pointed out, like the drunk who has lost his latchkey in the dark before his house door, but is searching for it under the lamp-post, because there he can see better. But although the biologist uses dead remains to study life, he has himself killed the living organisms a few moments before and by techniques known to him and selected to provide the least distorted picture. In fossils, on the other hand, the material of the former living organisms is replaced by "stone", that is by minerals, whereas even if some of the original organic material has been preserved, it has been thoroughly degraded and transformed. When this replacement has proceeded along orderly lines, molecule by molecule, the original structure of the organisms can be preserved in its minutest microscopic detail. This is the best the geologist can hope for. As we will see in Chapter 12 the geologist finds himself confronted with quite another set of difficulties in distinguishing "fossilized" non-living matter from fossilized remains of life.

For our subject we may therefore follow the most inclusive definition of life proposed by the biologists and define life as "macromolecular, hierarchically organized and characterized by replication, metabolic turnover and a regulation of its energy flow which is more or less independent of its surroundings" (GROBSTEIN, 1965).

3. CHEMICAL UNIFORMITY AND MORPHOLOGICAL DIVERSITY OF PRESENT MODE OF LIFE

From the biological studies of present life, one may draw as a salient conclusion the antithesis between the immense variation of its morphological expression and the relatively small number of chemical reactions on which it is based.

Morphologically there is an enormous number of separate forms of life, of species, genera, families and the higher systematic categories of microbes, plants and animals. Biochemically, in contrast, all present-day life in all its variations, is based on nucleic acids, proteins, carbohydrates and fats and on some minor compounds such as phosphoric esters. Although showing great variation in detail, these compounds are all interrelated and built upon a few scores only of basic biochemical reactions.

This basic biochemical unity, first stressed by Kluyver (cf. KLUYVER and DONKER, 1926; see also POSTGATE, 1968) is one of the main characteristics of the present mode of life. All plants and animals, whether marine or continental, from plankton to whale and from virus to elephant, whether aerobic or anaerobic, all of these diverse forms of life are based on this surprisingly small number of basic organic compounds.

The pun is often heard that nature is so organized that any living thing forms part of nature's food chain, that is, that it can be fed upon by other living things. This is due to the fact that a relatively small number of organic compounds is used in building up terrestrial life. So there is, always, in any living organism, something digestible for some other living organism. The pun rests, consequently, on the biochemical similarity of life. This can be interpreted, in a more scientific way, as indicating that all present life is somehow related. Which means, ultimately, that all of life has a common origin.

This does not mean to indicate that the chemistry of life on earth is not diversified, but only that it is not as diversified as it could have been. In fact, though life has used part of the inorganic world only, it has done so in an extremely subtle way, exploiting to full advantage the properties of the elements which go into its making. It is so well adapted, that there exists a unique relationship between life and its substratum, which again and again fills biochemists with wonder. A full tabulation of this adaption is found in a book by NEEDHAM (1965), aptly titled: *The Uniqueness of Biological Material*[1].

This unique relationship has led PIRIE (1966) to a renewed examination of the old question whether life and matter are not but two aspects of the same thing, or whether they are two independent entities, the one adapted to the other in such a remarkable way. The first viewpoint, which implies either that both life and matter have been designed together, or that organisms and matter could interact upon each other, has been most eloquently held by the philosopher Plotinus. The second viewpoint has been upheld facetiously, but not less eloquently, by Voltaire's philosopher Pangloss in *Candide*.

According to PIRIE (1966), adherence to the first view would lead to the necessity of overthrowing all our accepted scientific canons, a step so radical that it has as yet not been forced upon us by the observations. This leaves us the second view, which does not have to mean that the adaption of life, remarkable as it is, should already be at its acme. We may not yet live in Pangloss's best of all possible worlds, but life certainly has made a wonderful attempt towards it.

---

[1] There are, of course, deviations from the main stream, some of which have been discussed in the meeting on "Anomalous aspects of biochemistry of possible significance in discussing the origins and distribution of life", organized by PIRIE (1968).

As for this uniqueness of life in relation to its environment, we must realize that life has had all the time of the billions of years of organic evolution, to adapt itself in the main to the prevalent environment on earth. The overwhelming majority of life does at present use an environment which is rather narrowly bordered. That is, around 1 atm pressure, a salinity from 4% to zero, a temperature range from 0 to 40 °C. But even today, in exceptional environments, present life is able to withstand a much greater variety of environmental factors. This is demonstrated in Table IV, in which the minimum and maximum values for a number of such factors are given. Even many biologists might find this table revealing, not having realized that the possible range of present life is so wide.

TABLE IV

MINIMUM AND MAXIMUM VALUES OF SEVERAL ENVIRONMENTAL FACTORS OF PRESENT-DAY LIFE (From VALLENTYNE, 1965)

Factor	Lower limit	Upper limit
Temperature	-18 °C (fungi, bacteria)	104 °C (sulfate-reducing bacteria under 1,000-atm hydrostatic pressure
Eh	-450 mV at pH 9.5 (sulfate-reducing bacteria	$+850$ mV at pH 3 (iron bacteria)
pH	0 (*Acontium velatum*, fungus D, *Thiobacillus thiooxidans*)	13? (*Plectonema nostocorum*)
Hydrostatic pressure	essentially 0	1,400 atm (deep-sea bacteria)
Salinity	double-distilled water (heterotrophic bacteria)	saturated brines (*Dunaliella*, halophilic bacteria, etc.)
a_w*	0.65–0.70 (*Aspergillus glaucus*)	essentially 1.0

* a_w (activity of water) $= p/p_0$, where p is the vapor pressure of water in the material under study and p_0 is the vapor pressure of pure water at the same temperature.

The chemical compounds of present-day life together form the original subject of organic chemistry, as opposed to inorganic chemistry. If we now look more closely into these natural organic compounds, such as proteins, carbohydrates, fats, nucleic acids and some others, we find that they are mainly formed from the elements C, O, H, N and P. Typically they form very large molecules of an intricate structure, which is now in the process of being unravelled. The exact structure of these molecules is, however, less important for our subject, and it suffices if we keep in mind that most natural organic compounds are formed by large and complicated molecules.

4. IMPOSSIBILITY OF SYNTHESIS OF ORGANIC COMPOUNDS
THROUGH NATURAL CAUSES UNDER THE PRESENT ATMOSPHERE

The conclusion reached in the preceding section is of decisive importance for any evaluation of the origin of life. On the one hand the synthesis through natural inorganic processes of such large, complicated molecules happens to be wellnigh impossible under present environmental conditions. Whereas, on the other, such organic molecules are moreover unstable under present conditions of light, temperature and composition of the atmosphere.

Even if, by some extremely unlikely coincidence, such an organic molecule had been formed, it would be destroyed almost immediately either by inorganic oxidation, or by organic oxidation processes, such as rotting. So these large organic molecules cannot, at present, exist on their own, without a relation towards, or, to put it more strictly, without being incorporated in, living organisms. They do not form normally, nor even rarely, in natural inorganic chemistry, and if this happens, they are liable to immediate destruction. This statement does, of course, do no more than paraphrase the wide gulf that at present exists between organic and inorganic compounds. A gulf so wide that until it had been bridged by the synthesis of urea in 1824, it even seemed as if man-made chemistry would forever be incapable to synthesize organic molecules, an aptitude which up to that time was thought to be reserved exclusively to life processes.

In present-day chemistry this fundamental difference between inorganic and organic has faded away, owing to the extremely varied organic compounds which can now be synthesized. Very schematically, the organic chemist of these days occupies himself with compounds based on the carbon atom, the inorganic chemist with other compounds. Still, all these new compounds, whether dubbed organic or inorganic, or belonging to transitionary groups, and although made in vitro, in a laboratory or in a factory, are products of life also, i.e. of a chemist's brain.

In this book, in which we are concerned with the distinction of compounds formed either by life or by non-life in earlier days of the earth's history, when neither laboratories nor factories existed, it seems right to maintain the original distinction between organic and inorganic. Moreover, being personally indoctrinated with the rules of biological taxonomy, which favour the retention of the earliest meaning of a given term, I even think this is the only right thing to do. *Organic* then means compounds produced by life. It is the equivalent of terms such as "biogenic" or "biologically produced". *Organic processes* is used for life processes. *Inorganic* means compounds at present produced by non-life, or inanimate, processes, whereas it also applies to those processes leading to the formation of these compounds. Synonyms widely used are "abiogenic" and "abiologically produced".

For those compounds which at present purely belong to the organic world, but which, it is thought, were formerly synthesized inorganically under a primeval anoxygenic atmosphere the term *organic* is used. *Pseudoorganic* would be a possible synonym for this group of compounds, which is perhaps more euphonic and therefore preferable in lectures.

Returning from this by-path in semantics, we have to retain at this point that organic compounds at present can only be formed in nature by living matter already in existence. It follows that the origin of life from inorganic beginnings is at present impossible, because only living matter can synthesize organic compounds. *Only living matter can at present produce other living matter.*

5. THE OXYGENIC ATMOSPHERE OF THE PRESENT

The crucial point lies not, however, in the fact that the origin of natural organic compounds outside life is altogether impossible, but only that it is impossible under the present environmental circumstances. The most essential of these are all based on the fact that our atmosphere at present contains an appreciable amount of free oxygen, in other words that it is *oxygenic*.

The free oxygen of our atmosphere, and the free oxygen dissolved in our hydrosphere, leads to the oxidation processes mentioned in the preceding section, which will destroy any organic material, which is not protected, in one way or another, for instance by membranes of living matter. Free oxygen, moreover, enables animals and most plants to breathe, which is one of their important life processes. This aspect of the free oxygen of our atmosphere lies at the basis for poetic quotations about the "life-giving" oxygen. A third result of the free oxygen of our atmosphere is that through it we are shielded from most of the ultraviolet rays of the sun. In Chapter 15 we will see in more detail that this rests mainly on the fact that the free oxygen in the higher reaches of the atmosphere forms a thin layer of ozone, and that oxygen and ozone together trap the lethal part of the ultraviolet sunlight. This is important because, although a small amount of ultraviolet rays can be sustained, and will even produce a healthy sunburn, the full spectrum of the ultraviolet rays, in the intensity in which it is produced by the sun, is more deadly to present-day life than any radiation due to natural radioactive decay.

Of the three effects of free atmospheric oxygen, those of oxidation and rotting, of breathing and of the shielding from the ultraviolet sun rays, only the latter is a conditio sine qua non for life on earth. As to the first effect, we have seen already how life, being able to maintain an energy flow of its own, different from its surroundings, effectively shields its organic material from oxidation and rotting, which only make their debut after the death of the organism in question. Breathing, on the other hand, is only a requisite of part of life. We know a wide variety of microbes, which have quite a variation of metabolic cycles, in none of which breathing of free oxygen plays a part. Free oxygen is even deadly to most of them, and since our atmosphere contains free oxygen in large quantities, they can only subsist when they are so well shut off from the atmosphere that they are able to reduce the small amounts of oxygen which leak through. So there is at present life on earth which lives and propagates under the exclusion of free oxygen. Because this is at present only possible under exclusion from the air, this part of life is called *anaerobic* in contrast to the *aerobic* organisms, which live in free contact with the oxygen contained in the atmosphere and the hydrosphere.

However, both aerobic and anaerobic organisms would be killed off by the shorter ultraviolet sun rays, if these were not absorbed in the atmosphere by the combined effects of free oxygen and ozone. Therefore even the anaerobic life we know now would have difficulty in surviving in the absence of free atmospheric oxygen. Such an atmosphere would very well suit their metabolic processes, but they would not survive the lethal effect of the ultraviolet sun rays. Only when shielded by water or rock, that is in lakes or oceans, or in the pores of the soil, could early life similar to our present-day anaerobic organisms survive in such an environment.

The impossibility of an adequate synthesis of organic compounds through inorganic processes in our present-day environment has been fully portrayed by the French biologist and philosopher LECOMPTE DE NOÜY (1947). But he made the mistake often made when philosophy is applied to science, to think that in disproving one thing one proves the other. He disproved a natural evolution of life at present, and thought he had thus proved that life originated through creation.

The wide following he enjoyed rests in part on the distrust of church members against scientists "trying to do away with creation", cited already in Chapter 1. Here at last we had an honoured scientist, a biologist and member of the French Academy of Sciences, who lucidly proved how science taught us that life could not have been but created. But Lecompte de Noüy has not done more than prove that life could not have evolved from non-living under the present atmosphere. This does not constitute, however, proof in favour of creation. He seems to have been totally unaware of the possibilities offered by an anoxygenic primeval atmosphere.

6. ANOXYGENIC PRIMEVAL ATMOSPHERE

Consequently all present-day theories about a natural origin of life on earth, theories which all go back to the original ideas of Oparin, postulate an early or primeval atmosphere of reducing character, in which there is no, or almost no, free oxygen.

Oxygen will at that time only have been present in chemical compounds, of which water is thought to have been the most important. Apart from water this early atmosphere, and also the accompanying hydrosphere, the contempora neous oceans, lakes and rivers, will have carried carbon and nitrogen, and a host of other elements. Amongst the latter sulphur and phosphorus may have been important as catalysts in early energy cycles. The main difference, and the only important one at that, lies in the fact that there could not have been present, in that primeval atmosphere, an appreciable amount of free oxygen.

As an example of what is thought could have been the prevalent constituents of the atmosphere, the hydrosphere and the lithosphere at that time, Table V, taken from BERNAL (1961), is presented. But views on the relative abundance of the elements and their compounds present in the early atmosphere and hydrosphere vary strongly with the various authors who have presented hypothetic models on this matter. This will be further discussed in the next chapter.

TABLE V

THE PREVALENT CONSTITUENTS OF ATMOSPHERE, HYDROSPHERE AND LITHOSPHERE AT THE TIME OF THE PRIMEVAL ATMOSPHERE (From BERNAL, 1961)

Atmosphere	Hydrosphere		Lithosphere
CO_2 (or CH_4)	H_2O water		SiO_2 sand
N_2	NH_4HCO_3		$AlSiO(OH)Fe(OH)_2$ clay
NH_3 ⎫	H_2S ⎫		$CaCO_3$ inorganically precipitated
H_2S ⎬ very little	$NaCl$ ⎬ low concentrations		limestones
H_2O ⎭	KCl ⎪		
	KH_2PO_4 ⎭		

7. INORGANIC SYNTHESIS OF "ORGANIC" COMPOUNDS IN THE PRIMEVAL ATMOSPHERE

For our subject the important thing, however, is not the exact composition of the primeval atmosphere and hydrosphere. Important is the absence of free oxygen and the presence of inorganically formed relatively simple compounds of carbon, C, with elements such as O, H, N, S, P and others (see Fig.11), which now occur exclusively as natural organic compounds, together with other inorganically formed compounds such as CO, CO_2, H_2O, SiO_2, of silicates, sulphates and the like, which in our present-day environment are still formed inorganically.

The presence of these compounds results in a twofold way from the absence

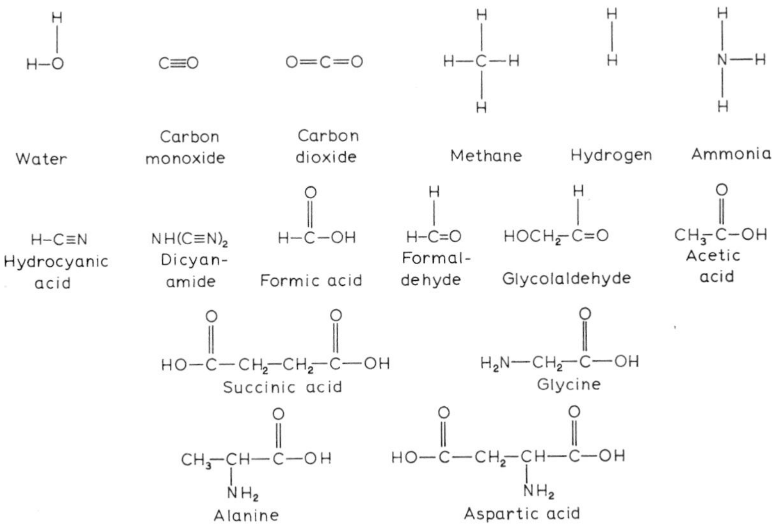

Fig.11. Schematic presentation of the smaller molecules which must be present in any stellar or planetary primeval hydrosphere, and how they combine to "organic" larger molecules, which in their turn are the building blocks of the larger molecules of life. (From CALVIN, 1965. For more information, consult Chapter 6.)

of free oxygen. First, these compounds can be built up by processes of inorganic chemistry in such an anoxygenic atmosphere, whereas under the present atmosphere only life processes supply the requisite energy for the formation of these compounds. And second, these compounds will be stable under an anoxygenic atmosphere, or they will be at least so stable that they are not destroyed at a rate comparable to that of their formation. This is due to the first effect of the presence of an atmosphere without free oxygen mentioned in the preceding section, i.e. the absence of present-day oxidation and decomposition processes.

The reason why in an anoxygenic atmosphere simple "organic" compounds can be formed by processes of inorganic chemistry lies in the absence of the shield of oxygen and ozone which in our atmosphere absorbs the shorter ultraviolet sun rays. In an anoxygenic atmosphere these rays consequently can freely reach the surface of the earth. Light of such short wavelength is of very high energy, which permits inorganic photochemical reactions which are not possible now, since that part of the sunlight which gets through the oxygenic atmosphere does not have a high enough energy. The shorter ultraviolet radiation of the sun is so rich in energy that it will exite elements of atmosphere and hydrosphere to form chemical bonds. That is, to form molecular compounds through the absorption of the energy of light quanta (see Fig.12). This is a thoroughly inorganic process, but as we will see in Chapter 6, it may produce typically "organic" molecules. Since the process is dependent upon the absorption of light quanta from the sun

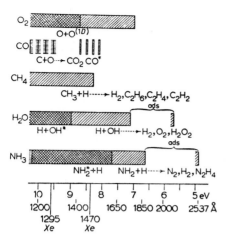

Fig.12. Spectral ranges of photochemical activity for gases. Oxygen is seen to absorb the ultra-violet rays from a wavelength of 1,850 Å downwards. The energy values of the light quanta, expressed in electronvolts, rises from about 7 eV for 1,800 Å to 10 eV for light of 1,200 Å. In this same region other gases, such as CO and CO_2, CH_4, H_2O and NH_3 are also chemically active, as is indicated by the hatching and cross hatching of the respective bars. The asterisks indicate at what wavelength the radicals in question start emitting light, a feature without significance for the synthesis of the molecular compounds, but important for their recognition in the experiment. The dashed arrows indicate the spectral ranges at which composite molecules, formed from atoms and radicals, are built up. The two xenon lines of 1,295 Å and 1,470 Å indicate the strong monochromatic radiation of the xenon lamp which Groth used as early as 1938 in experiments of this kind.

It follows that our present atmosphere is opaque in the wavelengths in which simple stable gases such as CO, CH_4, H_2O and NH_3 are chemically active. They cannot now, consequently, form compounds by using inorganic photosynthesis, because the sunlight of these wavelengths does not reach the surface of the earth. Hydrogen, on the other hand, is quite transparent for light of these wavelengths, so in an atmosphere composed principally of hydrogen these rays of the sunlight could penetrate through the atmosphere to the surface of the earth. (From TERENIN, 1959.)

rays it is a form of photosynthesis. But it is *inorganic photosynthesis*. As such it must not be confused with the organic photosynthesis which at present takes place in plants, mainly through the catalytic action of chlorophyll. This organic photosynthesis, which mainly uses light quanta from the red part of the sunlight, is able to form the same compounds synthesized inorganically by the shorter ultra-violet rays, because in plants the process is broken up into a large number of consecutive steps. Each step in itself requires much less energy than the process as a whole, and in this way plants are able to synthesize the same organic compounds, although they only use the much lower energy supplied by the red part of the sunlight.

Just as with the composition of the early atmosphere, for which, as mentioned in the preceding section, we have as yet no clear understanding of its exact composition, we also have no idea of the exact nature, nor of the absolute or the relative abundance, of the simple molecular compounds of carbon and other elements formed by inorganic processes in and under an anoxygenic atmosphere. From the side of the organic chemists in particular (cf. CALVIN, 1965), the presence of the highly reactive compounds of hydrogen cyanide or prussic acid and of hydrogen dicyanamide is postulated. The mixture of water with these compounds in the early hydrosphere is often referred to as the "thin soup", which obviates the necessity of making too strict an estimate of its composition.

From the literature we may gather that there is some difficulty in finding a good name for this forerunner of the present hydrosphere, characterized by its content of "organic" melocules formed through purely inorganic, or—if one likes that word better—through abiogenic processes. The reason why it has been widely compared to a soup is that these small "organic" molecules had nutritional value for the life which developed. But it is too colloquial a term for other authors, who have called it a "pre-biological", or a "probiological" or a "nutrient" soup. OPARIN (1964) has had his Russian term translated into "primeval broth", which seems to be on the heavy side. I believe that the original indication of "thin soup" has by now become so generally known, amongst the wider public too, that we may well continue to use it. The quotation marks will indicate that it is, as everybody knows, a very special type of soup.

How special its characters are will be set forth in more detail in the next chapters. At this point we might quote the philosopher HAWKINS (1964, p.270), who was so impressed by the fact that in this primeval ocean life found everything provided for by the previous stage of purely chemical development, that he was reminded of paradise, and called it "a sort of aquatic garden of Eden".

As we will see in Chapter 16, there are reasons to believe that the chemical reactions leading to the formation of the "thin soup" extended well into the time when early life had already been developed. Early life might have thus co-existed with pre-life, and enjoyed the privileges of this primeval garden of Eden, being able to browse on the "organic" molecules prepared for it by chemical inorganic processes, for a period of the order of two billion years.

8. GENERATIO SPONTANEA

In the literature the inorganic photosynthesis of "organic" compounds has sometimes been called a *generatio spontanea*. But generatio spontanea has for a long time been a sort of catch-all for any mode of origin of life, which was either not understood, or not possible to study, or something akin to, or just the opposite of, creation. At present, now that the possible modes of an origin of life through natural causes form the subject of intense scientific research, it seems better to drop the term of generatio spontanea altogether,

In this way we also get around discussions of the type such as: "Do you think generatio spontanea has occurred only once, or many times?" This is the type of discussion which has had its place in the era of the philosophical speculations on the origin of life. At that time it was quite feasible to ask the question —evolution having been proved by the paleontologists—whether life could have been created supernaturally only once, or several, or many times. That is, how many times were needed, before life created as such could further evolve on its own along natural lines.

Such discussions obscure the field of vision when the term generatio spontanea is applied to the processes studied now. The expression *inorganic synthesis of "organic" compounds* may be more cumbersome, but is states exactly what we are looking for[1]. It is not burdened with any metaphysical meaning, and it is therefore to be preferred.

It must be made quite clear, and if, perhaps, this statement is repeated too often, this is but the necessity of "frapper, frapper toujours", it must be quite clear that these processes have nothing divine, nothing extranatural, or mystic, or vitalistic, or even spurious. Once the physicochemical values of the environment are of the proper level, these processes just have to happen. This has by now been amply proved by the many and varied experimental checks described in Chapter 6.

To realize the normal character of these processes, we must be aware that they are of the same basic kind as all inorganic processes, those of the present day included. Of course, the physicochemical environment under a primeval atmosphere was different from that at present, and hence the products formed were different. But the nature of the processes involved in their formation was the same as, for instance, today in the formation of clouds in the skies, the deposition of salt crystals in lagoons drying out under the energy of the sun rays, or the rusting of iron.

Rain drops, snow flakes and hail will generate spontaneously in present day clouds, once the proper physicochemical level is reached, just as crystals of gypsum and salt will fall out spontaneously on the bottom of any lagoon, once there has been enough evaporation of sea water and iron will rust when not protected by a coat of paint. We are not at all curious whether a rain drop or a salt crystal will generate spontaneously only once, or if they will do so twice or many times, because we well know that they will just form in any number, when the environmental conditions have reached the proper values. Instead, we are interested in what these conditions are. Moreover, we do not only want to know the overall factors, but we are also interested in smaller details which govern this process. Such as the influence of dust in the atmosphere on the condensation of water

---

[1] It forms the first step of what BERNAL (cf. 1959) has called *biopoesis*, and what is nowadays often called *biogenesis* (see Chapter 6).

droplets, or that of the muddy bottom of a lagoon on the crystallization of gypsum and salt. We are quite indifferent to the question whether rain drops or salt crystals will generate spontaneously one hundred thousand or one hundred thousand and one times, but we want to know when and how they form. Will water drops of smaller or larger dimensions be formed? Will we find separate zones of crystals of gypsum and salt, or will we find a mixture of both? Such is the type of questions we are interested in, with respect to these inorganic processes.

To develop this parable a little further, we also want to have information on how the products once formed through these inorganic processes will fare with time. Will their life span be short or long? Will they remain separate, or will they combine with other compounds? Will, to take the example of a cloud, the water drops or the snow flakes evaporate again in the skies, or will they fall down on land or water? Will the crystals of gypsum and salt, to take the example of the lagoon, dissolve at the next high tide? Or perhaps only during the next monsoon or hurricane? Or what might be the conditions through which they may become buried by mud, and thus become preserved over much longer time spans?

So it is the physicochemical conditions governing such inorganic processes we are partial to. The conditions leading to the formation of certain products. And the variation of the physicochemical characters of such products with variations in the environment. And the later history of these products. That is the sort of environment in which they are either destroyed or preserved. But once the environment is of the right character, these processes will just happen, not once or twice, but countless times.

Paraphrasing these ideas in regard to our problem, the origin of life, we should like to unravel the nature of the conditions governing the processes leading to the accumulation or "organic" matter. We are of course keen on the question, whether such processes would lead to the formation of much, or very much, or very little of such material, but not on actual numbers. So strong has, however, been the stormy development of research into this problem over the last decade, that we do not stop any more at these basic questions. We are no longer interested solely in the possibility of a formation of such "organic" materials, but also in the processes leading to a concentration of such matter, to its preservation, to a possible separation according to their chemical constitution, and, finally, in the processes leading to the transition from such pre-life to life itself, and in the influence the transition from the primeval to the present atmosphere must have had on this evolution.

But, to return to the case in point, that is the misuse of the term generatio spontanea, there is no difference in the basic nature of the processes involved when compared with other inorganic processes. Moreover, just as in the case of the rain drop or the salt crystal, the nature of the processes involved during their later history does not differ basically from those which went into their making. In the example of the rain drop, the conditions of the environment may have changed, but there is no basic difference in the process of its condensation, that is in its *spontaneous generation*, or in its possible subsequent evaporation, or *spontaneous dissipation*, or even in its break up when it happens to strike the surface of the sea or the land. In the same way there is no basic difference in the nature of the processes leading to inorganic formation of "organic" material under the primeval atmosphere, and in those processes leading either to its subsequent destruction or to its preservation, or even to its further evolution, all dependent upon the nature of the environment.

9. CHEMICAL DIVERSITY OF PRE-LIFE AND CHEMICAL UNIFORMITY OF LIFE THE PIRIE (1959) DRAWING

Apart from the more common elements mentioned in the preceding section, many other elements may have taken part in these early processes of inorganic photosynthesis. In contrast to present life, with its almost infinite morphological variation based on a narrowly limited number of biochemical reactions, these early processes of inorganic photosynthesis, this *pre-life*, probably showed strong chemical variety, but was not coupled to any definite morphology.

The British chemist PIRIE (1959) has expressed this difference in a now famous drawing, reproduced here as Fig.13. In a simplified form the evolution of life can be presented in a double cone, reminiscent of an hourglass. The lower cone symbolizes pre-life with its large number of elements participating in inorganic photosynthesis and in allied processes under the primeval atmosphere without free oxygen. The upper cone represents the development of life, with a strong morphological variation and based on a group of narrowly limited biochemical reactions. As we shall see, life started already under the anoxygenic atmosphere, but has only come to full development under the oxygenic atmosphere.

From the first inorganic photosynthetic reactions, which led to the formation of the "thin soup", a long time must have elapsed before the appearance of anything resembling living matter. A stretch of time filled, of course, by a large number of different activities. An exact limit is difficult to draw. It is, perhaps, even irrelevant. It is less important where to draw the dividing line between non-living and living in a continuous evolutionary process, than to retain the fundamental difference existing between the primeval and the newer atmospheres, and, moreover, to realize that life must have existed already in the time of that primeval atmosphere. Real life, producing fossilized remains of organisms, co-existed with pre-life, with the *eobionts* of Pirie, and consequently also with the photosynthetic reactions which one should like to call inorganic by all means.

In general there will have been an early evolution towards bigger and more complicated molecules forming "organic" substances, but still formed by inorganic processes. These bigger molecules were often built up by smaller units, each of similar structure. They are repetitive and consist of chains of identical or related blocks. Such molecules are often able to incorporate similarly built new blocks into their structure, or in other words, to grow. Such growing will have been favoured, for certain kinds of compounds, by the nature of their substratum.

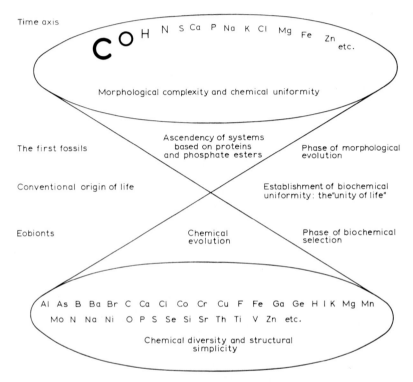

Fig.13. Simplified presentation of the origin and the development of life on earth, according to Pirie (1959). The lower cone shows the early inorganic chemical processes of pre-life, which are chemically diversified but without morphological expression. The upper cone represents the development of life, characterized by morphological diversification based on a restricted group of biochemical reactions. This idea is further discussed in Chapter 18.

For instance by adsorption on clays or quartz, then as now probably the two most common minerals on the surface of the earth. At that time sulphides too must have been abundant, for instance in the form of pyrite sands. Sulfur is known for its strong catalytic properties and the presence of sulfur must have been an important factor during those early days.

    A further step will have been that various of these bigger, "growing" molecules became dominant. From this stage to life-like compounds with specific metabolic reactions and capable of propagation still must have taken a great number of separate steps. It is, however, a development quite conceivable from our present knowledge of both inorganic and organic chemistry, and in Chapter 7 we will further analyze a number of these steps.

10. ADDENDUM

After the manuscript for this book was closed, the Third Conference on the Origin of Life was held at Pont-à-Mousson, France, and an International Conference on Biochemical Evolution at Liege, Belgium. In these conferences the present state of the art, mainly related to the topics treated in our Chapters 6 and 7, was given, but the results could no more be incorporated in this text. Those interested should consult the two volumes to be published under the title *Molecular Evolution* (BUVET and PONNAMPERUMA, 1971; SCHOFFENIELS, 1971). It will be seen that the classical experiments by MILLER (1959) have exploded into the study of a great variety of possible energetic pathways and possible geneses of abiotic compounds and subsequent polymerization and biopoesis.

Two of the main new insights developed in the first of these conferences were presented by Dr. G. Toupance of Paris and Professor G. N. Matthews of Chicago. Dr. Toupance showed that the energy needed to synthesize the small "organic" molecules inorganically is several orders of magnitude lower in an anoxygenic than in an oxygenic environment. This gives one more reason why these "organic" compounds synthesize so readily in all experiments using a simulated anoxygenic primeval atmospheric environment, whilst on the other hand it enhances the difficulty for the abiotic formation of such compounds in the present atmosphere. Professor Matthew's experimenting has been mainly concerned with the chemistry of HCN and its polymers. He showed that in a primeval atmosphere containing HCN and NH_3 formation of peptides would be far easier energetically, than that of animo acids. He submitted that formation of peptides is the primary reaction and that the animo acids found in the S. L. Miller-type of experiments are degradation products of the primary peptides, and not the other way round, as is commonly held. If his views were to be confirmed, the inorganic synthesis of (poly)peptides would become much easier to visualize and the inorganic synthesis of the building blocks of early life could have been accomplished in a far smaller number of steps.

REFERENCES

BERNAL, J. D., 1951. *The Physical Basis of Life*. Routledge and Paul, London, 80 pp.
BERNAL, J. D., 1959. The scale of structural units in biopoesis. In: A. J. OPARIN (Editor), *Origin of Life on the Earth*. Pergamon, London, pp.385–399.

BERNAL, J. D., 1961. Origin of life on the shores of the ocean. In: M. SEARS (Editor), *Oceanography*. Am. Assoc. Advan. Sci., Washington, D.C., pp.95–118.

BRYSON, V. and VOGEL, H. J. (Editors), 1965. *Evolving Genes and Proteins*. Academic Press, New York, N.Y., 629 pp.

BUVET, R. and PONNAMPERUMA, C. (Editors), 1971. *Molecular Evolution*, 1. North-Holland, Amsterdam, 560 pp.

CALVIN, M., 1965. Chemical evolution. *Proc. Roy. Soc. (London), Ser. A*, 288:441–466.

CALVIN, M., 1969. *Chemical Evolution*. Clarendon, Oxford, 278 pp.

FOX, S. W. (Editor), 1965. *The Origin of Prebiological Systems*. Academic Press, New York, N.Y., 482 pp.

GROBSTEIN, C., 1965. *The Strategy of Life*. Freeman, San Francisco, Calif., 118 pp.

HAWKINS, D., 1964. *The Language of Nature*. Freeman, San Francisco, Calif., 372 pp.

KLUYVER, A. J. and DONKER, H. J. L., 1926. Die Einheit in der Biochemie. In: *Chemie der Zelle und Gewebe*, 13 (134) (Also in: A. F. KAMP, J. W. M. LA RIVIERE and W. VERHOEVEN, 1959. *Albert Jan Kluyver*, North-Holland, Amsterdam, pp.211–267.)

LECOMPTE DU NOÜY, P., 1947. *Human Destiny*. Longmans, New York, N.Y., 289 pp.

MAMIKUNIAN, G. and BRIGGS, M. H. (Editors), 1966. *Current Aspects of Exobiology*. Pergamon, London, 420 pp.

MILLER, S. L., 1959. Formation of organic compounds on the primitive earth. In: A. I. OPARIN (Editor), *The Origin of Life on Earth*. Pergamon, London, pp.123–135.

NEEDHAM, A. E., 1965. *The Uniqueness of Biological Materials*. Pergamon, London, 593 pp.

OPARIN, A. I., 1938. *The Origin of Life*, 2nd ed. Dover, New York, N.Y., 270 pp.

OPARIN, A. I. (Editor), 1959. *The Origin of Life on Earth*. Pergamon, London, 436 pp.

OPARIN, A. I., 1961. *Life. Its Nature, Origin and Development*. Oliver and Boyd, Edinburgh, 207 pp.

OPARIN, A. I. (Editor), 1963. *Evolutionary Biochemistry—Proc. Intern. Congr. Biochem., 5th, Moscow, 1961*, 3:354 pp.

OPARIN, A. I., 1964. The chemical origin of life. *Am. Lecture Ser., Springfield*, 558: 124 pp.

PIRIE, N. W., 1959. Chemical diversity and the origins of life. In: A. I. OPARIN (Editor), *Origin of Life on Earth*. Pergamon, London, pp.76–83.

PIRIE, N. W., 1966. Pangloss or Plotinus. *Nature*, 209:230.

PIRIE, N. W. (Editor), 1968. A discussion on anomalous aspects of biochemistry of possible significance in discussing the origins and distribution of life. *Proc. Roy. Soc. (London), Ser. B*, 171:1–89.

PONNAMPERUMA, C. and GABEL, N. W., 1969. Prebiological synthesis of organic compounds. In: A. RENBAUM and R. F. LANDELL (Editors), *Chemistry in Space Research*, in press.

POSTGATE, J. R., 1968. Fringe biochemistry among microbes. *Proc. Roy. Soc. (London), Ser. B*, 171:67–76.

SCHOFFENIELS, E. (Editor), 1971. *Molecular Evolution*, 2. North-Holland, Amsterdam, in press.

TERENIN, A. N., 1959. Photosynthesis in the shortest ultraviolet. In: A. I. OPARIN (Editor), *Origin of Life on Earth*. Pergamon, London, pp.136–139.

VALLENTYNE, J. R., 1965. Why exobiology? In: G. MAMIKUNIAN and M. H. BRIGGS (Editors), *Current Aspects of Exobiology*. Pergamon, London, pp.1–12.

WINKLER, K. C., 1960. *Virus Mirabile. Leven en Dood*. Bohn, Haarlem, pp.86–99.

Chapter 5 | The Astronomer's View

1. DATA FROM ASTRONOMY

To follow up the introductory chapters on general aspects of geology and biology which are of importance to the study of the origin of life, it seems appropriate now to touch shortly on those aspects from astronomy which are relevant to our subject. We may then also use this chapter to incorporate some remarks on later events which took place on earth during the transition from its astronomical to its geological history. That is between the time the earth had been formed as an astronomical body and the time it had so far developed that an outer crust of rocks had formed, which could incorporate the data of the geologic history. This chapter will be a short one, because, although the major results of modern astronomy have strongly influenced the start of our modern insights in a possible origin of life through natural causes, the details of, for instance, the composition of planetary atmospheres or interstellar gases are not too relevant for our studies.

As indicated already in Chapter 1, it so happened that the recognition, in the world of biologists, that an abiogenic origin of life, i.e. its origin without the cooperation of life which existed already, was only possible in a reducing environment, more or less coincided with a development in the astronomer's world, where it was realized that, if not the whole universe, at least the atmospheres of comets and planets were predominantly in a reducing state.

Our knowledge about the composition of extra-terrestrial atmospheres or intrastellar clouds rests mainly on the data gathered by spectroscopy, which detects both emission and absorption lines and bands which are due to the influence of atoms and molecules within these extra-terrestrial gases. Spectroscopic measurements in which various compounds of hydrogen and carbon were detected in the atmospheres of planets and comets were already described in the early twenties. The realization that all of these extra-terrestrial atmospheres were fundamentally different from our own terrestrial atmosphere came, however, only with the classic treatise by H. C. Urey on the planets (UREY, 1952, 1959). Urey, a chemist by training, was the first astronomer who tried to combine the data obtained from the spectroscopic analysis of space with the thermodynamic equations which must govern the coexistence of the atoms and molecules recorded in the spectroanalytical data, and thus to try to arrive at a consistent history of the formation of the solar system.

As an example of the probable and possible constituents of the planetary atmosphere Table VI is given here. It will be seen that the elements C, H and N,

TABLE VI

THE ATMOSPHERES OF THE PLANETS

(From UREY, 1959)

Planet	Substance	Detected	Amount cm atm (NTP)	Basis of estimate
Mercury	–	–	–	–
Venus	CO_2	yes	10^5	spectroscopic
	H_2O	yes	oceans	polarization of clouds
	N_2	yes	?	spectroscopic
	CO	yes	<100	spectroscopic
Mars	CO_2	yes	3,600	spectroscopic
	H_2O	yes	?	polarization of clouds
	N_2	no	$1.8 \cdot 10^5$	total pressure measurement
Jupiter	CH_4	yes	$1.5 \cdot 10^4$	spectroscopic
	NH_3	yes	700	spectroscopic
	H_2	no	$2.7 \cdot 10^7$	
	He	no	$5.6 \cdot 10^6$	density of the planet
	N_2	no	$4 \cdot 10^3$	
	Ne	no	$1.7 \cdot 10^4$	
Saturn	CH_4	yes	35,000	spectroscopic
	NH_3	yes	200	spectroscopic
	H_2	no	$6.3 \cdot 10^7$	
	He	no	$1.3 \cdot 10^7$	density of the planet
	N_2	no	$9.5 \cdot 10^3$	
	Ne	no	$2.7 \cdot 10^4$	
Uranus	CH_4	yes	$2.2 \cdot 10^5$	spectroscopic
	H_2	yes	$9 \cdot 10^6$	calculated on Herzberg's
	He	no	$2.7 \cdot 10^7$	assumption
	H_2	yes	$4.2 \cdot 10^6$	calculated on assumptions
	He	no	$8.6 \cdot 10^5$	in UREY (1959)
	N_2	no	$4.2 \cdot 10^6$	
Neptune	CH_4	yes	$3.7 \cdot 10^5$	spectroscopic
	H_2	yes	larger	spectroscopic
	N_2	no	than in	
	He	no	Uranus	
Titan	CH_4	yes	$2 \cdot 10^4$	spectroscopic
	He	no		

either in their elementary state as atoms, or in combinations in simple molecules, have been most frequently detected. Oxygen is only known in combination with other elements; water, H_2O and carbon dioxide, CO_2, being the commonest, but no free oxygen seems to be present in the planetary atmospheres. This is, of course, in marked contrast with the present terrestrial atmosphere, which contains about 21% of free oxygen, as follows from Table VII.

TABLE VII

NON-VARIABLE CONSTITUENTS OF THE PRESENT ATMOSPHERE OF THE EARTH

(From UREY, 1959)

Constituents	Content	Stability	Spontaneous reactions
N_2	78.084%	stable	
O_2	20.946%	unstable	reacts slowly with FeO and carbon compounds
CO_2	0.033%	unstable	reacts slowly with silicates
A	0.934%	stable	
Ne	$18.18 \cdot 10^{-6}$	stable	
He^4	$5.24 \cdot 10^{-6}$	stable	
He^3	$6.55 \cdot 10^{-12}$	stable	
Kr	$1.14 \cdot 10^{-6}$	stable	
Xe	$0.087 \cdot 10^{-6}$	stable	
H_2	$0.5 \cdot 10^{-6}$	unstable	$2H_2 + O_2 = 2H_2O$
CH_4	$2 \quad \cdot 10^{-6}$	unstable	$CH_4 + 2O_2 = CO_2 + 2H_2O$
N_2O	$0.5 \cdot 10^{-6}$	unstable	$2N_2O = 2N_2 + O_2$

Although not of direct importance for the problem of the origin of life on earth, we might at this point remind ourselves that a difference exists between the so-called terrestrial planets and the so-called major planets, as follows from Table VIII. The terrestrial planets are nearer to the sun, whilst they are also smaller and denser than the major planets. The higher density of the terrestrial planets is thought to be a direct result of their position nearer to the sun, where the original solar nebula had a higher temperature than in its periphery. Accordingly, the more volatile components of the solar nebula, such as H_2, CH_4, NH_3, H_2O remained largely in the gas phase, so that the terrestrial planets mainly formed from the metallic oxide components of the primitive dust, which became reduced during its accretion. The major planets, on the other hand, seem more nearly to have retained the solar abundances of elements and consist dominantly of the more volatile components mentioned above. The distinction is, however, a fine one, because we know from the composition of the earth that an appreciable amount of these volatiles must have found its way, as liquids, into the terrestrial planets too, at the time of their accretion.

TABLE VIII

PHYSICAL PROPERTIES OF THE PLANETS OF THE SOLAR SYSTEM

(From RINGWOOD, 1966)

Planet	Mass (relative to earth)	Radius (relative to earth)	Density (g/cm³)	
Mercury	0.0543	0.383	5.33	
Venus	0.8136	0.9551	5.15	
Earth[1]	1.0000	1.000	5.52	terrestrial planets
(Moon)	0.0123	0.273	3.33	
Mars	0.1069	0.528	4.00	
Asteroids (chondritic)	⩽0.00013	⩽0.058	∼3.5	minor planets
Jupiter	318.35	10.97	1.35	
Saturn	95.3	9.03	0.71	
Uranus	14.54	3.72	1.56	major planets
Neptune	17.2	3.38	2.47	
Pluto	0.033?	0.45	2?	

[1] Mass of earth is $5.975 \cdot 10^{27}$ g (mean radius of earth is 6371.2 km).

2. THE PRIMARY AND SECONDARY ATMOSPHERES OF THE EARTH

We have seen how the realization that both space and planetary atmospheres are of reducing character was a major influence during the gestation period of the biological views about life originating in a reducing, that is, in an anoxygenic, environment. The data from astronomy brought home the fact that the terrestrial oxygenic atmosphere is an exception, at least in the solar system, but probably in all of the universe. So, what was more natural then to think of the possibility of an origin of life in a primeval, anoxygenic atmosphere which resembled in composition those of the planets and/or that of interstellar gases? Such an approach has, for instance, been taken already at an early date by BERNAL whose 1961 data have been given in Table V.

However, the story must have been, as always, more complicated. The history of the early stages of the earth, before it emerged as a full-fledged planet whose rocks could tell us something of its further geological history, was intricate and complex. There is still, at present, a considerable divergence in regard to the successive events which have led to the formation of the earth. But the pros and contras of this discussion need not bother us too much, as there is, on the other hand, considerable unanimousness as to the main events. As a concrete example of one of the proposed chains of events, we will here follow the outline suggested by UREY in 1962.

(*1*) Early in its history the sun acquired a nebula whose plane was that of the ecliptic.

(*2*) Within this solar nebula lunar-sized objects of the composition of the non-volatile fraction of cosmic matter were formed. The objects became heated to high temperatures and developed the characteristic iron-nickel phases of the achondritic meteorites (see Chapter 17).

(*3*) At least in that region where the meteorites originated, these objects were broken into small fragments by some crushing action. These processes occurred 4.5 billion years ago.

(*4*) There is some evidence that the lunar objects formed throughout the region of the terrestrial planets Mercury, Venus, Earth and Mars, and that they broke up into small fragments. The silicate fraction was selectively lost from the region of the terrestrial planets, resulting in an increased iron content of all these planets and a variable density due to the different degrees of silica loss.

(*5*) The gases were lost in the region of the terrestrial planets while the solid materials were at low temperature.

(*6*) The solid objects collected into the terrestrial planets probably so slowly that their general temperatures were low. It is likely that these planets never completely melted.

(*7–10*) Four more steps are given, but these are not of direct influence on the history of the earth.

For our problem the steps *2, 4, 5* and *6* are important. Although the validity of each step and moreover its succesive position is questioned by someone or other, we may still retain the overall picture of an earth formed by accretion from interstellar material[1] at relatively low temperatures, at least as far as the astronomer's scale goes. The temperatures nevertheless were so high that the earth has lost its primary components of the gases and part of its lighter solids. During this phase of accretion the earth was heated through the transformation of gravitational energy of the accreting particles, which at first were widely dispersed, into thermal energy. One of the uncertainties, touched upon in Urey's step number *6*, lies in the question whether the earth became molten all through in this period, or whether it remained at the stage of a red hot more or less solid ball.

This last question is an academic one in regard to the origin of life on earth, for it is generally understood that the earth must have been heated up so far that no previous life could possibly have survived the period of accretion. So the old assumption that the earth may have originated somewhere, complete with life, can now be answered most firmly in the negative. The origin of life on earth must

---

[1] An entirely different theory, in which it is postulated that the earth consists mainly of solar matter (KUHN and RITTMANN, 1941; RITTMANN, 1967) has several alluring points for the geologist, because it far easier explains the chemical composition of the crust and the differentiation of its magma, but it seems never to have been seriously considered by astronomers.

post-date the step number 6 of the sequence mentioned above, whereas a retention of a hypothetical earlier life is impossible.

There is also general agreement that, according to step 5, the original atmosphere of the earth has been lost. After the cooling posterior to step 6, the earth must consequently have known a period when it had no atmosphere to speak of, and a new atmosphere has since formed gradually. The building up of the new, the secondary, atmosphere was the result of the continuous outgassing of the earth, a process which is still in operation at the present time in at least part of the volcanic eruptions, solfataras and geysers. On top of this, continued accretion of materials from space, both in the form of meteorites and of interstellar dust, may have been active also.

The result of these considerations thus leads us to postulate an early, primary, atmosphere of the earth, which was, however, already lost during its pre-geological, astronomical history. The atmosphere we know of now is of a later formation. In the astronomical view it is a secondary atmosphere, but because it is the only atmosphere the earth has known during its geological history, this does not enter into the picture in the study of processes such as the origin of life which must have been limited to the geological history of the earth. So, when we, in treating this history from the geological aspects, speak about the primeval anoxygenic atmosphere as against the oxygenic present one, we have in mind two successive stages of what according to astronomy is the secondary atmosphere of the earth.

3. DIFFERENCES IN COMPOSITION BETWEEN THE PRIMARY AND SECONDARY ATMOSPHERES OF THE EARTH

One of the main arguments in favour of the loss of the primary atmosphere of the earth is that the relative amount of the noble gases present on earth (Tables VII and IX) is much less than in space. When the young earth lost its volatile materials those elements which do readily combine with other elements to form heavy, stable molecules, were of course retained more easily than volatile matter which remained on its own. It is easy to see why the noble gases, which lack chemical affinity and do not take part in the formation of larger molecules, have been lost preferentially and nowadays are found on earth in exceptionally low quantities.

Moreover, it follows from these considerations that in a general way the composition of the primary atmosphere of the earth will have been different from that of its secondary one. The composition of the primary atmosphere was dictated mainly by the composition of matter in space and by the temperature of accretion. The composition of the secondary atmosphere—in its primeval stage—,

TABLE IX

DEFICIENCY OF ELEMENTS IN THE EARTH[1]

(From RASOOL, 1967)

Element	Atoms/10,000 atoms of silicon		Deficiency factor (log b/a)
	whole earth (a)	solar system (b)	
H	250	$2.6 \cdot 10^8$	6.0
He	$3.5 \cdot 10^{-7}$	$2.1 \cdot 10^7$	13.8
C	14	135,000	4.0
N	0.21	24,400	5.1
O	35,000	236,000	0.8
Ne	$1.2 \cdot 10^{-4}$	23,000	10.3
Na	460	632	~0
Mg	8,900	10,500	~0
Al	940	851	~0
Si	10,000	10,000	0
^{36}A	$5.9 \cdot 10^{-4}$	2,280	6.6
Kr	$6 \cdot 10^{-8}$	0.69	7.1
Xe	$5 \cdot 10^{-9}$	0.07	7.1

[1] The deficiency of hydrogen and part of the very strong deficiency of helium can be attributed to outgassing. The remainder of the helium deficiency, like that of the other noble gases Ne, ^{36}A (not ^{40}A which is radiogenic and derived from the decay of ^{40}K), Kr and Xe, is attributed to the loss of the original atmosphere.

on the other hand, depended upon another set of environmental factors. The main influences in its formation were: (*1*) Which volatiles were retained by the primitive earth when it lost its primary atmosphere, for example by being bonded with other atoms in heavy molecules. (*2*) How these compounds broke down during the period of outgassing which led to the gradual build-up of the secondary atmosphere. And (*3*) the thermodynamical relations between these outgassing compounds in the primeval stage of the secondary atmosphere.

It stands to reason that a reconstruction of the composition of the secondary atmosphere in its primeval stage is quite difficult. As an example we might cite the calculations of ABELSON (1966), which clearly show the amount of uncertainties involved in such calculations. For one thing, use must be made of global geochemical inventories, which, as we will see in sections 4 and 5 of Chapter 14, are of limited reliability themselves. However, for our problem such disputes are again not so very important. It does not matter so much, whether the primeval atmosphere predominantly contained, for instance, CO or CO_2; CH_4 or HCN. For, as we will see in the next chapter, the inorganic synthesis of "organic" compounds is just as well feasible in an atmosphere containing, for instance, CO_2, CO, H_2O, N_2 and H_2, as in one formed by, say, H_2O, CH_4, NH_3 and H_2. As long as compounds of H, C, O and N are present, together with P, S and some other elements, the materials for an inorganic synthesis of "organic" compounds are available. Whereas, as long as this primeval atmosphere remains anoxygenic, the energy for the synthesis of the "organic" material is at hand in the light quanta of the shorter ultraviolet rays of the sunlight.

Consequently, although the composition of the secondary atmosphere of the earth—that is of our primeval atmosphere—forms in itself an intriguing subject for study (RASOOL, MCGOVERN, 1966; RASOOL, 1967; MCGOVERN, 1969), it doef not directly influence the origin of life. As long as the primeval atmosphere was anoxygenic, it may have varied widely in composition and still be at the origin os an inorganic synthesis of "organic" compounds.

Summarizing, we may state in the first place that in our studies we are only interested in what in the astronomer's viewpoint is the secondary atmosphere of the earth. Whilst, secondly, the composition of this atmosphere in its primeval stage was dictated by the composition of the early earth and not by that of matter in space. The main transition between the primeval stage of this secondary atmosphere and its present stage lies in the fact that the primeval stage was anoxygenic, the present one oxygenic, a distinction which will be further explored in Chapter 15.

REFERENCES

ABELSON, PH. H., 1966. Chemical events on the primitive earth. *Proc. Natl. Acad. Sci. U.S.'* 55:1365–1372.

BERNAL, J. D., 1961. Origin of life on the shores of the ocean. In: M. SEARS (Editor), *Oceanography.* Am. Assoc. Advan. Sci., Washington, D.C., pp.95–118.

CAMERON, A. G. W., 1966. Planetary atmospheres and the origin of the moon. In: B. G. MARSDEN and A. G. W. CAMERON (Editors), *The Earth-Moon System.* Plenum, New York, N.Y., pp.234–273.

HAYATSU, R., STUDIER, M. H., ODA, A., FUSE, K. and ANDERS, E., 1968. Origin of organic matter in early solar system. II. Nitrogen compounds. *Geochim. Cosmochim. Acta*, 32:175–190.

KUHN, W. and RITTMANN, A., 1941. Über den Zustand des Erdinnern und seine Entstehung aus einem homogenen Zustand. *Geol. Rundschau*, 32:215–256.

MCGOVERN, W. E., 1969. The primitive earth: Thermal models of the upper atmosphere for a methane dominated environment. *J. Atmospheric Sci.*, 26:623–635.

RASOOL, S. I., 1967. Evolution of the earth's atmosphere. *Science*, 157:1466–1467.

RASOOL, S. I. and MCGOVERN, W. E., 1966. Primitive atmosphere of the earth. *Nature*, 212: 1225–1226.

RINGWOOD, A. E., 1966. Chemical evolution of the terrestrial planets. *Geochim. Cosmochim. Acta*, 30:41–104.

RITTMANN, A., 1967. Die Bimodalität des Vulkanismus und die Herkunft der Magmen. *Geol. Rundschau*, 57:277–295.

STUDIER, M. H., HAYATSU, R. and ANDERS, E., 1968. Origin of organic matter in early solar system. I. Hydrocarbons. *Geochim. Cosmochim. Acta*, 32:151–174.

UREY, H. C., 1952. *The Planets.* Yale Univ. Press, New Haven, Conn., 425 pp.

UREY, H. C., 1959. The atmospheres of the planets. In: *Handbuch der Physik.* Springer, Berlin, 52:363–418.

UREY, H. C., 1962. Evidence regarding the origin of the earth. *Geochim. Cosmochim. Acta*, 26:1–14,

Chapter 6 | Experimental Checks

1. INTRODUCTION

In the preceding chapters we have seen it is theoretically possible that in a primeval anoxygenic atmosphere "organic" molecules can be formed by the energy of the shorter ultraviolet sunrays, or by that of electrical discharges or by still other sources of energy. Experiments concerning this hypothesis have now proved beyond all doubt the feasibility of such a process. A growing number of investigators have produced in vitro a large variety of "organic" materials of different chemical composition through the use of various kinds of energy. All these experiments have been carried out under laboratory conditions simulating the environment of an anoxygenic atmosphere.

In the literature reviewed in this chapter one will find the names of the many "organic" compounds thus synthesized under inorganic conditions. But, being an outsider in this field, I always have difficulty in remembering what some of these names, so commonly used in biochemistry, really stand for. Because we will need several of these notations time and again in order to be able to follow the amazing development of the experimental checks into the possible origin of life, a list of the more common names and formulas is given first. This might be helpful for those readers who also are outsiders in this field.

Apart from the narrative of the more striking of the various syntheses carried out so far, it seems appropriate to introduce at this point some description of the newer techniques with which in recent years it has become possible to analyse exactly the structure of these organic materials. That is, not only to ascertain their gross molecular weight and general chemical properties, but also their precise molecular structure. These techniques have developed rapidly over the past decades. Without them the analytical basis for distinguishing the various compounds synthesized in the experiments mentioned would be non-existent.

This chapter thus contains, first the notation of some of the more pertinent chemical formulas, second a description of the major experimental checks on a possible origin of life through natural causes; and third a discussion of some of the analytical techniques applied to this type of research. If this chapter is of too technical a nature it might just as well be skipped by anyone interested only in the general account. He might be satisfied by the statement already given above, that the experiments have proved such a possibility (cf. the summary of the state of the art in this field by J. M. Buchanan, cited on p. 58).

A much more detailed story of this subject may be found in the paper by PONNAMPERUMA and GABEL (1969) and in Professor CALVIN's latest book (1969).

2. SOME FORMULAS

Before we proceed it is necessary to discuss a few of the chemical formulas pertinent to our general narrative. Anyone wishing to avail himself of more information, should consult textbooks in this field such as FRUTON and SIMMONDS (1958), and specially the *Molecular Biology* of HAGGIS (1964) and co-authors.

Amongst the important groups of compounds essential to present life are the proteins, the polysaccharids, the lipids and the nucleic acids. As the experimental checks carried out so far are mainly concerned with the proteins and the nucleic acids, we will confine ourselves to these groups. In a general way, the proteins, and the polysaccharids, lipids and the like, take care of the daily processes of life, whereas the nucleic acids control reproduction and replacement by providing the means for the building of new molecules. The nucleic acids carry the genetic information, which sees to it that newly formed molecules are constructed in the same pattern as those they replace or complement. Apart from these two groups, there are other types of molecules serving special needs. Heme and chlorophyll (Fig.44), for instance, are examples of essential compounds serving respiration and photosynthesis respectively.

All *proteins* consist of a large number of building blocks formed by amino acid molecules. These are mainly selected from about twenty naturally occurring amino acids. Individual amino acid molecules, some 100 to 300, or even more, combine in single strands to form the so-called *peptide chains*. The latter again

---

Fig.14. Schematic presentation of protein structure.

A. Row of peptized amino-acid building blocks. Each block has a different composition, as indicated by the notation of the various radicals R_1-R_4. It is consequently a heteropolymer (see section 6 of Chapter 7).

B. Structural model of such a peptide chain. Due to the presence of hydrogen bonds (series of dots between the H and O atoms on different rungs of the coil) the chain will tend to coil. Both the diameter and the pitch of the coil are determined by the nature of the radicals R_1-R_4 in any specific peptide chain. (From CALVIN, 1965.)

Fig.15. A three-dimensional drawing of the amino acid, L-alanine. (From BERNAL, 1967.)

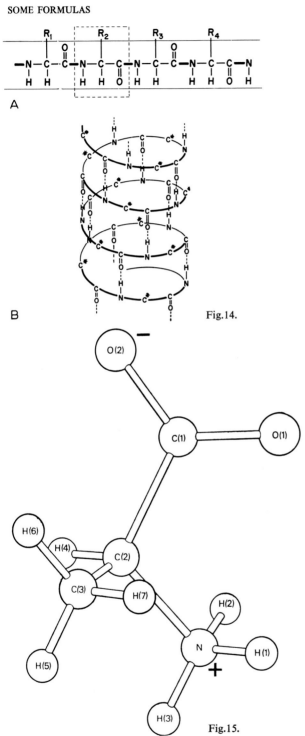

A

B   Fig.14.

Fig.15.

combine, the so-called polypeptization, to form the extremely large protein molecules (Fig.14). Each protein is composed of a number of repeating polypeptide chains. Whereas each peptide chain in itself contains a specific sequence of amino acids.

Glycine (gly)   L–Alanine (ala)   L–Valine (val)   L–Isoleucine (ileu)

L–Leucine (leu)   L–Serine (ser)   L–Threonine (thr)   L–Proline (pro)

L–Cysteine (cys)   L–Methionine (met)   L–Lysine (lys)   L–Arginine (arg)

L–Aspartic acid (asp)   L–Asparagine (asn)   L–Glutamic acid (glu)   L–Glutamine (gln)

L–Phenylalanine (phe)   L–Tyrosine (tyr)   L–Tryptophan (try)   L–Histidine (his)

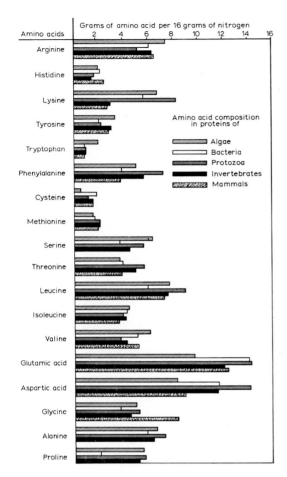

Fig.17. Distribution of eighteen of the twenty common amino acids in various main groups of present-day life. (After YOUNG and PONNAMPERUMA, 1964.) Glutamic and aspartic acids are at present more abundant in nature than the other amino acids.

Fig.16. The twenty common amino acids found in proteins. Together with a number of less common amino acids, these are the compounds found to occur in present-day proteins. It would have been easy, at the time of inorganic photosynthesis of amino acids, to form a much larger number of other, similarly built amino acids. One might assume that during the formation of prebiological systems many more amino acids have indeed been formed, only to be subsequently weeded out in one way or another during later evolution. It is also possible that amongst those amino acids which are now only found in minor quantities we have species which formerly have been much more common, or even dominant.

All common amino acids and most of the rarer species have *laevo* optical activity, that is they turn the plane of polarization of light which passes through them towards the left. A few of the less common amino acids do, however, show *dextro* optical activity (MEISTER, 1965). (From BERNAL, 1967.)

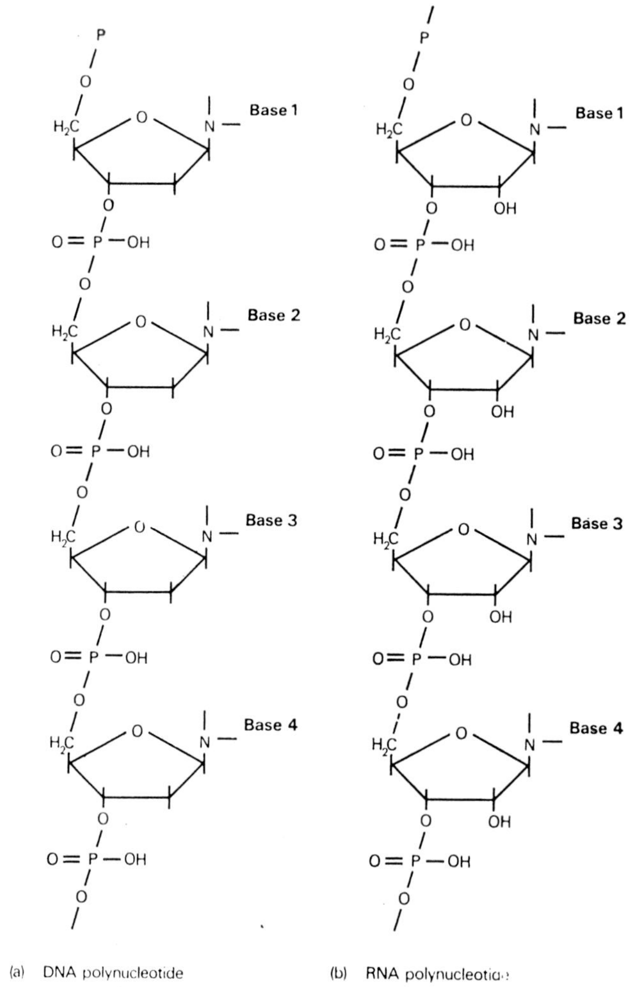

(a) DNA polynucleotide   (b) RNA polynucleotide

Fig.18. Schematic representation of the DNA and the RNA polynucleotides.

Fig.19. Schematic representation of the building of a nucleic acid chain. The base, in this case adenine, combines with the sugar 2-deoxyribose to form adenosine and with phosphoric acid to form adenylic acid. Two molecules of adenylic acid then combine to a first polymer, a dinucleic acid. Further polymerization, with nucleotides containing other bases, leads to the formation of a complete nucleic acid chain. (From CALVIN, 1965.)

Fig.20. The nitrogenous bases occurring in DNA and RNA. (From BERNAL, 1967.)

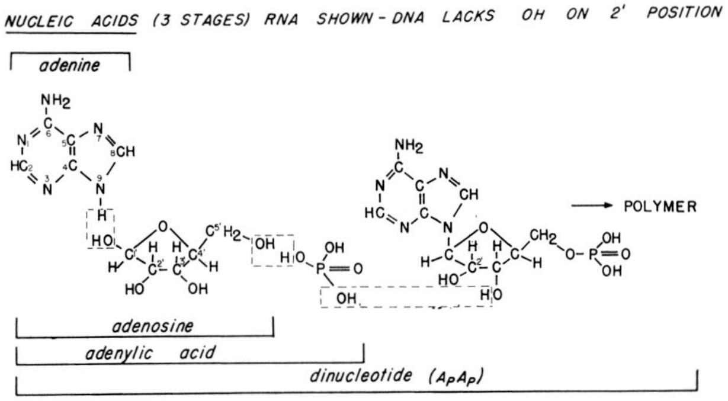
NUCLEIC ACIDS (3 STAGES) RNA SHOWN - DNA LACKS OH ON 2' POSITION
adenine
POLYMER
adenosine
adenylic acid
dinucleotide (ApAp)

Fig. 19.

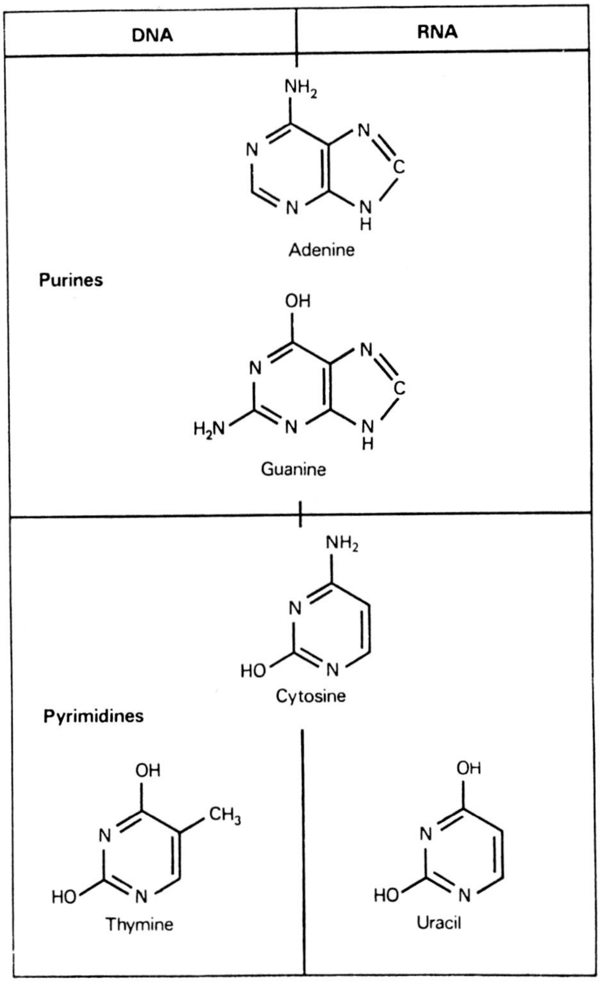
DNA
RNA
Purines
Adenine
Guanine
Pyrimidines
Cytosine
Thymine
Uracil

Fig. 20.

Amino acids are all based on the simple formula:

$$\text{R}$$
$$|$$
$$\text{NH}_2\text{—CH—COOH}$$

where R, for *radical*, represents various possible combinations of atoms (see Fig.15). The main amino acids which occur as the building blocks of proteins are listed in Fig.16. Their distribution in the main groups of present-day life is illustrated in Fig.17.

The *nucleic acids* can be classified in two general types. *Deoxyribonucleic acid*, always called DNA, is a constituent of cell nuclei, whereas *ribonucleic acid* (RNA) is chiefly located in the cytoplasm of the cell, outside the nucleus. Only in some simpler organisms, such as plant viruses, is all nucleic acid in the form of RNA.

Nucleic acids consist of two long chains arranged around each other in a helical structure, the *double helix*. Each chain consists of a series of similar sub-units, called nucleotides, and hence each chain itself is a polynucleotide structure. Individual nucleotides again consist each of three constituents: a molecule of phosphoric acid; a molecule of a sugar called deoxyribose; and a base formed by one out of four different nitrogen compounds. The partial structure of the DNA and RNA chains follows from Fig.18–21. The four nitrogen bases in DNA consist of two purines, adenine and guanine respectively, and of two pyrimidines, i.e.

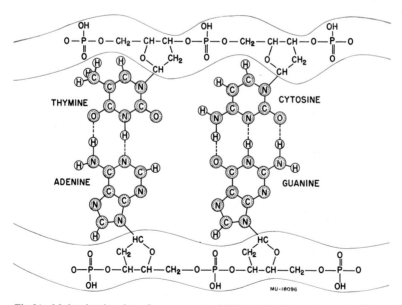

Fig.21. Molecular drawing of components of DNA. It indicates schematically how the two chains of the double helix of DNA are coupled by the bases belonging to each chain. (From CALVIN, 1965.)

thymine and cytosine. Their formulas follow from Fig.20. In DNA the two strands of the helix, that is the two polynucleotide chains, are mutually connected by hydrogen bonding between the purine molecules of one chain and the pyrimidine molecules of the other chain.

3. EXPERIMENTS BY S. L. MILLER

As stated already, the first experiments, which by now have become classic, as to the possibility of inorganic synthesis of "organic" material in a reducing environment were carried out by MILLER (1959), at that time a student of H. C. Urey.

Miller used a very simple apparatus, a spark-discharge flask, as illustrated schematically in Fig.22. It was filled with watery mixtures and by their gaseous counterparts. The main compounds used were hydrogen, methane and ammonia, in the absence of free oxygen. Electric sparks were set off continuously in the upper part of the flask, while in its lower part the water was kept boiling, to induce circulation (Fig.23).

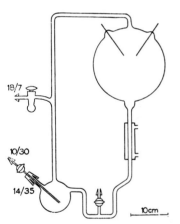 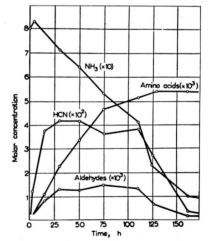

Fig.22. Diagram of a spark-discharge apparatus used to produce "organic" compounds from hydrogen, methane, ammonia and water in an anoxygenic environment. (From MILLER, 1959.)

Fig.23. Concentrations of ammonia, hydrogen cyanide and aldehydes in the U-tube, and amino acids in the 500-ml flask, when sparking a mixture of methane, ammonia, water and hydrogen for 175 hours in the apparatus of Fig.22.
During the first 25 hours mainly hydrogen cyanide and aldehydes form, at the expense of the ammonia. After this time the concentration of the hydrogen cyanide and the aldehydes becomes stabilized, whilst that of the amino acids continues to rise, still at the expense of the ammonia, up to about 125 hours. Their concentration remains constant thereafter, although the amounts of ammonia, hydrogen cyanide and aldehydes now all diminish rapidly. It looks as if at this stage synthesis of amino acids equals consumption. Later experiments have since indicated that peptization of amino acids, not analyzed for by S. L. Miller, is responsible for this development during the later part of the experiment. (From MILLER, 1959.)

Electric sparking instead of irradiation by ultraviolet rays was used in these first experiments for convenience, because it presented fewer experimental difficulties. The energy produced by the sparks is smaller than that of ultraviolet irradiation, so it was assumed that syntheses which are realized by sparking will also be possible by ultraviolet irradiation.

The original experiments by Miller have excited much interest, and similar experiments have been taken up by a large number of other scientists throughout the world. Among them we find Americans (ORÓ, 1965a,b; PONNAMPERUMA, 1965; CALVIN, 1965), Germans (GROTH and VON WEYSSENHOFF, 1959) and Russians (OPARIN, 1965).

The important result of all these experiments, to my mind, is not only that the experiments of Miller were confirmed, but that our horizon has become greatly widened. In fact these newer experiments show not only a great variety in the starting material used, but also in the experimental environment, in the source of energy, and in the compounds synthesized.

It is impossible to review in this text all of these newer experiments on the synthesis of "prebiological" systems. We must be content to choose a number of examples which highlight the stormy development in this field.

4. EXPERIMENTS BY WILSON AND BY PONNAMPERUMA

Already in 1960 Wilson succeeded in producing much larger polymeric molecules, each built up by 20 or more carbon atoms, by adding sulphur to watery mixtures similar to those used by S. L. Miller. In the spark-discharge flask these formed sheetlike solids, of the order of 1 cm across (Fig.24). It is thought that in this case surface-active molecules were formed, which built up films on the gas-liquid surfaces. This agrees well with the idea that films of sheetlike molecules, forming either on the gas-liquid, the gas-solid or the liquid-solid interfaces, must have been of importance in the early stages of the development of life.

The presence of sulphur is believed to have had a catalytic action on the formation of these films. The result of the addition of sulphur is also important in relation to the origin of life, because sulphur will have been commonly present in the form of sulphide grains (e.g. in pyrite sands) under the primeval anoxygenic atmosphere (see section 8 of Chapter 13).

Experiments similar to the original ones by Miller, but using ultraviolet irradiation as the energy source, have been reported by PONNAMPERUMA (1965) and collaborators. Although it was conceded, on theoretical grounds, that the syntheses resulting from electric sparking and from ultraviolet irradiation ought to be similar, it is nevertheless gratifying to know that this really is the case. For we are convinced that ultraviolet irradiation was available in much larger quantities than electric sparking in the primeval atmosphere.

Ponnamperuma not only managed to synthesize amino acids and purines, buildings blocks of protein and of nucleic acids respectively (Fig.25), but under certain conditions he even managed to synthesize combinations of these building blocks. In the presence of hydrogen cyanide, it was, for instance, found that amino acids polymerized to peptide chains. Whilst when phosphoric acid was added, various nucleotides were formed (PONNAMPERUMA, 1965; PONNAMPERUMA and PETERSON, 1965; PONNAMPERUMA and MACK, 1965; SCHWARTZ and PONNAMPERUMA, 1968; RABINOWITZ et al., 1968; STEINMAN et al., 1968).

Fig.24. Sheet-like films of "organic" macromolecules, produced by sparking a mixture of water, ammonia, hydrogen sulphide and ashes of baker's yeast. (From WILSON, 1960.)

We will return to the influence of hydrogen cyanide on the polymerization of the building blocks into larger "organic" molecules in section 7 of this chapter, in relation with the experiments performed by M. Calvin and collaborators.

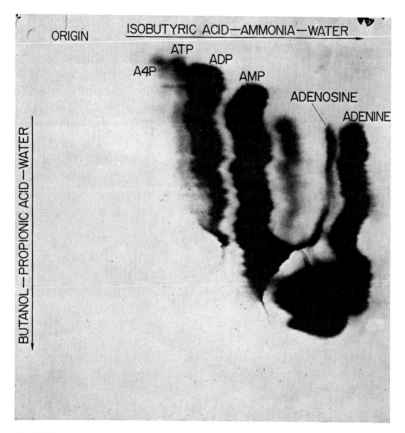

Fig.25. Paper chromatogram (see p. 113) of the mixture of compounds resulting from the irradiation of adenosine monophosphate (AMP) by ultraviolet light.

The products synthesized are adenosine tetraphosphate (A4P), adenosine triphosphate (ATP) and adenosine diphosphate (ADP). In earlier experiments the inorganic synthesis of adenosine monophosphate (AMP) from adenosine and adenine by ultraviolet irradiation had been proved to be possible, whereas the latter two compounds can be formed in watery solutions containing hydrogen cyanide, either by applying a somewhat elevated temperature (ORÓ, 1965a,b), or by ultraviolet irradiation.

Inorganic synthesis of many of the smaller organic molecules upon which present life is based, therefore appears quite feasible under the conditions of the primeval atmosphere. (From PONNAMPERUMA, 1965.)

5. EXPERIMENTS BY ORÓ, USING SOMEWHAT ELEVATED TEMPERATURES IN A WATERY ENVIRONMENT

An important breakthrough has been accomplished by Oró and collaborators (ORÓ, 1965a,b), who showed that it is possible to synthesize larger "organic" molecules by using the ambient energy of an elevated temperature, and without the help of irradiation by ultraviolet rays.

Once it has been established that it is possible for smaller "organic" molecules to be synthesized by the energy of ultraviolet sunrays in a reducing atmosphere, we inevitably come to the next question, that of the transition of this pre-life towards early life. We will go further into this question later in this text, and have only to retain at this point that the conditions in the primeval atmosphere were just as lethal for early life as they would have been for present life. Early life, although it was anoxygenic under the primeval atmosphere, was not aradiatic for the shorter ultraviolet light.

It is therefore important to know that in the transition from pre-life to life, and during the further development of early life, a change in the energy sources utilized was possible. Once the free radicals and the smaller "organic" molecules were formed by the influence of the high energy ultraviolet sun rays, other, more complicated molecules could be synthesized through lower level sorts of energy.

In the experiments of Oró and collaborators watery mixtures of simple "organic" molecules were allowed to stand for several days at temperatures varying from room temperature to 150 °C. Thus, as far as temperature is concerned, these experiments in most cases ran their course in conditions well within the boundaries of present life, as listed in Table IV. The compounds synthesized by Oró are listed in Table X. Some illustrations of several of the analyses of the products formed

TABLE X

AMINO ACIDS SYNTHESIZED IN AQUEOUS SOLUTIONS FROM VARIOUS MIXTURES OF SMALLER "ORGANIC" MOLECULES

(From ORÓ, 1965b)

Amino acid or amino amide[1]	Approximate relative amount
A. Glycinamide	+ + +
Glycine	+ + +
Alanine	+ +
Aspartic acid	+ +
Serine	+ +
Glutamic acid	+
Threonine	+
Leucine	+
Isoleucine	+
Arginine	+
α-Amino-n-butyric acid	+
β-Alanine	+ +
α,β-Diaminopropionic acid	+ +
B. Glycine	+ +
Valine	+ +
Lysine	+ +

[1] A: from $HCN + NH_3 + H_2O$; or $CH_2O + NH_2OH + H_2O$; B: from $CH_2O + N_2H_4 + H_2O$.

are given in Fig.26–33. When starting from a formaldehyde-hydroxylamine solution, a formaldehyde-hydrazine solution, or from a solutions containing hydrogen cyanamide, amino acids were formed. In other experiments these products were polymerized to peptide chains, a big stride towards the inorganic synthesis

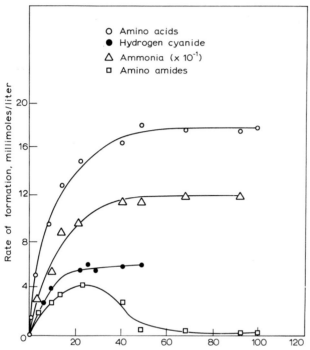

Fig.26. Rate of formation per hour of amino acids and other products from watery formaldehyde-hydroxylamine mixtures. Amino amides clearly are the intermediate product in the formation of amino acids. (From ORÓ, 1965a.)

$$H_2C = O + H_2NOH \longrightarrow H_2C = NOH + H_2O$$

$$H_2C = NOH \longrightarrow HCN + H_2O$$

$$H_2C = O + NH_3 \longrightarrow H_2C = NH + H_2O$$

$$H_2C = NH + HCN \longrightarrow CH_2(NH_2)CN$$

$$CH_2(NH_2)CN + H_2O \longrightarrow CH_2(NH_2)CONH_2$$

$$CH_2(NH_2)CONH_2 + H_2O \longrightarrow CH_2(NH_2)COOH + NH_3$$

COMPLEMENTARY REACTION:
$$2H_2C = NOH + 2H_2O \longrightarrow H_2C = O + NH_2OH + HCOOH + NH_3$$

Fig.27. Proposed mechanism of inorganic formation of the amino-acid glycine from an aqueous formaldehyde-hydroxylamine mixture. (From ORÓ, 1965a.)

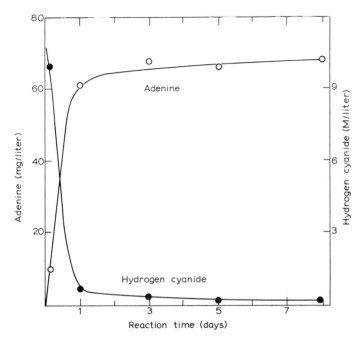

Fig.28. Rate of formation at 90 °C of adenine from a watery mixture of ammonia and hydrogen cyanide. The adenine, which in fact contains 5 HCN molecules, forms the hydrogen cyanide. (From Oró, 1965b.)

of protein. When starting from a hydrogen cyanide solution in an aqueous ammonia system even the more complicated purines and pyrimidines, the nitrogen bases of nucleic acids, appeared.

As stated already, these experiments consequently point the way towards the possibility of a transition from the synthesis of small "organic" molecules formed by the energy of the shorter ultraviolet rays of the sunlight, to more complex "organic" molecules formed under less drastic conditions. For a fuller account of these important results the reader is referred to Oró (1965a,b).

6. EXPERIMENTS BY FOX, USING HIGH TEMPERATURE IN A DRY ENVIRONMENT

An altogether different approach was followed by Fox and collaborators (Fox, 1965a,b; Fox and Waehneldt, 1968) in an attempt to synthesize inorganically the huge molecules of proteins. As we saw, proteins are formed by a large number of so-called polypeptide chains, which in turn are built up by a large, or even a very large, number of amino acids of various kinds.

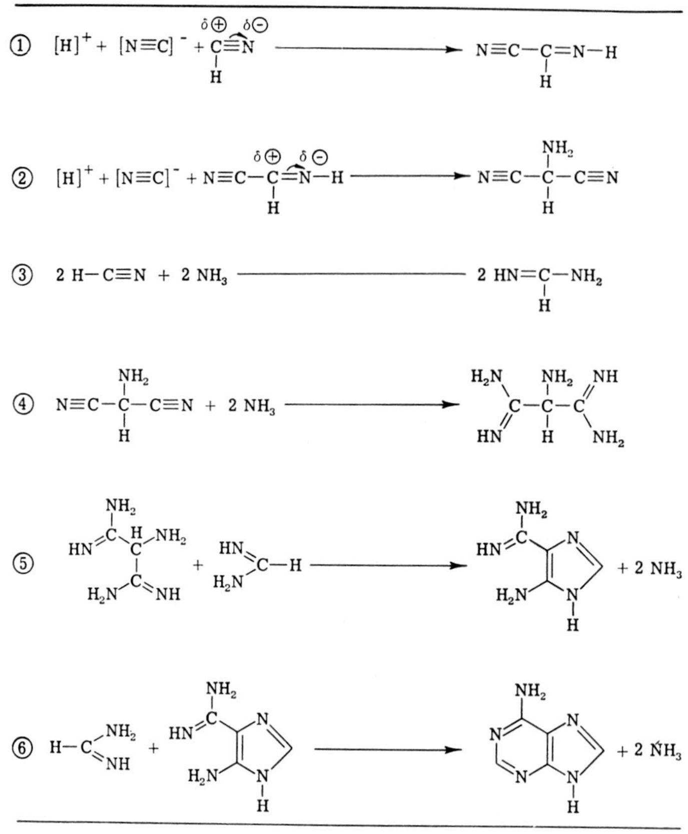

Over-all reaction: 5 HCN = Adenine

Fig.29. Proposed mechanism for a direct synthesis of the purine adenine from a watery mixture of ammonia and hydrogen cyanide. The notations δ^+ and δ^- in formulas *1* and *2* indicate positive and negative partial charges. (From Oró, 1965b.)

Thus, once amino acids have been formed, the next step in the formation of a protein is their combination into peptide chains, a process called peptization. This process can be described as the combination of two amino acids of the same general formula, and with the same or with different radicals, R and R′ (see Fig. 14 and 39). The general formula of peptization is:

$$\begin{array}{cccccccccccccc}
R & O & & & R' & O & & & R & O & R' & O & & \\
| & \| & & & | & \| & & & | & \| & | & \| & & \\
HC- & C- & OH & + & HC- & C- & OH & = & HC- & C- & HC- & C- & OH & +H_2O \\
| & & & & | & & & & | & & | & & & \\
NH_2 & & & & NH_2 & & & & NH_2 & & NH_2 & & &
\end{array}$$

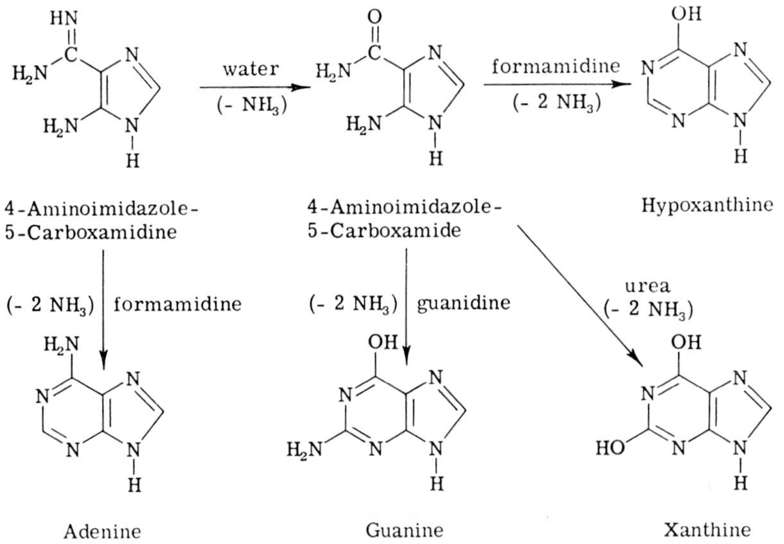

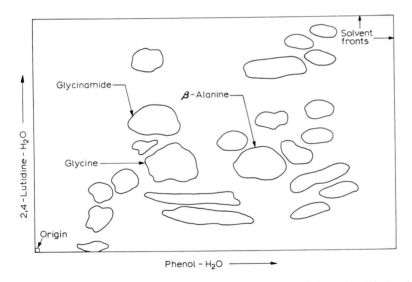

Fig.30. Proposed mechanism of the formation of purines from a watery mixture of ammonia and hydrogen cyanide, via the intermediate products 4-aminoimidazole–5-carboxamidine (AICAI) and 4-aminoimidazole–5-carboxamide (AICA), which were actually isolated from the reaction products. (From ORÓ, 1965b.)

Fig.31. Paper chromatogram (see section 9 of this chapter) of the ampholitic fraction from the formaldehyde–hydroxylamine reaction. Among the various smudges on the paper chromatogram the amino acids glycine and β-alamine could be identified directly. Identification of the other smudges required further analysis, which eventually led to the recognition of all amino acids listed under *A* in Table X. (Whatman no. 1 paper, ninhydrin spraying. From ORÓ, 1965a.)

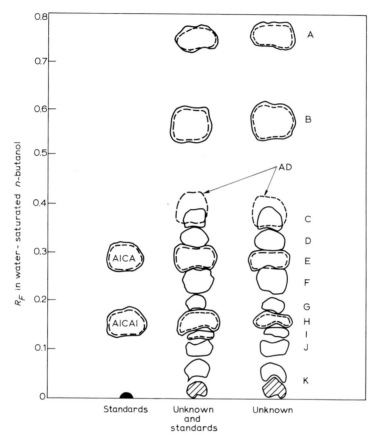

Fig.32. Reproduction of a paper-chromatographic analysis of the diazotizable amines formed from hydrogen cyanide.

The amino acid adenine and the intermediate products AICA and AICAI (cf. Fig.30) could be directly identified from the paper chromatogram. The other smudges had to be further analyzed by other techniques. Capitals refer to these further analyses, as reported in the original paper. (From Oró, 1965a.)

According to Fox, the main point lies in the fact that during peptization a molecule of water is produced by the combination of two molecules of amino acids. In other words, the reaction involves a dehydration. Accordingly it will evolve better, either more rapidly, or more completely, or both, in the absence of water. Fox therefore postulated that the early development of life must have taken place in a volcanic environment, an idea which agrees with a generally expressed opinion that the earth originally must have been more highly volcanic than during its later years.

Fox devised experiments in which a dry mixture of amino acids was subjected

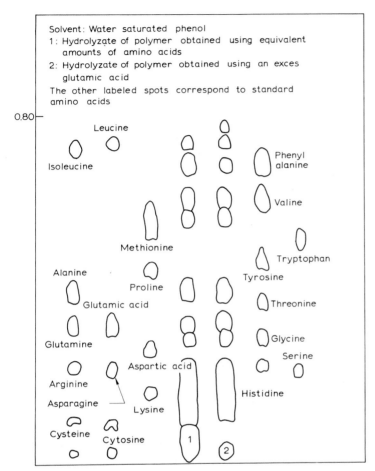

Fig.33. Reproduction of a chromatogram of the hydrolysates of two polymers obtained in aqueous ammonia systems from amino acid mixtures. This shows that the inorganically synthe-sized peptides are indeed built up by a surprisingly large number of different amino acids. (From Oró, 1965a.)

to temperatures of up to 170 °C. It was found out that whenever the amino acids aspartic and glutamic acid were present in these mixtures, the results were truly amazing. The reason why these amino acids have to be present to get the significant results described below is not yet well understood. It is, however, striking that these same two amino acids are the most important amino acids found in present life.

During the experiments mentioned above, compounds were formed which in all important aspects resemble natural protein. That is, they are of large mole-cular size with molecular weights of up to 300,000 and are composed of the same molecular building blocks found in natural protein. That is, they contain 18 out of the 23 amino acids common to present life, thereby fulfilling the common definition of a protein. A number of other characteristics are also similar to natural protein, notably the binding of polynucleotides (WAEHNELDT and FOX, 1968), the fact that the synthetic compounds have nutritive value for bacteria and rats; and that they are able to fulfil reactions similar to those performed by enzymes in living matter. One such function is the catalytic decomposition of glucose by these artificially synthesized "organic" molecules. Admittedly the activity is weak, but to quote from the report: "...no more than a weak activity was needed in the first proteins" (FOX and KRAMPITZ, 1965). Another function resembles that of a melanocyte-stimulating hormone (FOX and WANG, 1968).

Research into such activities has been actively pursued since, and Fox (1968a–c) has been able to report a far larger number (see Fig.34). The semantics have become more subtle too, over the years, and instead of "enzymatic" activities, the words "catalytic", or even "rate-enhancing" activities are used now. This is in keeping with the careful distinction between the activities of the artificially

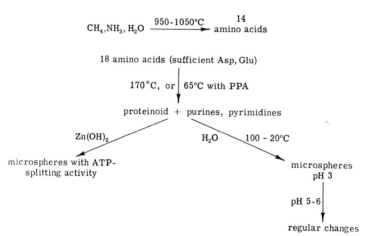

Fig.34. Catalytic or "rate-enhancing" activities of inorganically synthesized proteinoids. (From Fox, 1968a.)

synthesized proteinoid compounds and those of present day life. Although one could maintain that the basic character of an enzymatic action is a catalytic, or a rate enhancing one, still the difference between the rather weak activities executed by the proteinoids and the strong and very specific activities performed by present day enzymes might lead some scientists to doubt the correlation proposed between these two groups of activities. But even if one would tend to call the activities reported from the proteinoid compounds "pre-enzymatic", the importance lies not in the difference with present day enzymatic actions, but in their existence during the period of transition from pre-life to early life, when strong and specifically reacting enzymes were not yet available.

Another important character of the proteinoid compounds, to which we will return in section 6 of the next chapter, is their "limited heterogeneity". That is that their peptide chains are not formed by a random sequence of their building blocks of amino acids, but that these follow each other in a more or less orderly sequence.

According to Fox, a rigorous comparison between these artificially formed compounds and natural proteins cannot yet be carried out, because proteins form such complex molecules that most of them have not as yet been defined rigorously themselves. To indicate the many points of similarity existing between the artificial compounds and natural proteins, Fox uses the term *proteinoid* for the artificial compounds. And, because they have been formed by thermal activity, they have been further designated as "thermal proteinoids".

The main steps in the formation of these thermal proteinoids, according to Fox, are given below[1]. A complete, though condensed report on these experiments can be found in Fox, 1965a.

There is one aspect to the experiments of Fox and collaborators which is so spectacular that it must be related somewhat more fully. This is the formation of microspheres from the thermal proteinoids. The process is simple. Washing the hot mixture of the artificially synthesized polymers with water or aqueous salt solutions of different compositions produces microspheres in very large numbers. Their size is microscopic and their diameter varies around 2μ. Fig.35–38 give an idea of their morphology.

The microspheres are quite stable. Moreover, they react when brought into contact with solutions of a concentration different from that in which they were formed. In solutions of a higher concentration they shrink, in those of a lower concentration they swell. They consequently react to osmotic pressures in a way similar to living cells. This is due to the fact that they have evidently developed

---

[1] Synopsis of the "thermal theory of the origin of life" (From Fox, 1965b):

An early formation of amino acids is supposed to have taken place at very high temperatures. This is still another possibility for inorganic formation of amino acids, and can be compared with the sparking (S. L. Miller), irradiation (C. Ponnamperuma) and heat (J. Oró) models described earlier. The feasibility of this synthesis has been experimentally demonstrated by HARADA and Fox (1964).

The next step is the peptization of the 18 amino acids formed experimentally, either at 170 °C or at 65° in the presence of certain phosphates (PPA). This peptization succeeds when enough aspartic acid and glutamic acid (Asp, Glu) are present.

As stated more fully in the text, the proteinoid mixture will form microspheres when washed out by water or by acidic aqueous solutions (rain) (Fig.35–38).

The faculty of the proteinoids to perform life-like functions is indicated by their capacity to hydrolize the nucleotide ATP in the presence of zinc hydroxide. This represents a weak enzymatic, or "rate-enhancing" action.

an outer wall, a sort of membrane, which is semipermeable. As Fig.38 shows, electron-microphotos have disclosed that this wall may even be a double one.

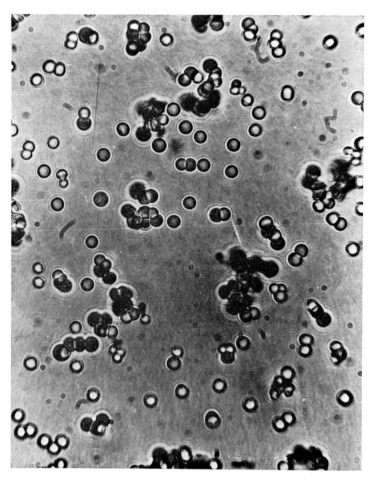

Fig.35. Proteinoid microspheres formed by washing out hot proteinoid mixtures. Average diameter is 2μ. (From Fox, 1965b.)

    The formation of microspheres from the thermal proteinoids is important, because it might show the way in which yet another step in the evolution of life might have taken place. That is the step from non-organized "organic" molecules into groups, each forming a definite unit, and separated from the exterior world by some sort of a membrane. We will return to this question in the next chapter in the section describing coacervation and the formation of membranes.

    For me as a geologist there remains one aspect of the "thermal theory" that is difficult to visualize, that is the high temperatures envolved in the model. Fox stresses the fact that such temperatures are readily available in volcanoes during an eruption, both in the craters and on top of hot lava flows. But it seems difficult to combine the idea of a first development of smaller molecules in the "thin soup" with a further evolution in and around volcanoes if and when there is an eruption. We will see, notably in Chapter 16, that pre-life and early life must have been contemporary over very long time spans, and I would prefer a more prevalent and regularly occurring process than the chance eruptions of volcanoes.

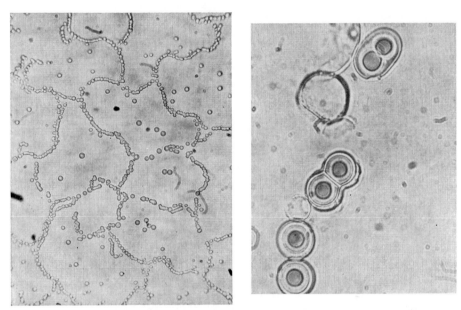

Fig.36. Proteinoid microspheres associated in algae-like chains, a form they take when produced under slight pressure. (From Fox, 1965b.)

Fig.37. Proteinoid microspheres which show twinning as a reaction to a rise of the ambient pH to 6.0. (From Fox, 1965b.)

However, in the natural processes, the compounds formed are always more varied than in simplified model experiments. So, in nature, comparable reactions could perhaps have taken place at lower temperatures, with the help of some catalyst. This possibility has in fact already been indicated in the experiments of Fox, where phosphates were added to the amino acid mixture, and peptization occurred already at 65 °C (see Fig.34).

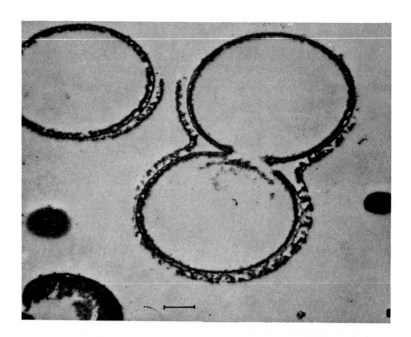

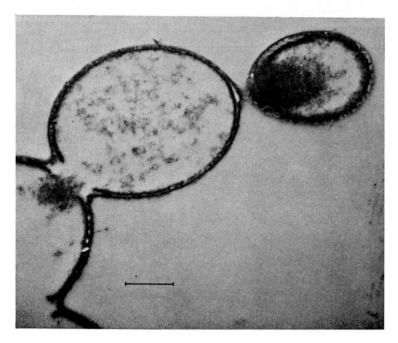

Fig.38. Electron micrographs of sectioned microspheres, showing double walls; magnification about 1400 ×. (From Fox, 1965b.)

7. EXPERIMENTS BY CALVIN. CONDENSATION OF SMALLER MOLECULAR
BUILDING BLOCKS INTO LARGER UNITS IN AQUEOUS SOLUTIONS IN THE
PRESENCE OF HYDROGEN CYANIDE

With regard to the assumption underlying the "thermal theory" of an origin
of life, as proposed by Fox, CALVIN (1965) has remarked that such a dehydratation
occurs not only in the peptization of amino acids but also in the combination of
other building blocks into the larger "organic" molecules. Such a combination
will always entail some condensation, which in most cases occurs through the
expulsion of a hydrogen atom of one block and a hydroxyl group of the next
block (Fig.39).

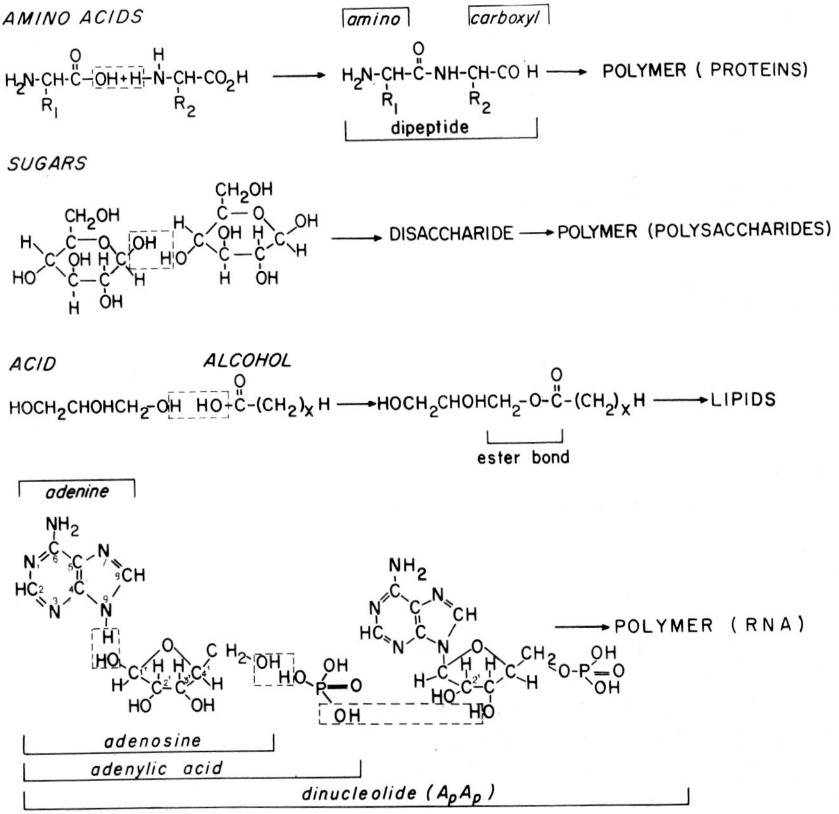

Fig.39. Schematic representation of the condensation through dehydratation of individual
building blocks into larger "organic" molecules.
    Upper three equations represent the condensation and further polymerization of amino
acids to proteins (cf. p. 99ff.); sugars to polysaccharids; and acids and alcohols to lipids
respectively. Lower equation represents the condensation of adenine with oxyribose and
phosphoric acid, to form a nucleotide (p. 92). Further water molecules are then lost through
the condensation of separate nucleotide molecules into a nucleic acid chain. (From CALVIN, 1965.)

Instead of following Fox's reasoning that these condensations must have taken place in a dry—in an anhydrous—environment, Calvin searched for a possibility through which, paradoxically, such a dehydration could take place in the watery environment of the "thin soup". The answer was found in the presence of HCN, which has the faculty to bind water molecules in the "thin soup". The presence of HCN as one of the most common compounds of the "thin soup" has not only been assumed on theoretical grounds, but it was found already in the first experiments by Miller (Fig.23).

Besides HCN it was found that cyanamide and dicyanamide—$HN(C \equiv N)_2$— two somewhat more complicated compounds, were found to possess this dehydrating faculty in an even stronger way. The reactions are somewhat more complicated and the exact reaction pathways have not yet been detected (HUNTRESS et al., 1969). However, it has been demonstrated that condensation of building blocks through dehydration will occur at normal temperatures in very dilute aqueous solutions, equivalent to the thinnest of "thin soups" in the presence of HCN and related cyanamides (MATHEWS and MOSER, 1968).

A further sophistication of this line of evidence has been supplied by ABELSON (1966), who pointed out that the reactions with HCN are strongly influenced by the acidity of the watery solutions in which they occur. They are not found in acid environments, but are favoured by alkaline conditions with a pH of 8–9. Even if it may seem improbable that the primeval oceans had this composition (see Chapter 14, sections 7 and 8), it is well possible that larger bodies of fresh water in lakes, which were in contact with basalts, would have had this pH. Hence it seems quite possible that such reactions could indeed have taken place.

8. GENERAL RESULTS OF THE EXPERIMENTAL CHECKS

When we now try to review the general aspects of the experimental checks into a possible origin of life through natural causes, we find two important answers. The first relates to the many different pathways in which "organic" molecules have experimentally been formed by inorganic reactions under conditions simulating an anoxygenic primeval atmosphere. The second is that such experiments are entirely insufficient for rigorous geological application, because they have not, and are not able to, take into account the long time spans of geologic history.

To take the positive result first, it is amazing to see in how many ways, and under what variety of environmental conditions "organic" molecules will be formed by various sources of energy, all of them presumably available under the conditions of the primeval atmosphere during the early history of the earth. Such experiments have, indeed, become so commonplace, that *Scientific American*, in its January 1970 issue gave complete instructions for do-it-yourselves scientists under the title "Experiments in generating the constituents of living matter from inorganic substances".

The experiments reviewed earlier in this chapter form by no means a complete record of all the experiments carried out in this field. But I have tried to stress the environments tested and sources of energy used, under which these reactions have taken place. Conditions tested include cold, warm and even hot aqueous, as well as dry environments, to which different energy sources such as heat, ultraviolet radiation and electric sparks have been applied.

It follows that for the formation of pre-life, and thus for the initial conditions of the formation of life through natural causes, two conditions only must be met. First the atmosphere must be anoxygenic, and second the building blocks of the "organic" molecules—that is the atoms of carbon, nitrogen, inorganic catalysts, water and such—must be present. As soon as these conditions are fulfilled inorganic formation of "organic" compounds will start forthwith.

It follows that the formation of pre-life cannot be a process limited to our own earth. On every planet, either within or without our solar system, similar reactions will occur as soon as the two above-named conditions are met. We have seen in Chapter 5 that an anoxygenic atmosphere, containing the necessary atoms or molecules to synthesize inorganically "organic" molecules, seems to be the normal case for both the planetary and the stellar environment. The main condition for the formation of pre-life on a planet being that the existence of liquid water is possible (see section 6 of Chapter 17).

This far-reaching conclusion drawn from the experiments divests pre-life, and by implication life, from one of its most cherished characters, that is its terrestrial uniqueness. *Inorganic formation of "organic" compounds must instead be assumed to be a normal cosmic process.*

The other general aspect of the experimental checks into the possibility of an origin of life through natural causes, the difficulty to extrapolate the results to periods of geologic significance, is inherent upon all experimenting in earth history questions. It is always nice to know how certain things can be formed by a certain process, but it is just as important to know if these things will be conserved over any length of time. It is important to know not only the possibility of a certain process, but also that of the conservation and in many cases the concentration of its products. Already in Miller's classical experiments (Fig.23), it was seen that the results of the reactions after a day were quite different from those formed after a week.

A model consistent with the time involved in all earth historic processes could only be arrived at, if we knew what compounds would be formed under a given set of experimental conditions in which a sort of steady state existed between formation and destruction of these "organic" compounds, obtained over a period of, say, a thousand years. As the atomic properties of the atoms which build up the "organic" compounds are quite well known, it might even be that a computer calculation as to such an equilibrium or equilibria might yield a better insight, than mere experimenting.

As an attempt to lay the groundwork for such calculations we may cite the paper by STEINMAN (1967). He presented evidence that during abiotic synthesis of peptides the interaction of one amino acid with another follows well-defined statistics based on the relative reactivity of each amino acid. The physicochemical characters of the environment, the pH, the sort of the side chain of the amino acid, and the characters of the polymer already formed, were all found to have their effect. It follows that, dependent upon the relative reactivities of the amino acids and upon the environmental circumstances, populations of polypeptides of specific sequences could be produced abiotically, without involvement of nucleic acids. If only the characteristic rates for each of these steps could be determined for various sets of circumstances, appropriate calculations could lead to estimates of the relative amounts of the various polymers produced. It follows that the sequence of amino acids in a peptide chain growing under primitive abiotic conditions would be anything but random.

In an earlier report ECK et al. (1966) investigated the thermodynamic equilibrium of inorganically formed "organic" compounds. In such cases there remains, however, always the doubt whether conditions on the primitive earth are really simulated. For it is probable that the "organic" compounds formed at that time never reached equilibrium, because of the continuous influx of solar energy.

9. ANALYTICAL TECHNIQUES

All of these experimental checks would not have been possible without the truly remarkable development of analytical methods in organic chemistry. For only now has it become possible to find out exactly what compounds were formed during the experiments.

Earlier methods of analysis in organic chemistry were extensions, more or less, of those used by the inorganic chemists. There, the time-honoured way had been to "attack" an unknown compound, by degrading it into smaller fragments which could be determined more easily. By analyzing the nature and the relative amount of these fragments one could in many cases arrive at the correct composition of the original compound by a process of recombination. Although what has been said above is an oversimplification, it still gives the basic approach to early organic chemistry. The information that can be gained in this way is limited. It is possible to determine whether one is dealing with, say, a hydrocarbon, an ester, a fat or a carbohydrate. But within these groups there exist various compounds of single atoms or groups of atoms. Such compounds, having the same molecular composition, but differing only in spatial arrangement of atoms are called *isomers*.

Even in a simple molecular series, such as the saturated hydrocarbons or alkanes, which only consist of atoms of carbon and hydrogen, the number of possible isomers of the same molecular composition rises astronomically with the number of atoms in each molecule. Butane (C_4H_{10}) is the first hydrocarbon with two isomers, followed by pentane (C_5H_{12}) with three; hexane (C_6H_{14}) with five; heptane (C_7H_{16}) with nine isomers, etc. Whereas there are 366.319 different isomeric alkanes with the formula $C_{20}H_{42}$ (PAULING, 1964).

One could not with the tools available to the early organic chemist reliably analyze mixtures of such complex isomeric structures. Most newer techniques, on the other hand, rely on separating and analyzing the compounds as they are, instead of first breaking them up into their component fragments. Only further refinements in structure identification might then be dependent upon the analysis of various fragments of a given compound, such as the application of the mass spectrograph, as referred to in section 11 of Chapter 12.

For full information on these newer techniques we must refer the reader to textbooks such as WILLARD et al. (1958), or handbooks such as the *Comprehensive Analytical Chemistry* series of Elsevier, or *Ullmann's Encyclopädie der technischen Chemie* of Foerst. Moreover, several of these newer techniques have been described on an intermediate level in *Scientific American* (see GRAY, 1951; STEIN and MOORE, 1951; CRAFORD, 1953; KELLER 1961). However, there are three methods which stand out so prominently in this type of research that they merit a special mention. These are the liquid adsorbent chromatography, or chromatography for short; the vapour phase or gas chromatography; and the newer applications of mass spectrometry. These techniques will be described here from examples cited earlier in this chapter and from the description of the chemical fossils in Chapter 12.

The use of *liquid adsorbent column chromatography* was applied already many years ago as a means of separating individual compounds from complex mixtures. The mixture, dissolved in a solvent, is applied to a glass column filled with a specific adsorbent material. Large quantities of pure solvent are then passed through the column and the individual compounds of the original mixture migrate according to the extent of their interaction with the adsorbent and hence the ease with which they are replaced from the adsorbent surface by the solvent molecule determine their retention time on the column.

Recent modifications of liquid adsorbent chromatography have led to the use of the more efficient *paper chromatography* and *thin-layer chromatography*. In these methods either a specially prepared paper, or a thin layer of a specific adsorbent spread upon a glass or metal plate, is employed. A drop of the mixture to be analyzed is placed on one corner of the paper or plate, whereupon the bottom edge is placed in a solvent mixture. The migration of the solvent through capillary action to the top of the chromatogram effects the desired separation. Examples of such separations are found in Fig.25 and 31. Not all of the smudges appearing in this way on a chromatogram represent single compounds. A further separation can be arrived at by turning the chromatogram over 90° and attacking single smudges with a new solvent mixture. This is the so-called *two-dimensional chromatography*.

Recent sophisticated use of the proper solvent with various adsorbent materials has effected many isolations of non-volatile compounds that were previously considered impossible. Carefully controlled conditions allow the identifications of unknown compounds by comparing their relative migration rates to those of known compounds.

Gas chromatography, also called *vapour phase chromatography* and *gas liquid chromatography*, differs from liquid adsorbent column chromatography in that a gaseous mobile phase is led through a column filled with solid particles each coated with a film of stationary liquid. The gas is a mixture of a carrier gas and the product which must be analyzed. The liquid film, coated on the particles, exerts a retarding action on the gaseous compounds which pass through the column, the amounts of interaction depending on their nature. The individual compounds of a given mixture thus become separated and can be detected or collected at the exit of the column.

In gas chromatography minute quantities of organic compounds, of only 10^{-12} g, and occasionally down to the order of 10^{-22} gram, can be detected (SUPINA and HENLEY, 1964). These authors therefore speak of gas chromatography as the "snooper par excellence". It has played a prominent part in the discovery that pesticides form a common element of our daily food, a topic triggered by Rachel Carson's *Silent Spring*. In our subject gas chromatography has been successfully applied to the detection of molecular fossils in very old Precambrian rocks (see sections 10 to 12 of Chapter 12).

The detectors placed at the exit of a gas chromatography column measure some physical constant of the gas flowing by on a moving strip of paper and print

Fig.40. Flowchart of a combined gas-liquid chromatograph and mass-spectrometer. The sample is mixed with the carrier gas and separated in the gas-chromatography column (top). After passing a further separator and an ionizer, it is split in two. One part goes straight to a recorder of the ion current, which in this instrument is used instead of a hydrogen flame detector for instrumental reasons (right). The other part passes through a mass-spectrometer (left), where the original molecules break up into a number of constituent parts, which are specific for a certain type of molecule. Different species of molecules therefore give different records, as is schematically indicated in the four mass-spectra. Variations in the magnetic force of the mass-spectrometer alter the beam of the ionized molecules and their break-up products. This is effected by the magnetic scan control unit, whereby the various beams of ionized molecules of different weight one after the other fall on the slot in the electron multiplier and can be recorded separately. (From EGLINTON and CALVIN, 1967.)

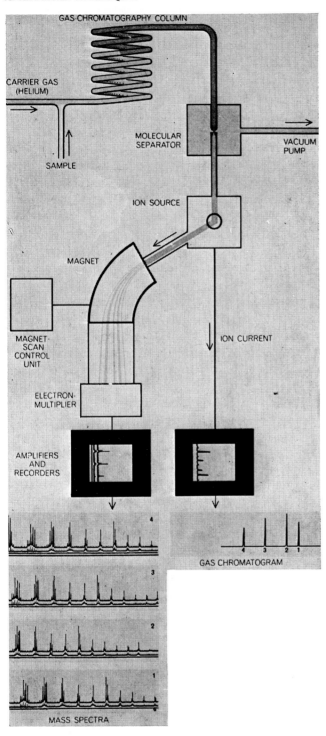

GAS CHROMATOGRAPHY COLUMN

CARRIER GAS
(HELIUM)

SAMPLE

MOLECULAR
SEPARATOR

VACUUM
PUMP

ION SOURCE

MAGNET

MAGNET-
SCAN
CONTROL
UNIT

ION CURRENT

ELECTRON-
MULTIPLIER

AMPLIFIERS
AND
RECORDERS

GAS CHROMATOGRAM

MASS SPECTRA

out the information. So instead of the smudges of liquid chromatography, the record of gas chromatography is a chart in which the various peaks represent different gases (see Fig.69–71). These gaseous compounds, now nicely separated, can also be collected for further analysis, such as by a mass spectrograph (Fig.75, 77).

A hydrogen flame detector is, for instance, often used as detector. This measures the ionized particles formed when the compound mixed with—or *eluted by*, as it is called—the carrier gas is burned. When only the carrier gas leaves the separation column zero conductivity shows on the chart. If, however, a mixture of the carrier gas and some other compound passes through the detector, a peak shows on the recorder chart. The time elapsed before such a peak appears is a very sensitive parameter that often suffices to identify the compound, whereas the area on the chart covered by the peak is a quantitative measure of the amount present.

We met already with the *mass-spectrometer*, or the *mass-spectrograph*, as the case may be, in section 10 of Chapter 3. It is used to determine the exact weight of a given atom or molecule. In the newer techniques of organic chemistry it does, however, play a further reaching role, because it can also be applied to determine the molecular weights of the individual building blocks of larger organic molecules. This faculty resides in the fact that many larger molecules are unstable when ionized. They dissociate into smaller fragments, which in turn may become ionized. These particles are accelerated in the normal way by the magnetic field of the instrument and are separated and identified according to their mass-to-charge ratio. The recorder moreover plots out the mass number of the ionized fragments against their concentration, as indicated by the heights of the various peaks on the charts (see Fig.75, 77). Thus identified, the many fragments can be pieced together mentally, providing additional evidence towards the structure and identification of the parent molecule.

REFERENCES

ABELSON, PH. H., 1966. Chemical events on the primitive earth. *Proc. Natl. Acad. Sci. U. S.* 55:1365–1372.

BERNAL, J. D., 1967. *The Origin of Life*. Weidenfeld and Nicolson, London, 345 pp.

CALVIN, M., 1965. Chemical evolution. *Proc. Roy. Soc. (London)*, Ser. A, 288:441–466.

CALVIN, M., 1969. *Chemical Evolution*. Clarendon, Oxford, 278 pp.

CRAFORD, JR., B., 1953. Chemical analysis by infrared. *Sci. Am.*, 189:8 pp.

ECK, R. V., LIPINCOTT, E. R., DAYHOFF, M. O. and PRATT, Y. T., 1966. Thermodynamic equilibrium and the inorganic origin of organic compounds. *Science*, 153:628–633.

EGLINTON, G. and CALVIN, M., 1967. Chemical fossils. *Sci. Am.*, 216 (1):32–43.

FOX, S. W., 1965a. A theory of macromolecular and cellular origins. *Nature*, 205:328–340.

FOX, S. W., 1965b. Simulated natural experiments in spontaneous organization of morphological units from proteinoid In: S. W. Fox (Editor), *The Origins of Prebiological Systems*. Academic Press, New York, N. Y., pp.361–373.

FOX, S. W., 1965c. Experiments suggesting evolution to protein. In: V. BRYSON and H. J. VOGEL (Editors), *Evolving Genes and Proteins*. Academic Press, New York, N. Y., pp.359–370.

Fox, S. W., 1968a. Self-assembly of the protocell from a self-ordered polymer. *J. Sci. Ind. Res.*, 27:267–274.

Fox, S. W., 1968b. A new view of the "synthesis of life". *Quart. J. Florida Acad. Sci.*, 31:1–15.

Fox, S. W., 1968c. Spontaneous generation, the origin of life and self assembly. *Currents Modern Biol.*, 2:235–240.

Fox, S. W. and Krampitz, G., 1964. Catalytic decomposition of glucose in aqueous solution by thermal proteinoids. *Nature*, 203:1362–1364.

Fox, S. W. and Waehneldt, T. V., 1968. The thermal synthesis of neutral and basic proteinoids. *Biochim. Biophys. Acta*, 160:246–249.

Fox, S. W. and Wang, C. T., 1968. Melanocyte-stimulating hormone: Activity in thermal polymers of alpha-amino acids. *Science*, 160:547–548.

Fruton, J. S. and Simmonds, S., 1958. *General Biochemistry*, 2nd ed. Wiley, New York, N. Y., 1077 pp.

Gray, G. W., 1951. Electrophoresis. *Sci. Am.*, 185: 11 pp.

Groth, W. and Von Weyssenhoff, H., 1959. Photochemische Bildung organischer Verbindungen aus Mischungen einfacher Gase. *Ann. Phys.*, 4:69–77.

Haggis, G. H. (Editor), 1964. *Molecular Biology*. Wiley, New York, N. Y., 401 pp.

Harada, K. and Fox, S. W., 1964. Thermal synthesis of natural amino-acids from a postulated primitive terrestrial atmosphere. *Nature*, 201:335–336.

Huntress Jr., W. T., Baldeschwieler, J. D. and Ponnamperuma, C., 1969. Ion-molecule reactions in hydrogen cyanide. *Nature*, 223:468–471.

Keller, R. A., 1961. Gas chromatography. *Sci. Am.*, 205:11 pp.

Mathews, C. N. and Moser, R. E., 1968. Peptide synthesis from hydrogen-cyanide and water. *Nature*, 215:1230–1234.

Miller, S. L., 1959. Formation of Organic Compounds on the Primitive Earth. In: A. I. Oparin (Editor), *The Origin of Life on Earth*. Pergamon, London, pp.123–135.

Meister, A., 1965. *Biochemistry of the Amino Acids*, 2nd ed. Academic Press, New York, N. Y., 1:119 pp.

Oparin, A. I., 1965. The pathways of the primary development of metabolism and artificial modelling of this development in coacervate drops. In: S. W. Fox (Editor), *Origins of Prebiological Systems*. Academic Press, New York, N. Y., pp. 331–341.

Oró, J., 1965a. Investigation of organo-chemical evolution. In: G. Mamikunian and M. H. Briggs (Editors), *Current Aspects of Exobiology*. Pergamon, London, pp.13–76.

Oró, J., 1965b. Prebiological organic systems. In: S. W. Fox (Editor), *The Origin of Prebiological Systems*. Academic Press, New York, N. Y., pp.137–162.

Pauling, L., 1964. *College Chemistry*, 3rd ed. Freeman, San Francisco, Calif., 832 pp.

Ponnamperuma, C., 1965. A biological synthesis of some nucleic acid constituents. In: S. W. Fox (Editor), *The Origin of Prebiological Systems*. Academic Press, New York, N. Y., pp.221–236.

Ponnamperuma, C. and Gabel, N. W., 1969. Prebiological synthesis of organic compounds. In: A. Renbaum and R. F. Landell (Editors), *Chemistry in Space Research*. In press.

Ponnamperuma, C. and Mack, R., 1965. Nucleotide synthesis under possible primitive earth conditions. *Science*, 148:1221–1223.

Ponnamperuma, C. and Peterson, E., 1965. Peptide synthesis from amino acids in aqueous solution. *Science*, 147:1572–1573.

Rabinowitz, J. S., Chang, S. and Ponnamperuma, C., 1968. Phosphorylation of inorganic phosphate as a potential prebiotic process. *Nature*, 218:442–443.

Schwartz, A. and Ponnamperuma, C., 1968. Phosphorylation of adenosine with linear polyphosphate salts in aqueous solution. *Nature*, 218:443.

Stein, W. H. and Moore, S., 1951. Chromatography. *Sci. Am.*, 184:9 pp.

Steinman, G., 1967. Sequence generation in prebiological peptide systems. *Arch. Biochem. Biophys.*, 121: 533–539.

Steinman, G., Smith, A. E. and Silver, J. J., 1968. Synthesis of a sulfur-containing amino acid under simulated prebiotic conditions. *Science*, 159:1108–1109.

SUPINA, W. R. and HENLEY, R. S., 1964. Gas chromatography- snooper par excellence. *Chemistry*, 37,12–17.

WAEHNELDT, T. V. and FOX, S. W., 1968. The binding of basic proteinoids with organismic or thermally synthesized polynucleotides. *Biochem. Biophys. Acta*, 160:239–245.

WILLARD, H. H., MERRIT, L. L. and DEAN, J. A., 1958. *Instrumental methods of analysis*, 3rd ed. Van Nostrand, Princeton, N. J., 626 pp.

WILSON, A. T., 1960. Synthesis of macromolecules. *Nature*, 188:1007–1009.

YOUNG, R. S. and PONNAMPERUMA, C., 1964. Early evolution of life. *B.S.C.S. Pamphlets, 11.* Heath, Boston, 29 pp.

Chapter 7 | Stages in Biopoesis

1. INTRODUCTION

Biopoesis is the term used by BERNAL (cf. 1959a, b; 1961) for the process of evolution from non-living to living. Although it is not widely used by other authors, and has been more or less superseded by the term *biogenesis*, I prefer to retain Bernal's term. At least in this chapter, in which we will enquire into the various steps which must have occurred, at one place or another, in the chain of events leading from non-living to living. In the preceding chapter information has been offered on the many experimental checks which exemplified the possibility of a formation of "organic" molecules by inorganic reactions under the primeval atmosphere. The overall result of these investigations was that there are many possible ways along which such reactions can take place.

The very fact that a number of possible pathways has already been demonstrated to exist even indicates that under an anoxygenic atmosphere such reactions would not be single, or rare events. Quite to the contrary. Although we do not yet know at present which would have been the preferred pathway in any set of circumstances, or if there were perhaps a number of different pathways operating simultaneously, such reactions must have been very common indeed under an anoxygenic atmosphere. To return to the parable of the rain and the salt crystal in section 8 of Chapter 4, the formation of "organic" molecules under the primeval atmosphere must have been just as normal an event as the formation of rain drops and salt crystals, or the rusting of iron is under the present atmosphere.

The next step in biopoesis would then be the formation of polymers, such as peptide chains and nucleic acids, from the building blocks produced by the inorganic reactions of the first step. Again, the experimental checks described in the preceding chapter indicate that there are various possible pathways in which inorganic reactions can produce such results. Even a further polymerization to proteins and other larger molecules seems to be well within the possibility of inorganic synthesis.

Accordingly, we come to a further step in the evolution from non-living, that is the accumulation of such "organic" molecules into definite particulate groups of larger—though often still microscopic—dimensions. For this step a process has been invoked which long has retained a certain nebulous character, i.e. *coacervation*. We will see, however, that in this field too modern research has opened up promising vistas.

Another step in biopoesis rests on a fundamental property of life, that is the presence in every living organism of a wall, isolating the organisms from the pernicious influence of the outer world. Consequently the formation of walls, or *membranes* around the particles accumulated earlier will be a further step in biopoesis.

At present such walls are often extremely sophisticated in structure and in function. But membranes developed during the early evolution of life were probably of a much simpler nature than the present ones. We will see further in this chapter how even models have been proposed for the development of simple membranes through inorganic processes. So the first membranes could possibly have developed already very early in biopoesis.

Another stage in biopoesis must have been *metabolism*. Again, as we will see in the section dealing with this property, metabolism need not have been a very efficient process at the start, only gradually developing in its present-day intricacy and efficiency.

Once organized and surrounded by a membrane, such prebiological material still had to find a way to *duplicate* itself, before it could be called truly living. So this is another stage in biopoesis. This does not mean that individual duplicating mechanisms could not have existed before such an advanced stage was achieved. Such duplication may even have occurred in chemical systems outside the membranes. However, a coordination of duplicating systems toward a common goal can only be accomplished within a membrane and as such forms one of the advanced stages in the transition from non-living to living.

This breaking up into a number of stages does not mean that Bernal visualized the transition from non-living to living as an intermittent process. But although the process is a fluid one, and even though the development of various stages might have taken place concurrently, there is a certain hierarchy in this evolution. As we have seen, it is not before atoms of hydrogen, carbon, oxygen, nitrogen, etc. are present, that they can start to combine into the simplest molecules. Only then could the smaller "organic" molecules be synthesized inorganically. Whereas only after condensation and subsequent polymerization, could the formation of larger units have taken place. Only from then onward were coacervation, the forming of membranes, duplication and metabolism possible. Moreover, as we found already indicated in the preceding chapter, the environment leading to the formation of each stage will have been different from stage to stage.

This will accordingly be the theme of this chapter, and an illustration of Bernal's ideas may be used to close this introduction (Table XI).

TABLE XI

SOME STAGES OF AGGREGATION IN BIOLOGICAL SYSTEMS

(From BERNAL, 1959b)

Name of particle	Nature of last stage binding	Order of magnitude of molecular weight	Order of magnitude of particle dimensions	Examples
Simple molecule (monomer)	homopolar bonds	50–200	10 Å^3	amino acids, purines, porphyrins, sugars, lipids
Chain polymer (homo or hetero)	the same	1000–100,000	5 × 10 × 1000 Å	silk fibroin, β-type, denatured proteins, cellulose, rubber
Coiled polymers	hydrogen bonds or S–S links	the same	10 × 10 × 500 Å	coiled fibrous protein, α-type, deoxyribosenucleic acid
Folded or coiled-coil polymers Globular particles	the same	10,000–100,000	(50 Å)^3	smaller globular proteins, ribonuclease
Homogeneous agglomerated particles Twinned fibres	ionic or cryo-hydric forces	50,000–1,000,000	(100 Å)^3 20 × 20 × 1000 Å^3	larger globular proteins, haemoglobin, seed globulins, haemocyanin, fibrous insulin, collagen
Heterogeneous agglomerated particles or fibre aggregates	the same	10,000,000	(200 Å)^3 100 × 100 × 5000 Å	nucleoproteins, lipoproteins, mucoproteins, etc., smaller viruses

2. INORGANIC SYNTHESIS OF SMALLER AND LARGER "ORGANIC" MOLECULES

The first two steps of biopoesis, the formation of smaller "organic" molecules by inorganic reactions, and their condensation and polymerization to larger molecules, have already been treated in Chapter 6, and will not be further discussed here.

Although inorganic synthesis of "organic" compounds has been shown to be possible in a simulated anoxygenic environment, similar experiments in a simulated atmosphere containing oxygen have only given negative results (GETOFF, 1962). This might be due to destruction by oxydation which overrides the synthesis, or to other factors, but it underlines the importance of the anoxygenic primeval atmosphere for the development of life. In this atmosphere the influence of the shorter ultraviolet sunlight is, as we saw, of predominant importance. So, although the other forms of energy certainly will have played their part, I will, for shortness' sake, in this text use the "shorter ultraviolet sunlight" as an indicator for the various kinds of energy available in the primeval anoxygenic atmosphere.

Another point to retain is that the development of the larger "organic" molecules, which must have played their part during the transition from pre-life to life, will have been a step-by-step process. We have no reason to postulate the sudden appearance of a naked gen or an isolated protein molecule. Only the compounds formed in large numbers could have stood a chance of becoming abundant enough to form the basis for the next step.

In these early stages the chemical processes of pre-life will have been almost ridiculously simple, when compared with the biochemical processes of present-day life, which have had some three billion years to attain perfection through the influence of mutations and the weeding-out-of-the-weaker. It does not matter whether the chemical processes of pre-life have no survival value at present, but whether they did have so in the much less complicated environment prevailing on the surface of the earth some three billion years or more ago.

An example of such a situation can be found in the decomposition of hydrogen peroxide, H_2O_2—a common product of the irradiation of water by ultraviolet sunlight—by the enzyme catalase, as studied by Calvin in 1962 (cf. CALVIN, 1969). The hydrated ferric ion is capable of splitting up H_2O_2, which is a purely inorganic process. The yield of this process is, however, very low, when compared to similar actions by heme or by the enzyme. The ferric ion is only capable of splitting up 10^{-5} mol/sec. When incorporated in an organic porphyrin molecule (Fig.45) it already has an activity one thousand times as large, whereas if the porphyrin molecule is included in the protein component of the enzyme catalase, the ferric ion's activity is ten billion times as strong. But, although at present the influence on the degradation of H_2O_2 through the inorganic action of the hydrated ferric ion is negligibly small, it might well have been of relative, or even of dominant, importance during pre-life, when organic substances such as heme and enzymes did not yet exist.

Before proceeding with our study of biopoesis, we must at this point still stress one aspect of the great variation in energy sources capable of inorganically synthesizing "organic" compounds, which has up to now not yet received enough attention. This is that, whereas the synthesis of the smaller "organic" molecules seems to be dependent on energy-rich sources such as ultraviolet irradiation and/or electric sparking (read:lightning), their combination into the larger "organic" molecules is possible through the influence of less destructive sources of energy, such as heat and chemical energy (compare the influence of cyanide and dicyandiamide, reported in section 7 of Chapter 6). It seems probable that, whereas the inorganic synthesis of smaller "organic" compounds took place at the surface of the earth under the influence of ultraviolet irradiation or of lightning, the further condensation into larger "organic" molecules has taken place in entirely different environments, from which the influences of the shorter ultraviolet light and of lightning were excluded. It stands to reason we may postulate that it is in these latter environments that the transition from pre-life to life has taken place.

To conclude, we might quote RAUWS, who already in 1964, in a review article on prebiotic organic chemistry, a paper which has not become as widely known as it deserves because it is written in Dutch, states that the experimental checks have proved beyond doubt that carbon compounds, whether "organic" or organic, are not different in principle from other chemical compounds. According to Rauws, further experimenting will supply us with much more detail but not with much more fundamental insight, because the possibility of inorganic synthesis of "organic" material has already been amply proved. We will now have to direct our research into the complicated interplay of these compounds, into the dynamics of structure and of metabolic—or, as long as it applies to pre-life—of "metabolic" activities.

3. COACERVATION

As has been indicated in the introduction to this chapter, the next step in biopoesis is less well understood than the earlier steps in this chain of events. It is the combination of "organic" macromolecules, formed earlier by inorganic reactions, into particles. The Dutch chemist BUNGENBERG DE JONG (1936) has already at an early date proposed the term *coacervation* for this process. But although the term has since been widely used, it has as yet scarcely acquired a meaningful background.

Perhaps the main reason for this lies in the complex nature of coacervation particles. No two coacervate particles need to be of similar size or weight or composition. This is a horrible situation for a chemist, whose whole research is centered upon finding out by analysis the exact characters of a given compound. Moreover, coacervates often form a sort of bluish or black, gluey or rubbery material of a repulsive sort. PIRIE (1965) referred to them as "The black gunk which unskilled chemists get when they are trying to make something the textbooks say they ought to get,... the stuff they throw down the sink". "Goo" is often used instead of "gunk" in English literature, and it seems as if the mere sound of these words already indicates the chemist's distaste. BLOIS' (1965) paper on random polymers and OPARIN's (1965) on metabolism in coacervate drops may serve as an illustration of the difficulties encountered in experimenting in this field.

A new light has since been thrown on this matter from an unexpected angle. Oceanographers have in the last decade been reporting on the unsuspectedly wide

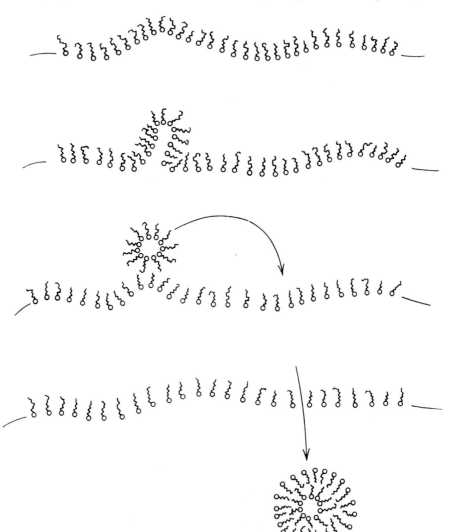

Fig.41. Formation of coacervate droplets through wave action upon a monomolecular film of polar molecules at an air–water interface. The polar end of each molecule is represented by the open circle, the non-polar end by the zig-zag line.

The molecules depicted here in chemical shorthand are lipids which are strongly polar in that they possess a polar and a non-polar end. Any molecule with similar properties will, however, tend towards the same sort of reactions. (From HAGGIS, 1964a.)

dissemination of organic particulate matter in sea water. First detected in Long Island Sound, such aggregates are found to occur not only in coastal waters, but also in the open oceans. RILEY (1963; see also WANGERSKY, 1965) described such aggregates as being delicate and plate-like, ranging in size from 5 μ to several mm in diameter. It has been found experimentally that by leading a stream of air bubbles through sea water such particles can be formed. Hence bubbling is thought to be the main factor in the formation of such particles, and in this type of coacervation. Bubbling is of course omnipresent in nature in waterbodies of all sizes due to wave action.

It is thought that molecules which exhibit a special sort of polarity in that they contain a polar end and a non-polar end, such as soaps, steroids, etc., can become involved in this type of coacervation. Such molecules are surface-active and will tend to form thin films of a layer one molecule thick at the water surface. Or, more technically expressed, monomolecular films at an air–liquid interface. Such films will of course be broken up by wave action. As has been schematically expressed in Fig.41, the molecules in the droplets will tend to rearrange in a

Oil

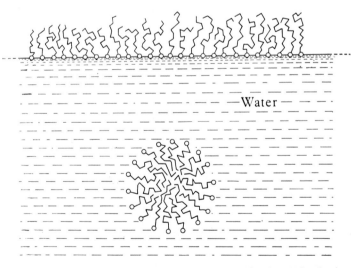

—Water— — — —

Fig.42. Schematic illustration of the arrangement of polar molecules in coacervate droplets in water and in oil and of a monomolecular layer at the oil–water interface. (From HAGGIS, 1964a.)

globular pattern, placing themselves radially, with the polar end of the molecules at the surface. When such droplets fall back into the water, they may break through the surface film, and remain immersed in the waterbody as small, separate, coacervate droplets.

4. FORMATION OF MEMBRANES

Another, possibly a further, but perhaps a step narrowly related in time, is the formation of membranes. The main function of membranes is to allow living matter to have both a composition and an energy level different from its environment. In present life these membranes are well developed, and make it possible for life to have a composition and energy level strongly different from its surroundings. In vertebrate animals, for instance, most species have an inside fluid concentration, which has properties similar to sea water, regardless of the fact whether the animal lives in the sea, in fresh water, or in the air. In the early days membranes will not have been as sophisticated as those of present life. Some sort of membrane activity must, however, already have functioned at the earliest stages of metabolism. For metabolism always results in a change of concentration of one or more substances, and hence in a difference in composition "within", as against "without".

In order to be able to maintain internal conditions different from the exterior world, membranes are normally so constructed that they let pass some molecules, but obstruct the passage of others. Size and/or electric charge of the molecules forming the membranes in question are the main factors by which membranes operate in this selective way. Such membranes are called *semi-permeable* and their activity results in a difference in osmotic pressure on either side of the membrane.

Just as in formation of coacervate particles, it is thought that the formation of primitive membranes could have started by an alignment of polar molecules on any air-water or oil–water interface. The membranes of present life are formed by double monomolecular layers of lipids. Lipids play a predominant part in present membranes, but non-lipid membranes are also known. A lipid membrane has been schematically illustrated in Fig.43.

5. METABOLISM

Once the stage of formation of membranes has been passed, metabolism may follow. Metabolism in its simplest form is no more than ingestion and egestion plus the intervening reaction or chain of reactions which transform the absorbed material into the waste product to be egested. For instance, in digestion it is the ingestion of food and the egestion of wastes. In respiration it is the ingestion of oxygen and the egestion of carbon dioxide.

The processes intercalated between ingestion and egestion are, in present life, performed by rather complicated chain reactions. But basically similar, very primitive reactions could already occur in non-biological compounds at the molecular level. A glance at Fig.43 will show, very schematically, how such a possibility would work. The larger, globular molecules, adsorbed by the membrane, could easily lose some component to a nearby polar molecule, which in turn could transfer this component to another molecule.

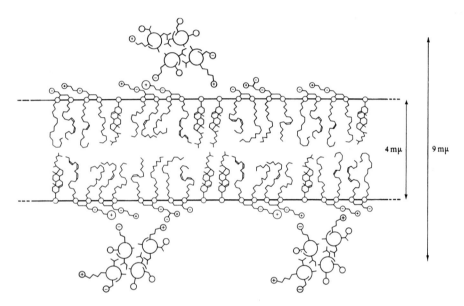

Fig.43. Schematic illustration of a double monomolecular layer of polar molecules, serving as a membrane, and of globular protein molecules, adsorbed to this membrane without any structural change. The monomolecular layers which form the membranes are built up by a repetition of building blocks each formed by two molecules of a lipid and one of cholesterol. This is a sophistication of present life, and not basic to the functioning of a membrane. Any film of polar molecules will be able to serve as such, provided they allow some molecules to diffuse through them. (From HAGGIS, 1964a.)

A more sophisticated model has been proposed by GRANICK (1965) which might indicate that processes akin to organic photosynthesis and to respiration may already occur at the molecular level. The model proposed by Granick, and called by him a "photovoltaic mineral unit", uses elements which should be readily available in the "thin soup" and in its substratum, the crust of the earth, during the early geologic history. Moreover, it is thought it would accomplish these simple reactions comparable to either photosynthesis or respiration, dependent upon the nature of the "organic" compounds with which it was surrounded. It could in this way serve as an inorganic powerhouse, preceding the more powerful and more specialized organic reactions of early life (Fig.44).

Although the energy involved in such transfers is much smaller than that used by present-day metabolism, such transfers would become meaningful at the moment the reactions are not random but show some direction.

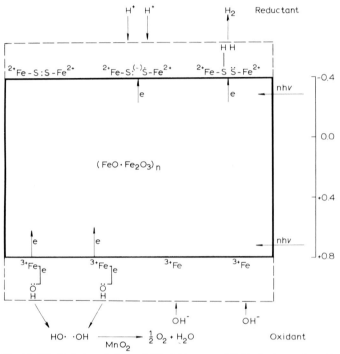

Fig.44. The "photovoltaic model" of GRANICK (1965). It shows how the basic reactions leading to photosynthesis and respiration, that is reduction and oxidation respectively, can be performed by minerals available on the earth's surface.

The mineral postulated is an impure magnetite, $FeO \cdot Fe_2O_3$, of quite common occurrence. Due to the complex juxtaposition of ferric and ferrous ions in the crystal lattice it may serve as an electron trap. Moreover, it easily absorbs light, and thus energy, and also acts as a semiconductor. If such a mineral would become coated with a thin layer of other elements, again a quite common feature in the inorganic mineral chemistry on the earth's surface, it would upon the absorption of sunlight, act either as a reductant, or as an absorbant.

The first possibility is indicated at the upper surface of the model, where the magnetite is thought to be covered by a monolayer of FeS and FeS_2. Light quanta striking the surface will cause electrons (e) to flow upwards, towards the surface of the magnetite. As is schematically indicated, an $S–S$ bond would be reduced to SH, whereas H_2 would ultimately escape.

The lower surface of the magnetite is thought to be covered by a monolayer of ferric hydroxide and some manganese oxide, MnO_2. Absorption of light quanta by this black coloured layer will cause electrons to flow from the hydroxyl ions, via the iron ions, and ultimately upwards, that is into the interior of the magnetite. This will lead to oxidation of the hydroxyl ions, and water will be formed eventually.

So, although simple molecular reactions of ingestion and egestion seem to be the normal state of affairs at the molecular level, once larger polymers and polar molecules have been formed, we must still have some kind of coordination. We may assume that the environment exerted a selection pressure upon such simple molecular reactions, so that some reactions were favoured, just as functions the environmental selection pressure in modern life. But we have, as yet, no idea how such a selection pressure would function.

Consequently, although the development of metabolism seems quite feasible to theorists like Oparin and Bernal, it still remains, as far as the experimental checks are concerned, one of the great stumbling blocks in the experimental, in vitro studies of the origin of life.

6. DUPLICATION

The ultimate stage in biopoesis then would be the acquisition of the faculty of duplication, through which life can perpetuate itself. As before, the faculty of duplication in present life is a process so complicated that one could hardly believe it has evolved gradually, step by step, from non-living simple precursory reactions. Still, if we want to arrive at an understanding of the origin of life through natural causes, it is exactly these simple steps to which we have to turn our attention.

BERNAL (1965) rather pointedly stressed his impression that many biochemists consider life to begin with ribonucleic and deoxyribonucleic acids, the nucleic acids that at present somehow carry the heriditary information in reproduction, whilst most of the biochemical literature is concerned with the processes involved in duplication at this stage, in the "genetic code" and such. To Bernal, a structural crystallographer, the earlier development is the more interesting aspect. He stresses that reproduction is a much more general process than is normally thought, and may possibly have been reached in the early development of life by a number of pathways. According to him, the essential principle of reproduction is already found in crystallization, where it is a consequence of the interaction between molecules of the crystal and the atoms and ions in the fluid.

To achieve reproduction it is necessary for the object that is to be reproduced to be accessible to the atoms or the smaller molecules which must unite to form a copy of it. Therefore the object must be linear or planar. We may confine ourselves to the most simple model, that of a molecule built up by a number of smaller units all arranged one after the other in a single row. Such a molecule is called a *linear polymer*. Since life must be able to execute several processes, such as growth and metabolism, it must convey information about these processes during duplication. Now a polymer consisting of a repetition of the same building blocks does not normally convey this information. Just as words, to be able to convey any meaning at all, must consist of a sequence of various different smaller units, the letters, the polymers active in duplication must also be built by a sequence of different building blocks. Such a molecule is called a *linear heteropolymer* (CRICK, 1968; ORGEL, 1968).

BERNAL points out that our knowledge of crystallography makes it probable that such a structure would be more stable in a coiled form than if it was built in a straight line. Consequently it is probable that even the most primitive linear heteropolymers used in early processes of duplication had already the coiled form so characteristic for the much larger and more sophisticated nucleic acids which transfer information during duplication in present life.

A confirmation of these ideas can be found in CALVIN (1965), who reported upon the coiled spirals of the peptide chains forming the proteins (Fig.14). It was found that peptide chains, when uncoiled artificially, through changes of either temperature or pH, will coil again in exactly the same form when the original conditions have been restored. It follows that the coiling of linear polymers is not something akin to life, but a character dependent upon the thermodynamic properties of the compounds themselves.

Many more examples of such a reassembly of degraded organic macro-molecules have since been cited, not only of proteins, but also of pigments and enzymes. A full treatment can be found in Chapter X of Professor CALVIN's (1969) latest book. This does, however, only apply to the three-dimensional structure of these polymers. Quite another question is whether such polymers, which are always heteropolymers, are built up by a random series of smaller molecules, or if there is a sort of orderly sequence of these building blocks. In present-day organic polymers an ordered sequence is always found, as everybody will know since the double helix and the genetic code have become household terms. How-ever, Fox and collaborators (FOX and NAKASHIMA, 1967; FOX, 1968, 1969) have shown that already in their proteinoids, which have been synthesized inorganically (see section 6 of Chapter 6) a well-expressed orderly sequence of amino acid and amide building blocks is found.

I will not go into the chemical details of this question, referring the interested reader to the publications cited, but retain only that the proof of such an orderly structure can be found by decomposing the proteinoids once synthesized, and analysing their components. The important feature is that these ordered polymers are synthesized inorganically, without the presence of nucleic acids. Consequently already in pre-life ordered polymers, rightly called *self-ordered polymers*, can synthesize in protein-like compounds. Neither life, nor the presence of nucleic acids form a necessary prerequisite for their formation. Such an orderly sequence appears explainable on the basis of the three-dimensional structure and the electro-nic configuration of the individual amino acids (FOX, 1965).

This latter consideration is important for those people who do not readily accept Professor Fox's proposed environment of rainfall on eruptive volcanoes as very propituous (see section 7 of Chapter 6). For it does not matter in what environment such heteropolymers have grown, in the hot environment proposed by Fox, or in the aqueous environment in the presence of hydrogen cyanide or dicyanamide, as favoured by Calvin, the influence of the three-dimensional structure and the electronical configuration of individual amino acids remains similar. So we may well use the results obtained by Fox from an environment which evidently presents less experimental difficulties than those used by the other organic chemists, and transpose them to other environments which might seem more attractive from a geological standpoint.

Eventually the inorganically synthesized proteinoid or "Urprotein" will react with inorganically formed nucleotides and the present overriding influence of the nucleic acids will have come about. This is a field in which several preliminary investigations have been reported so far, which are summarized in Fox (1969; see also WAEHNELDT and Fox, 1968; YUKI and Fox, 1969). A first experimental check as to the possibility of such a "pre-DNA" evolution has recently been presented by LACEY and PRUITT (1969) who, upon setting up an interaction between mono-nucleotides and poly-amino-acids, found that the glycine codon GGG of the present-day genetic code formed inorganically. It can be safely predicted that more of these "pre-DNA" syntheses will be found in the future.

7. MUTANTS ACQUIRING A NEW SKILL: ORGANIC PHOTOSYNTHESIS

Even the earliest life, and also pre-life, must have produced mutants. For mutation is something akin to both the "organic" and organic macromolecules. Although still not well understood, mutation seems to be the result of some minor switchover during growth, due to some energy flux, probably an external one, which might be either of thermal or of radiative character.

In present life mutants seem to develop mainly from a faulty rearrangement of a complete nitrogen base, somewhere along the line of a nucleic acid (cf. SONNEBORN, 1965; FREESE and YOSHIDA, 1965; CRICK, 1966). In nucleic acids the various bases follow each other in a rigid sequence, which is different for different organisms. This sequence forms the basis for the "genetic code", which, as is well known, ensures exact duplication upon reproduction. When, through some mishap, exact duplication fails to be achieved, a mutant occurs. But in early life, when the reproductive process had not yet reached its present-day perfection, mutants may well have developed from other, more simple, mistakes at the molecular level.

Organic photosynthesis, or assimilation, is at present effectuated by an extremely complex series of biochemical reactions (CALVIN and BASHAM, 1962).

But we must not forget that the plant kingdom has had around three billion years in which to weed out the less complex and consequently less successful pathways. Of the early stages of photosynthesis, the "primary act" of WASSINK (1963), in which we are especially interested here, we know as yet nothing.

Somewhere, sometime, mutations of early life will have produced compounds capable of organic photosynthesis, i.e. of organic assimilation. They acquired the unique capability of dissociating carbon dioxide (CO_2) into carbon (C) and to liberate free oxygen (O_2). The carbon could be used to build more organic matter for the living organism, whilst the oxygen could excape into the atmosphere. As we have indicated already in Chapter 4, and as will follow from Chapters 8 to 15, the results of this new capability have in the end been overwhelming. We must, however, make a clear distinction between the effects of this new mutation upon the living forms themselves, and the general result of organic photosynthesis as a whole.

The benefit of organic photosynthesis for the organisms themselves which had acquired this capability lies in the much greater amount of energy freed by this process, than by other sorts of metabolism. Any reactions in which free oxygen is not involved, such as fermentation, or the nitrate-, sulphate- and carbonate-reducing reactions, have a much lower energy level than that of the assimilation of carbon dioxide. This new way of life consequently gave an immediate advantage in the struggle for life over those contemporary organisms which were still based on other types of metabolism.

The general result of organic photosynthesis will have lain in production of free oxygen for the atmosphere. Although its effect will have been slow and gradual, and will have presumably taken a couple of billions of years (Chapter 16), it will, in the end, have dominated over the benefits acquired by the photosynthetic organisms themselves. For through organic photosynthesis free oxygen will have become a more and more important part of the atmosphere.

8. RESPIRATION

Once enough oxygen had appeared in the atmosphere, respiration could start; probably as the result of another mutation. Respiration is, in a way, the opposite of organic photosynthesis and produces carbon dioxide. On the other hand it has much in common with photosynthesis. On the one hand, it is a metabolic process involving a much higher energy field than that of other processes such as fermentation. On the other hand, both processes, so opposite in their results, are based in present life upon composite molecules which are very much alike, i.e. heme and chlorophyll (Fig.45).

All present aerobic life, both animals and plants, uses respiration for its

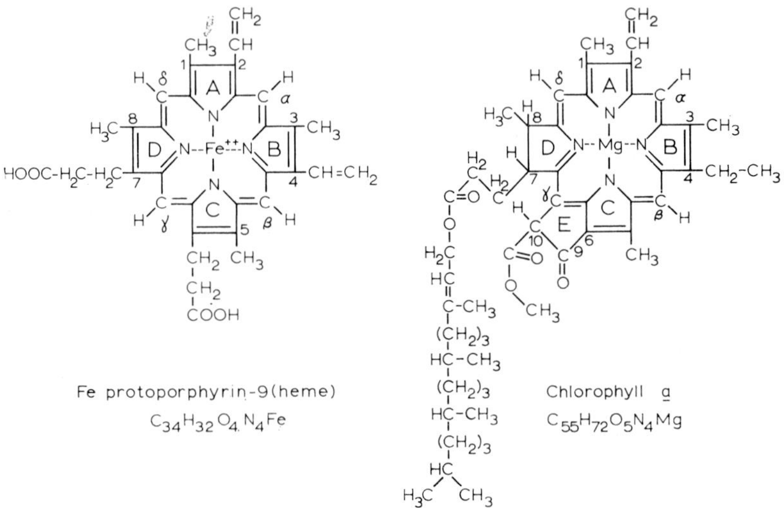

Fig.45. Structure of heme and chlorophyll. The complex molecules of heme and chlorophyll are both built around a single metal atom surrounded by four atoms of nitrogen. In heme this is iron, whilst in chlorophyll it is magnesium.

The long "tail" of the chlorophyll molecule is not easily destroyed. It is this tail, more or less completely preserved over billions of years, which, under the names of pristane, etc., forms the basis for the molecular fossils described in sections 10–12 of Chapter 12. (From GRANICK, 1965, fig.1.)

life processes[1]. The difference being that animals have respiration only, whereas plants have both photosynthesis and respiration. In plants organic photosynthesis is the more important of the two processes. This permits the plant kingdom not only to produce the atmospheric oxygen needed for its own respiration, but also for that of the animal kingdom.

We will see in the next chapter that there is a definite level below which animals of various classes cannot freely respiration; beneath which they cannot live. Bacteria are able to respire at a much lower level than animals, i.e. at about 1% of the present level of 21% of free atmospheric oxygen. Moreover, there are many bacteria, called the *facultative aerobes*, and which could perhaps better be called *facultative respirators*, which change their metabolism from fermentation to respiration at this level. Although it is possible that respiration has started at an already much lower level of free oxygen—and GRANICK's (1965) "photovoltaic powerhouse" model (Fig.44) points strongly to such a possibility—there seems to be a certain advantage of fermentation and related metabolic processes over respiration

---

[1] Another type of oxydation, called *oxygenase*, has not been taken in account here. It is not yet clear what its importance is, relative to respiration; whilst it possibly does not use molecular oxygen but derives its oxygen from oxides such as ferrioxide (HAYASHI and NOZAKI, 1969),

below a level of 1% of the present free atmospheric oxygen. We will therefore schematically assume in Chapter 15 that respiration started at the attainment through organic photosynthesis of this level of free oxygen.

REFERENCES

BERNAL, J. D., 1959a. The problem of stages in biopoesis. In: A. I. OPARIN (Editor), *Origin of Life on the Earth*. Pergamon, London, pp.38–53.
BERNAL, J. D., 1959b. The scale of structural units in biopoesis. In: A. I. OPARIN (Editor), *Origin of Life on the Earth*. Pergamon, London, pp.385–399.
BERNAL, J. D., 1961. Origin of life on the shores of the ocean. In: M. SEARS (Editor), *Oceanography*. Am. Assoc. Advan. Sci., Washington, D. C., pp.95–118.
BERNAL, J. D., 1965. Molecular matrices for living systems. In: S. W. FOX (Editor), *The Origin of Prebiological Systems*. Academic Press, New York, N.Y., pp.65–88.
BLOIS, M. S., 1965. Random polymers as a matrix for chemical evolution. In: S. W. FOX (Editor), *The Origin of Prebiological Systems*. Academic Press, New York, N.Y., pp.19–33.
BUNGENBERG DE JONG, H. G., 1936. La coacervation, les coacervats et leur importance en biologie. *Actualités Sci., Ind.*, 397, 398—Herman, Paris, 52 + 56 pp.
CALVIN, M., 1965. Chemical evolution. *Proc. Roy. Soc. (London), Ser. A*, 288:441–466.
CALVIN, M., 1969. *Chemical Evolution*. Clarendon, Oxford, 278 pp.
CALVIN, M. and BASSHAM, J. A., 1962. *The Photosynthesis of Carbon Compounds*. Benjamin, New York, N.Y., 127 pp.
CRICK, F. H. C., 1966. The genetic code: III. *Sci. Am.*, 215 (4):55–62.
CRICK, F. H. C., 1968. The origin of the genetic code. *J. Mol. Biol.*, 38:367–379.
FOX, S. W. (Editor), 1965. *The Origin of Prebiological Systems*. Academic Press, New York, N.Y., 482 pp.
FOX, S. W., 1968. Self-assembly of the proto-cell from a self-ordered polymer. *J. Sci. Ind. Res.*, 27:267–274.
FOX, S. W., 1969. Self-ordered polymers and propagative cell-like systems. *Naturwissenschaften*, 56:1–9.
FOX, S. W. and NAKASHIMA, T., 1967. Fractionation and characterization of an amidated thermal 1:1:1-proteinoid. *Biochim. Biophys. Acta*, 140:155–167.
FREESE, E. and YOSHIDA, A., 1965. The role of mutations in evolution. In: V. BRYSON and H. J. VOGEL (Editors), *Evolving Genes and Proteins*. Academic Press, New York, N.Y., pp.341–355.
GETOFF, N., 1962. Über die Bildung organischer Substanzen aus Kohlensäure in wässeriger Lösung mittels 60Co-Gamma-Strahlung. *Z. Naturforsch.*, 17 b:751 –757.
GRANICK, S., 1965. Evolution of heme and chlorophyll. In: V. BRYSON and H. J. VOGEL (Editors), *Evolving Genes and Proteins*. Academic Press, New York, N.Y., pp.67–88.
HAGGIS, G. H., 1964a. The structure and function of membranes. In: G. H. HAGGIS (Editor), *Molecular Biology*. Wiley, New York, N.Y., pp.151–192.
HAGGIS, G. H., 1964b. The origin of life. In: G. H. HAGGIS (Editor), *Molecular Biology*. Wiley, New York, N. Y., pp.315–354.
HAYASHI, O. and NOZAKI, M., 1969. Nature and mechanisms of oxygenases. *Science*, 164: 389–396.
LACEY JR., J. M. and PRUITT, K. M., 1969. Origin of the genetic code. *Nature*, 223:766–804.
OPARIN, A. I., 1965. The pathways of the primary development of metabolism and artificial modelling of this development in coacervate drops. In: S. W. FOX (Editor), *The Origin of Prebiological Systems*. Academic Press, New York, N.Y., pp.331–341.
ORGEL, L. E., 1968. Evolution in the genetic apparatus. *J. Mol. Biol.*, 38:381–393.
PIRIE, N. W., 1965. Discussion. In: S. W. FOX (Editor), *The Origin of Prebiological Systems*. Academic Press, New York, N.Y., p.33.

RAUWS, A. G., 1964. Prebiotische organische chemie, een kritisch overzicht. *Chem. Weekblad*, 60:361–369.

RILEY, G. A., 1963. Organic aggregates in seawater and the dynamics of their formation and utilization. *Limnol. Oceanog.*, 8:372–381.

SONNEBORN, T. M., 1965. Degeneracy of the genetic code: extent, nature and genetic implications. In: V. BRYSON and H. J. VOGEL (Editors), *Evolving Genes and Proteins*. Academic Press, New York, N.Y., pp.377–397.

WAEHNELDT, T. W. and FOX, S. W., 1968. The binding of basic proteinoids with organismic or thermally synthesized polynucleotides. *Biochim. Biophys. Acta*, 160:239–245.

WANGERSKY, P. J., 1965. The organic chemistry of seawater. *Am. Scientist*, 53:358–374.

WASSINK, E. C., 1963. The nature of the primary act in photosynthesis. In: H. TAMIYA (Editor), *Mechanism of Photosynthesis—Proc. Intern. Biochem. Congr., 5th, Moscow, 1961*, 3: 100–121.

YUKI, A. and FOX, S. W., 1969. Selective formation of particles by binding of pyrimidine polyribonucleotides or purine polyribonucleotides with lysine-rich or arginine-rich proteinoids. *Biochem. Biophys. Res. Commun.*, 36:657–663.

Chapter 8 | Stages in the Early Evolution of Life

1. HYPOTHETICAL CHARACTER OF ASSUMPTIONS ABOUT THE EARLY EVOLUTION OF LIFE

We will in this chapter consider some aspects of the early evolution of life, which, at some stage or other must have played their part during the early evolution of life. It is, however, impossible to tell in which sequence they might have occurred. For in the whole chain of events which eventually led to life as we know it, the stages of the early evolution of life remain the most obscure. Consequently all of our deductions about these early stages of life tend to be even more hypothetical than those on the preceding or the subsequent stages.

The reason for this is that life at that time had outgrown the simple molecules, whereas it had not yet developed discrete morphological forms. In the simple molecules, which, as we saw in the preceding chapters, form the basis of life, the properties of the individual atoms making up the molecules play a discernible part. Physico-chemical considerations, drawn from the known properties of the elements, are able to indicate within certain limits, what might and what might not have happened. Whereas, on the other hand, fossils may, again within certain limits, tell us what has happened in the later stages of life.

So, during the early stages of the evolution of life the molecules going into living organisms have already outgrown the possibility of physico-chemical predictions, whereas the absence of discrete morphological forms in these early stages of life precludes the possibility of providing recognizable fossils. Nevertheless, there are certain considerations to be drawn as to what must have happened anyway sometime during the early evolution of life, whereas certain basic facts of present-day life also impose their restrictions on what may have happened during its early evolution. Accordingly, although we cannot tell exactly what happened when, we had better take these aspects into consideration, because, to a certain extent, they help to clarify the early evolution of life.

In a general way, we must be concerned with the environment of the primeval atmosphere. That is with the difference between what is now normally called *anaerobic* and *aerobic*, and what in this context had better be called *anoxygenic* and *oxygenic*. An important aspect of this early environment is that life probably was always *aradiatic* for the shorter ultraviolet sunlight. We must also take note of an important property of present-day bacteria in connection with the so-called *Pasteur Point*. This indicates the oxygen level above which some bacteria may live through respiration, whereas below this level they can have recourse to fermentation or to still other metabolic pathways. Another major fact of present-day life, that has only recently transpired through the wide application of electron microscopy, is that all present life forms are characterized by the possession of either one or the other of two cell types. These are called the *procaryotic* and the *eucaryotic cells*. The procaryotic cells are the more simply built, and as the name implies, are thought to be the more primitive.

2. AEROBIC AND ANAEROBIC, OXYGENIC AND ANOXYGENIC

The difference between the environment of early life, under the primeval anoxygenic atmosphere, and that of present life, under the actual oxygenic atmosphere, is as clear-cut as between black and white. Admittedly, there will have been a period of transition, as we will see in Chapter 15, but for the time before and the time after, we find complete antithesis. This difference has, notably in the usage of the words "aerobic" and "anaerobic", not always been sufficiently stressed in the literature. In our present life aerobic organisms live in contact with the air, and, because of that fact, they are in free contact with oxygen. The present-day anaerobic forms live excluded from fresh air, and consequently excluded from oxygen.

The metabolism of early life forms, under the anoxygenic atmosphere, may well have been more or less similar, biochemically, to that of our present anaerobic organisms. Most authors therefore speak of "anaerobic conditions" prevailing in the primeval atmosphere, without taking into account the question whether life at that time was, or was not, in contact with the atmosphere.

We should therefore better distinguish between anoxygenic and oxygenic forms of life, which then might be either anaerobic or aerobic, as the case might be. In the time of the primeval atmosphere all life will have been anoxygenic, but it might well have been aerobic for its larger part. At present all aerobic life is forcibly oxygenic, whereas present-day anoxygenic life occupies an exceptional environmental niche, that of anaerobic conditions. Things will probably have been even more complicated during the period of transition from the primeval anoxygenic to the present oxygenic atmosphere.

3. ARADIATIC FOR SHORTER ULTRAVIOLET LIGHT

Apart from the properties just mentioned, aerobic or anaerobic, oxygenic or anoxygenic, we must take into consideration still another set of environmental conditions, that between radiatic or aradiatic for shorter ultraviolet sunlight.

The early development of pre-life, the formation of the "thin-soup", must have been radiatic, as we saw earlier. In fact the inorganic synthesis or "organic" molecules took place by using the energy of these shorter ultraviolet rays of the sunlight. Once the complicated molecules of early life had been formed, and probably even already during the later stages of biopoesis, these were, on the other hand, aradiatic for the shorter ultraviolet sunlight.

One of the many paradoxes encountered in the early history of life lies in the fact that the same rays of the sun which formed the building blocks of the molecules of life were lethal for life. Early life had therefore only limited environmental possibilities. It could survive only when shielded by a thick layer of water, or when living in pores in the soil, or in natural caves. Even so it will probably have been aerobic. For not only does the air filter freely into caves and also into pores of the soil, but the upper part of most water bodies is well aerated too and therefore belongs to the aerobic environment.

Concluding, we may summarize the various environmental possibilities of life in relation to the atmosphere as follows:

aerobic or anaerobic = exposed or not to the atmosphere,
oxygenic or anoxygenic = exposed or not to free oxygen,
radiatic or aradiatic = exposed or not to direct sunlight.

4. THE IMPORTANCE OF MICROBES

Before further pursuing our investigations into the circumstances that may have been surrounding the early evolution of life, it seems well to interpolate at this point a statement as to the importance of microbes.

Microbes are not only important themselves, but also in providing the right environment for all other life. As to the first aspect, microbes make up three quarters of present living matter by sheer weight, surely a claim to importance. But it is found, moreover, that many of the processes of the life cycle of microbes form the basis for the daily functions of the higher plants and animals. The possibilities for these higher organisms would be very much reduced, if there were no microbes present. And one could even facetiously ask with Kluyver, whether such a life could really be called life. As he once put it: "Is this to be called life? Is such life, without bread and wine, cheese or beer, really worth living?" (KLUYVER, 1937, 1955; see also KLUYVER and VAN NIEL, 1956.)

Another important feature of microbe life is that it shows a much greater variety in its metabolic processes, than do plants and animals. The latter have in a way arrived at a certain set of patterns in their life cycles, which they exploit to the full, but to the exclusion of other possible metabolic pathways. The microbes, on the other hand, can still make use of all possible ways of energy transfer. As a result, questions such as about aerobic and anaerobic metabolism, or about autotrophic (plants) and heterotrophic (animals) ways of feeding, are much less black and white in the microbial world, than in that of the higher organisms.

The importance of microbes for the early evolution of life is, of course, still much larger than for the present life. One could even call it all-important, because microbes also are the evolutionary basis of all higher organisms. It is therefore imperative at this point to take a look at some aspects of the microbial world.

Taxonomically microbes form part of a larger group, called the Protista. This is commonly set up as a third separate kingdom of life in addition to the plant and animal kingdoms. An exact definition of the boundaries between the Protista and either the plants or the animals is difficult to give. Consequently there is not much agreement between authors as to what is to be called a protist and what is to be called a plant or an animal. As a general rule, one may, however, state that the Protista have a relatively simple organization, without tissue differentiation. A classification in which the protist kingdom has been extended as far as possible upwards into the plant and animal kingdoms is given in Table XII.

TABLE XII

THE COMPONENT GROUPS OF THE THREE KINGDOMS OF LIFE

(From STANIER et al., 1963)

	Plants	*Animals*
Multicellular, showing extensive tissue differentiation	seed plants ferns mosses and liverworts	vertebrates invertebrates
	Protists	
Unicellular, coenocytic or multicellular; without tissue differentiation	Algea Protozoa Fungi Bacteria	

It follows that in our search for the remnants of early life we must turn away from normal paleontology. Normal paleontology is mainly interested in well-preserved hard parts of high organisms, which can be nicely classified and interpreted into evolutionary schemata. We, on the other hand, must look for remains of early microbes and other forms of life with a low morphological organization, and deduce from their extremely scarce remains the development and history of early life on earth. The scarcity of these remains obliges us to tell our story in but a very general way.

5. THE TRANSITION FROM FERMENTATION TO RESPIRATION: THE PASTEUR POINT

Returning now to the question of aerobic versus anaerobic, of oxygenic versus anoxygenic ways of living, we saw that microbes have a greater variety of ways of acquiring energy, than the higher organisms. We always more or less automatically think only of either organic photosynthesis, or of respiration. Both are aerobic in the true sense of the word, in that they are only possible in contact with the atmospheric air. But photosynthesis can be both anoxygenic as well as oxygenic, and it is only respiration which is an exclusively oxygenic process.

But microbes have still other means of acquiring energy from their surroundings. Amongst these other possibilities the process of fermentation is the most widely known. The individual pathways of these processes show a surprisingly wide variation, but they all have the same final result. Always an electron is eventually transferred by the organism from an electron donor to an electron acceptor. The yield of energy which accompanies this transfer is used by the organism in its life cycle[1].

With STANIER at al. (1963) we may summarize the principal types of microbial metabolism as follows:

(1) Using radiant energy (sunlight): *organic photosynthesis.*

(2) Using energy-yielding oxidation-reduction (so-called "black" reactions):

(a) *respiration*, in which free oxygen is the ultimate electron acceptor;

(b) *anaerobic respiration*, in which some other inorganic compound is the ultimate electron acceptor;

(c) *fermentation*, in which an organic compound serves as the ultimate electron acceptor.

This wide variety of metabolic processes in microbes leads to the existence of a broad zone of transition between anoxygenic and oxygenic life. Instead of only the two groups of the strictly oxygenic (or as they are called in present life

[1] This entails oxydation of the compound which donates the electron and reduction of the receiving compound. Such reactions are therefore called "energy-yielding oxidation-reduction reactions". Because they do not use "radiant energy", that is energy from the rays of the sunlight, such reactions are moreover called "black". In respiration the ultimate electron receiver is free oxygen, which is reduced and combines with hydrogen to form water. Microbes are, however, able to use a surprising variation of other compounds, both inorganic and organic, as ultimate electron acceptors.

the *obligately aerobic*, or better *obligately oxygenic*) and of the strictly anoxygenic (or *obligately anaerobic*, or *obligately anoxygenic*) organisms, we find that many microbes may live both in an oxygenic or in an anoxygenic environment. In present life these are called the *facultative anaerobes*. In early life they had better be called *facultative respirators*.

To this group of facultative anaerobes in present life belong yeast and a great many bacteria. Below a certain oxygen level they acquire their energy through fermentation, above this level through respiration. The point of transition is called the *Pasteur Point*.

The transition from fermentation to respiration in the facultative anaerobes is a sharp one. The Pasteur Point is situated at a level of oxygen of about 0.01 PAL (= Present Atmospheric Level; L. V. Berkner and L. C. Marshall, personal communication, 1966).[1]

The Pasteur Point must not be confused with the level of free oxygen below which respiration becomes impossible in purely aerobic organisms. This level is written as P_{50}. For the vertebrates it is indicated in Table XIII.

TABLE XIII

PARTIAL PRESSURES OF FREE OXYGEN BELOW WHICH RESPIRATION IS OBSTRUCTED (P_{50})

(From L. V. BERKNER and L. C. MARSHALL, personal communication, 1966)

Class	P_{50} (mm Hg)	P_{50} in fraction of PAL
Man	50	0.31
Mammals	22–50	0.14 –0.31
Birds	35–50	0.22 –0.31
Reptiles	13–31	0.1 –0.2
Amphibians	5–18	0.031 –0.11
Fish	4–17	0.025 –0.11

The transition from fermentation to respiration gives a powerful advantage to the facultative respirators over the obligately anoxygenic organisms, whenever the Pasteur Point is reached, because the energy released by respiration process is much greater than that yielded by fermentation. The reason for this lies in the circumstance that in fermentation the end product that serves as electron acceptor is an organic compound, whereas in respiration this function is taken over by free oxygen. The energy released in breaking up the organic compound which forms the end product of fermentation thus is added to the process in respiration. As follows from the schematic equations cited below, the energy yielded by respiration is even more than ten times greater than that liberated by fermentation.

[1] As the partial pressure of free oxygen in the present atmosphere is 159 mm Hg, a level of 0.01 PAL corresponds to a partial pressure of 1.59 mm Hg.

fermentation: $C_6H_{12}O_6 \rightarrow 2\ CH_3CH_2OH + 2\ CO_2 + 50$ cal./mol.
(*ethyl alcohol*)
respiration: $C_6H_{12}O_6 + 6\ O_2 \rightarrow 6\ CO_2 + 6\ H_2O + 686$ cal./mol.

6. IMPORTANCE OF FACULTATIVE RESPIRATION FOR THE EARLY EVOLUTION
OF LIFE

The widespread existence in the microbial world of today of organisms capable of both fermentation and respiration, dependent upon the oxygen level of their environment, is of the greatest importance for our understanding of the early evolution of life. In relation to the history of the oxygen in the atmosphere, a topic which will be more fully treated in Chapter 15, this leads us to an important consideration. The transition from fermentation to respiration by the precursors of the present facultative anaerobes would have capitalized on the free oxygen produced by the early photosynthesizing organisms of those days.

A feedback mechanism consequently will have operated, when the oxygen level reached 0.01 PAL. For at that level, that is at the Pasteur Point, any facultative respirator present at that time will have switched from fermentation to respiration. That is, they began to use oxygen, a thing which they could not do, at least in such quantities, below that level. If the total volume of the facultative respirators of that time would have been anywhere comparable to that of the contemporary photosynthesizing organisms, this will tend to regulate the oxygen level of the atmosphere at 0.01 PAL. When it should again drop below this level, the facultative respirators could revert to fermentation, whereas any anoxygen produced in excess of this level through photosynthesis would immediately have been consumed by a more vigorous respiration of the facultative respirators.

Of course, there will have been a certain lag when the oxygen level reached 0.01 PAL for the very first time. For at that time the respiratory mechanism, with its requisite enzymes, could not yet have developed. But with BERKNER and MARSHALL (1965, 1966) we may postulate that the possibility of respiration offered such an immense advantage to early life that it must have been developed by evolution in what was—geologically speaking—a short time. Once respiration had developed, the organic feedback mechanism regulating the oxygen content of the primeval atmosphere at 0.01 PAL may be thought to have operated almost instantaneously.

Little is known about the "why" of the transition from fermentation to respiration in the present-day microbial world. It seems as if there have been no serious investigations. It is felt, but only in a general and vague way, that the kinetics of fermentation and respiration at low levels of free oxygen determine the threshold at which respiration becomes the more rewarding of the two. But whether this is perhaps only the rate of diffusion of oxygen in water, or—more likely—the kinetics of the various enzymatic reactions employed in the step-by-step processes involved in respiration, seems to be unknown at present. It seems as if an attractive field for investigation has been overlooked. For our topic, we may retain, however, the overall picture which tells us that the transition at the Pasteur Point is a generalized event, which comes at about the same level of free oxygen in various, otherwise unrelated, microbial organisms of the present-day life. This is the reason why we may assume that it has already been operative in early life also.

7. MICROBIAL FEEDING HABITS

Microbes do not only show much more variation than plants and animals in the processes of acquiring their necessary energy, but also in the types of food used during their life cycle. As early life has been microbial, it seems well to establish this point now, as it leads to a clearer understanding of the nutritional requirements of early life.

With the higher organisms we do again meet with only two clear-cut cases. These are the *autotrophic* feeders, in which the organism feeds on inorganic compounds and the *heterotrophic* feeders, which require organic compounds to feed on. Plants, with a few exceptions, are autotrophs, whilst animals are heterotrophs.

But in the study of the life cycle of microbes other possibilities are met with. With STANIER et al. (1963), we may now follow a classification, which takes into account both the type of energy and the type of food used during metabolism. A first division, already given in section 5 of Chapter 7, distinguishes between the *phototrophic* (or photosynthetic) organisms, that use radiant energy, and the *chemotrophic* organisms that use the energy of dark oxidation-reduction processes for their life cycle. A further distinction is then made according to the kind of electron donor an organism normally employs. Organisms using inorganic electron donors are called *lithotrophic*, organisms using organic compounds are called *organotrophic*.

A combination of both classifications leads to the following system:

Photolithotrophic organisms use radiant energy and inorganic electron donors. Not only the green plants, but also the blue-green algae and some bacteria belong to this group.

Photoorganotrophic organisms use radiant energy and organic electron donors. To this group, which in the old classification would have to be called "photosynthetic-heterotrophic" belong some sulphur bacteria, and also some of the non-sulphur bacteria, notably the Athiorhodacea (see section 9 of Chapter 12 and GAST et al., 1963).

Chemolithotrophic organisms use oxidation–reduction reactions to acquire their energy and inorganic electron donors. Representatives of this group also are found only amongst the bacteria.

Chemoorganotrophic organisms use oxidation–reduction reactions for energy and organic compounds for electron donors. This group is heterotrophic, according to the earlier definition. But it not only embraces the animals, but also the majority of the bacteria.

It is necessary to stress this more refined classification, because all too often heterotrophy is still equated with animal life, whilst the much wider spectrum of possibilities in the microbial feeding habits is slurred over.

Extrapolating this classification backwards in time, one could, at least theoretically, distinguish two more groups, those of the *photo"organo"trophic* and the *chemo"organo"trophic* organisms. The use of the quotation marks would indicate those organisms of early life, which would be able to accept as electron donors the "organic" molecules, formed through inorganic photosynthesis.

The earliest life will have been chemo"organo"trophic and chemolithotrophic. With photosynthesis phototrophic organisms developed. In our present life the photolithotrophic organisms dominate strongly over the photoorganotrophic organisms. But in early life the photolithotrophic organisms were perhaps at a disadvantage. For life was at that time, in the anoxygenic primeval atmosphere, aradiatic for direct sunlight, whilst the photoorganotrophic bacteria are at present more diversified in their metabolic ways and means than the photolithotrophic organisms. They do not require direct sunlight, and can even grow in the dark. So it is quite possible that in early life photo"organo"trophic microbial forms have first dominated, and have only been replaced by photolithotrophic organisms as the dominant group at a later time.

An important factor in this question, already mentioned before, is that organic photosynthesis, although it produces oxygen, itself is not an oxygenic process. It only depends on the presence of carbon dioxide in the atmosphere and of molecules of chlorophyll or related compounds which are able to dissociate the atmospheric carbon dioxide and liberate oxygen.

8. PROCARYOTIC AND EUCARYOTIC CELLS

Another important aspect of the microbial world is the prevalence of a simple cell type, called *procaryotic*.

When cells were studied by the light microscope it was already found that most cells show a complex inner structure, and contain smaller, discrete subunits. But it was not possible to say for certain, whether, in those cells in which such subunits were not seen, this was due to the fact that there were no subunits at all, or to the fact that the subunits were too small in size to be detected by the light microscope. The advent of the electron microscope, which allows of a much stronger magnification, has since proved the existence of cells that show no further internal division into discrete subunits. Moreover, it has been found that all organisms are built up entirely by either one or the other of these cell types. One either finds the simple, non-divided *procaryotic*, or the complex *eucaryotic* cells.

Another difference between procaryotic and eucaryotic cells lies in the outer cell membrane. All procaryotic cell membranes are formed by a single molecule of a compound which is never found in the membranes of eucaryotic cells, whilst it also does not occur in viruses. This is called *mucopeptide* or the *mucocomplex* and consists of at least three to four amino acids, two sugars, one of which is acetylmuramic acid, and a relatively large amount of lipids (STANIER, 1961; STANIER et al., 1963; STROMINGER and GHUYSEN, 1967). The single molecule of this compound, which forms the membrane, sack-wise envelops the cell and expands with its growth.

It is only a limited part of the Protista, composed of the bacteria and of a peculiar group of the Algae, the blue-green Algae, which is characterized by procaryotic cells (Table XIV). They are classified as the Lower Protista, whilst all other Protista, that is most of the Algae, and all of the Fungi and the Protozoa,

TABLE XIV

PRINCIPAL DIFFERENCES BETWEEN PROCARYOTIC AND EUCARYOTIC CELLS

(From STANIER et al., 1963)

Groups where found as unit of structure:	Procaryotic cell	Eucaryotic cell
	Bacteria, blue-green Algae	most Algae, Fungi, Protozoa, higher plants and animals
Nuclear membrane	−	+
Mitotic division	−	+
Chromosome number	1 (?)	always greater than one
Cytoplasmic streaming	−	+ or −
Mitochondria	−	+
Chloroplasts	−	+ or −
Contractile locomotor	bacterial flagella axial filaments } in some	multistranded flagella or cilia } in some
Ameboid movement		+ or −

are eucaryotic and classified together as the Higher Protista. It follows that the procaryotic Lower Protista are basically different from the eucaryotic Higher Protista, the plants and the animals taken together (for a proposed division of all organismus into four or five kingdoms, see WHITTAKER, 1969). They really form the most primitive organisms of present-day life, a fact unfortunately obscured by the taxonomic inclusion of Lower and Higher Protista in one kingdom of life. A fact which has, moreover, become more strongly garbled because most botanists consider the blue-green algae to belong to the plant kingdom.

The eucaryotic cell is the cell with which biologists are most often concerned. In the more popular scientific treatment of the cell one even always meets only with the eucaryotic cell. All the spectacular electron microscopic photographs of cells, exhibiting their diversified and complicated internal anatomy, are pictures of eucaryotic cells. More often than not the existence of the procaryotic cell type in nature is not even mentioned.

The various subunits of the eucaryotic cell are surrounded by membranes. These are comparable to the outer cell membrane, in that they are semi-permeable, and therefore are able to maintain a difference in chemical composition within the subunits, as compared to that of the general cell matter outside these units. The best known subunit is the nucleus, which in turn possesses even smaller subunits, the chromosomes. These carry the nucleic acid (DNA), and are responsible for the transmission of the hereditary characters of the organism during reproduction. But other groups of discrete subunits also exist within a single eucaryotic cell, such as the chloroplasts and the mitochondria. The chloroplasts contain the pigments and the enzymes which, together, permit the conversion of radiant energy of the sunlight into the energy of chemical bonds. They contain, moreover, a whole series of enzymes which transform the original compounds into more complex carbon compounds which can be used by the cells. The mitochondria, on the other hand, contain the enzymatic system for the absorption of oxygen and the formation of ATP. The functions of chloroplasts and mitochondria are exclusive. Chloroplasts cannot serve as a respiratory complex, and vice versa. The same applies to their respective enzymes, which are either found in chloroplasts, or in mitochondria, but never in both.

In procaryotic cells of photosynthetic microbes small discs consisting of double flat lamellae, similar to those found within the chloroplasts of the eucaryotic cells, occur scattered throughout the cell cytoplasm. It is therefore supposed that photosynthesis is not a localized process in procaryotic cells, but occurs throughout the protoplasm. Respiration, in those microbes which are able to execute this process, is, on the other hand, localized in the cell membrane.

Turning now to the geologic history of procaryotic and eucaryotic cells, we have no evidence as to the time when eucaryotic cells were first present on earth (see section 10 of Chapter 14).

As we will see more fully in Chapter 11, actual cells are only preserved in the very best ways of fossilization. The finer internal structure of cells is, however, almost always destroyed, and therewith normally goes the possibility to distinguish in fossils between a procaryotic and a eucaryotic cell.

To biologists the transition from the simple procaryotic cell to the complex eucaryotic cell, with its numerous and diversified subunits, still presents an enigma. The most elegant of the theories proposed to explain this transition, is by SAGAN (1967). She postulates that eucaryotic cells developed from a symbiosis of two and more different procaryotic cells, to form the independent structure of an eucaryotic cell. The basis for this assumption lies in the fact that procaryotic cells have no inner membranes that are able to protect the general cell matter against lethal products formed by the cell during its metabolism. The most important of such products is the free oxygen developed by early photosynthesizing organisms, the paradox being that the oxygen liberated during photosynthesis by part of the cell, is able to oxidize other cell matter, and as such is a lethal product. In an eucaryotic cell the photosynthesis is carried on in the chloroplasts, and various internal membranes see to it that the oxygen is liberated from the cell without harming its other constituents. But in the procaryotic photosynthesizing organisms, oxygen defences had to develop concurrently with the ability to photosynthesize.

In the theory of Sagan the first step towards the development of eucaryotic cells originated from a symbiosis of two different types of procaryotic organisms. One would be a heterotrophic anaerobe—that is a photoorganotroph in the classification given in the preceding section—which normally browsed upon organic matter. When this availed—"ingested" is the proper term, which indicates that in the microbial world life does not necessarily end directly upon being eaten— a respiratory organism it would then find that the surplus of free oxygen which it developed itself, but which threatened its very existence, would be taken up by the ingested oxygenic procaryotic organism. In this way a symbiosis between an oxygen-producing procaryotic microbe and an oxygen user could be established, which was so successful that it developed into an obligate symbiosis.

To explain the existence of the various groups of other subunits in the present eucaryotic cells, Sagan proposes a number of other, but fundamentally similar, symbiotic steps. The ancestral eucaryote, as figured in Fig. 46 would then carry in a single large procaryotic host cell a number of smaller procaryotic guest

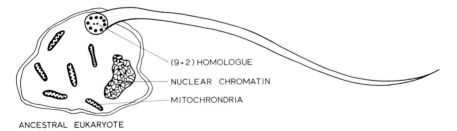

ANCESTRAL EUKARYOTE

Fig.46. Schematic picture of an ancestral eucaryotic cell. A larger procaryotic cell is thought to have "ingested" various other procaryotic cells of diverse life functions. They will now together proceed as nature's first symbiont, that is, as a eucaryotic cell. (From SAGAN, 1967.)

cells, which, when still individual cells, already had the same life functions as they now accomplish as units within the complex single cell. The eucaryotic cell would thus represent the earliest, and also the most widespread, symbiosis on earth.

Another possibility would, of course, be that procaryotic and eucaryotic cells developed separately and independently from pre-life during the long time of the order of two billions of years that pre-life and early life have co-existed on earth (see section 2 of Chapter 18). In this case we might even postulate that there have been originally two main strains of procaryotic cells—those with and those without the mucocomplex—and that only the latter of these groups developed into eucaryotic cells, possibly in the way proposed by Sagan. The original pro-caryotic cells of the latter strain, one might further postulate, have become extinct because their eucaryotic descendants were so much better adapted to life that the survival value of these procaryotes became nil. The other strain of procaryotes, with their differently built cell wall and possessing the mucocomplex, in that case may have thought to have a survival value, which enabled them to live on, and to occupy, in present-day life, a large and varied number of ecological niches which are not open to the more choosy eucaryotes.

9. OXYGEN DEFENCES AND OXYGEN OASES

In the preceding section we have come across the paradox that the oxygen produced by early organic photosynthesis, although it has eventually led to the full development of life as we know it now, was at that time lethal to that selfsame life. As we have seen, the origin of life has been accompanied by many a paradox. But the one between anoxygenic and oxygenic, of which the above-mentioned case is one of the many expressions, is perhaps the most fundamental. It must have accompanied life all the way during the transition from the anoxygenic to the oxygenic atmosphere and further during the oxygenic atmosphere.

The production of oxygen by organic photosynthesis will have led to quite contradictory reactions and results. On the one hand, it must have led, as we saw in the last paragraph, to the development of oxygen defences. On the other hand, local production of oxygen may have led to local accumulation and to the early local development of oxygenic organisms at a time when most organisms were still in the anoxygenic way of life. It is not yet possible to state how these things have developed. But in the vein of this chapter, we will mention one example of a possible oxygen defence mechanism, and give an indication of how the opposite mechanism, that of oxygen oases, might have developed.

A mechanism which has been thought to represent a left over from an early oxygen defence is the phenomenon of *bioluminescence* (McELROY and SELIGER, 1962). Bioluminescence—the faculty of emitting light by living organisms—is widespread throughout the living world today. It does not only occur in insects, such as the well-known fireflies, but also in fish, crustaceans, clams, worms, protozoans and in plants. It is so widespread, without any apparent relation to the normal groups of either the plant of the animal kingdoms, that it is thought to represent a very old faculty, which has only been retained in a few groups scattered throughout the multitude of plants and animals. Retention of such a faculty may occur when it has a secondary value for the living organism, which might not be related at all to its primary function. As such bioluminescence might be of value for actual lighting, such as in deep-sea fish, or in signalling in mating, such as in fireflies.

The process of bioluminescence in all of these varied groups of plants and animals is basically similar in that some substrate—called luciferin—is oxidized through the influence of an enzyme called luciferase. Although the structure of these compounds varies with the organisms capable of bioluminescence, this does not distract from the fact that in all cases a direct and vigorous oxidation is involved, in which free oxygen is used up. This is the reason why McElroy and Seliger have postulated that bioluminescence is an early oxygen defence, developed during the early days of organic photosynthesis, and which has been retained only by those organisms to which it has proved advantageous for other reasons, even in the oxygenic atmosphere.

Oxygen oases are the counterpart to oxygen defences. With this term, coined by FISCHER (1965) are indicated narrowly restricted areas around photosynthesizing organisms. In Fig.47 they have been represented as being occupied by early animals living around marine algae. But the concept might just as well be applied to respiratory microbes, obligate or facultative surrounding early photosynthetic microbes.

We will hardly ever have an exact idea about the actual structure, nor about

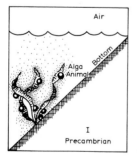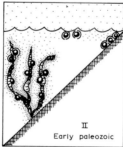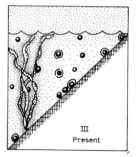

Fig.47. Oxygen oases, such as might have existed during the development of the oxygenic atmosphere. In stage I "animals", or other oxygenic organisms, are shown to live in complete respiratory dependence on photosynthesizing algae. In the further development, of which neither the exact nature nor its dating, has, contrary to the opinion of Fischer, as yet been clarified, the "animals" may leave the plants when the sea water becomes oxygenated. (From FISCHER, 1965.)

the importance of such oxygen oases. But the theoretical concept is helpful in picturing the many intermediate stages which must have existed somehow during the development of the oxygenic atmosphere.

10. THE BLUE-GREEN ALGAE

It seems appropriate to close this chapter, in which various events which might have happened during the early evolution of life have rather loosely been brought together, with a short note on the blue-green algae. The blue-green algae form a group of procaryotic organisms full of contradictory qualities. Individual species show a great variation in their ways of metabolism. For our topic, the most important aspect of the blue-green algae is, however, that they do possess chlorophyll, and as such are able of photosynthesis (ECHLIN and MORRIS, 1965; ECHLIN, 1966). In our present life they are by far the most important photosynthesizing procaryotic organisms, and moreover they are photolithotrophic. Although, as we saw in section 7 of this chapter, early life will probably have been photo-"organo"trophic, the production of oxygen will have proceeded at a much faster rate, once photolithotrophic organisms had developed. It is therefore now generally assumed that blue-green algae, or earlier organisms, either related or non-related, but possessing a similar metabolism, have played an important part in the early oxygen production of our atmosphere. It was only through the development of the eucaryotic cell and of our present eucaryotic green plants, that a further step in the production of oxygen could be taken. As we will see in section 11 of Chapter 14 the superficial resemblance of algae-like fossils about two billion years old, is no proof of a relationship between those organisms and the present-day blue-green algae, a fact often postulated in the literature. The latter are so primitive that they might just as well represent a young retrograde offshoot of more complicated algae. However, what matters is not whether the ancient and modern organisms were related, but whether they had a similar metabolism, a process about which we have, as yet, no data.

REFERENCES

BERKNER, L. V. and MARSHALL, L. C., 1965. On the origin and rise of oxygen concentration in the earth's atmosphere. *J. Atmospheric Sci.*, 22:225–261.
BERKNER, L. V. and MARSHALL, L. C., 1966. Limitation of oxygen concentration in a primitive planetary atmosphere. *J. Atmospheric Sci.*, 23:1133–1143.
ECHLIN, P., 1966. The blue-green algae. *Sci. Am.*, 214:75–81.
ECHLIN, P. and MORRIS, I., 1965. The relationship between blue-green algae and bacteria. *Biol. Rev.*, 40:143–187.
FISCHER, A., 1965. Fossils, early life and atmospheric history. *Proc. Natl. Acad. Sci. U. S.*, 53: 1205–1215.

GAST, H., SAN PIETRO, A. and VERNON, L. (Editors), 1963. *Bacterial photosynthesis*. Antioch Press, Yellow Springs, Ohio, 523 pp.

KLUYVER, A. J., 1937. 's Levens nevels. *Handel. 26e Natuurwetensch. Congr.*, pp. 82–106. (Translated as "Life's fringes" in: A. F. KAMP, J. W. M. LA RIVIÈRE and W. VERHOEVEN, *Albert Jan Kluyver*. North-Holland, Amsterdam, pp.329–348).

KLUYVER, A. J., 1955. Microbe en leven. *Jaarb. Koninkl. Ned. Akad. Wetenschap.*, 1954/55:27.

KLUYVER, A. J. and VAN NIEL C. B., 1956. *The Microbe's Contribution to Biology*. Harvard Univ. Press, Cambridge, Mass., 182 pp.

MCELROY, W. D. and SELIGER, H. H., 1962. Origin and evolution of bioluminescence. In: M. KASHA and B. PULLMAN (Editors), *Horizons in Biochemistry*. Academic Press, New York, N. Y., pp.91–101.

SAGAN, L., 1967. On the origin of mitosing cells. *J. Theoret. Biol.*, 14:225–274.

STANIER, R. Y., 1961. La place des bactéries dans le monde vivant. *Ann. Inst. Pasteur*, 101: 297–312.

STANIER, R., DOUDEROFF, M. and ADELBERG, E., 1963. *The Microbial World*, 2nd ed. Prentice Hall, New York, N. Y., 753 pp.

STROMINGER, J. L. and GHUYSEN, J. M., 1967. Mechanisms of enzymatic bacteriolysis. *Science*, 156:213–221.

WHITTAKER, R. H., 1969. New concepts of kingdoms of organisms. *Science*, 163:150–160.

Chapter 9 | Further Evolution of Life

1. THE THREE MAIN STAGES IN THE HISTORY OF LIFE

To round off the preceding chapters, in which attention has been drawn to the biological and chemical aspects of the origin of life, we will now shortly review some aspects of life's further evolution, in so far as they are of importance for an understanding of its origin. In the next chapters we will then return to the origin of life with an evaluation of the geological data, followed by a review of the pertinent data from atmospheric physics.

Notwithstanding the fact that every schematization of natural processes entails some falsification, it seems appropriate for a better understanding to divide the history of life into three main stages. These are: (*1*) pre-life, (*2*) early life, by which in this text is meant life under the anoxygenic atmosphere, and (*3*) later life under the oxygenic atmosphere. Although courting the danger of an even stronger schematization, we may state that the development of pre-life is dependent upon the anoxygenic atmosphere and its corollary, the shorter ultraviolet sunlight penetrating all the way through the atmosphere. Biopoesis, the transition from pre-life to early life, is also dependent upon the anoxygenic atmosphere, because here life forms may develop step by step from pre-life without the danger of immediate destruction by oxydation. Moreover, although the shorter ultraviolet sunlight itself is lethal to at least a part of the products formed during biopoesis, it is dependent upon pre-life because it uses the "organic" compounds formed by the shorter ultraviolet sunlight. Once biopoesis had resulted in the transition to early life, the latter was still dependent on the anoxygenic atmosphere and only when oxygen defences (see section 9 of Chapter 8) had become fully developed did it stand a chance of survival in the oxygenic atmosphere. In later life, on the other hand, all respiratory organisms, that is not only the animals, but also the plants which do not only photosynthesize, but also respire, are dependent upon the oxygenic atmosphere. Only the anaerobes, which might either be a left-over from early life, or a younger evolutionary off-shoot from later life, are independent of the present oxygenic atmosphere.

As already indicated, there has probably existed a major overlap between the two first stages thus delineated. Pre-life was probably still present when early life was already fully established. In the later chapters we will see that this probably has been the case over most of the history of the anoxygenic atmosphere, from the first development of early life unto the transition to the oxygenic atmosphere, a time span of some two billion years. On the other hand, there has probably been a transition, or a gradual evolution from early life to later life, but it is more difficult to acquire relevant data on this period. It seems, however, entirely within the possibilities that an, albeit primitive, but well-developed flora like that of the Gunflint Formation (see section 14 of Chapter 12) could have developed oxygen defences at a rate which enabled it to follow the transition from anoxygenic to oxygenic.

In contrast to the gradualness of the evolution of life itself, there is a serious gap in the fossil record of life, if we look at the transition from early life to later life. Since primitive plants may have lived as well under the anoxygenic as under the oxygenic atmosphere, the only fossils which without any doubt proclaim the existence of a contemporaneous oxygenic atmosphere are those of the continental plants and the animals. The earlier of these are the animals, and their fossils have so far only been found in the latest Precambrian (see sections 3 and 4 of Chapter 12). Moreover, the abundant fossil record is limited to the time from the Cambrian onward when animals began to wear shells, a fashion probably not related to any major evolution of life, but to a temporary depletion of atmospheric carbon dioxide (see sections 3 and 4 of Chapter 16). These factors combine in bringing about a rather sharp distinction in the study of the history of life on earth. As we saw already, all normal studies on the evolution of life are only concerned with that part of the evolution which took place after the beginning of the Cambrian, that is during the Phanerozoic.

2. THE EVOLUTION OF LATER LIFE

The gap in our data between early life and later life is not only a gap in time, as explained in the preceding section, but also a difference in the character of our data. In early life, when animals had not yet developed, all records rest on fossil remains of plants and—perhaps—of microbes. Our records of later life, on the other hand, are based mainly on the fossils of animals and fossilized plants play only a minor role.

This distinction rests mainly on two factors. The first is that in later life animals possessing hard parts, such as shells, skeletons and teeth, stood a much better chance in fossilization than did other organisms. But moreover, it so happens that in most groups of animals these same hard parts are also used in the systematic description of present-day life. It is therefore relatively easy to integrate the taxonomy of fossilized remains of animals with those of the present. With plants the situation is normally different. For even with those plants which do possess hard parts, such as trees, the systematic description of the present-day flora is not based on the structure of the wood, but on its flowers, which only rarely become fossilized. Moreover, most higher plants grow on land, whereas most fossils stem from a marine environment, which is but another factor leading to the fact that the description of the evolution of later life predominantly rests upon the fauna.

This distinction in our approach to the data of early life and of later life is not always recognized and the findings of the study of the evolution of later life are often extrapolated backwards in an impermissible way. As a case in point of such a mixing of the record of early life and later life we may cite KERKUT (1960) who postulates that the theory of evolution is based on seven assumptions, which together form the "general theory of evolution". The assumptions are:

(1) Non-living things gave rise to living material, i.e. spontaneous generation[1] occurred.

(2) Spontaneous generation occurred only once.

The other assumptions, according to Kerkut, all follow from the second one. They are:

(3) Viruses, bacteria, plants and animals are all interrelated.

(4) Protozoa gave rise to Metazoa.

(5) The various invertebrate phyla are interrelated.

(6) The invertebrates gave rise to the vertebrates.

(7) Within the vertebrates the fish gave rise to the amphibia, the amphibia to the reptiles and the reptiles to the birds and mammals.

The first two assumptions quoted deal with the origin of life, the last five with its further evolution and it is not, to my mind, permissible to gather them together in a "general theory of evolution". The last five assumptions deal with facts, i.e. the evolution of life such as it has become known from the study of later life, whereas the first two assumptions, as we have repeatedly seen in this text, still are hypothetical.

It is, however, permissible to use relevant aspects of the evolution of later life in hypothetical approximations of the history of pre-life and of early life. We will proceed to do so in three cases, i.e. (1) the question of monophyletic versus polyphyletic origin, (2) the speed of occupation of new ecological areas, and (3) the dangers of comparative biochemistry.

---

[1] As already explained in section 8 of Chapter 4, I do not think the use of the term "spontaneous generation" a happy one in this context. Terms like "biopoesis" or "biogenesis" are preferable in relation to the transition from non-living to living.

3. MONO- OR POLYPHYLETIC DEVELOPMENT

The question of mono- or polyphyletic development has been widely discussed in the study of the evolution of later life, but the amount of ink it has caused to flow, is, as is so often the case, inversely proportional to the amount of data relevant to this question.

If we take as an example the origin of man, we can easily see that even a definition of the terms involved is difficult. Purely monophyletic origin would mean that man had descended from a single pair of higher primates, which is, of course, nonsense. The strictest possible form of monophyletic origin would be if man has descended from a single interbreeding population of higher primates. But would we still call it monophyletic if man had descended from a number of closely related, but not normally interbreeding, primate populations? Most people would, I think, still call this a monophyletic origin, but it is obvious that even if we know all about the breeding habits of the populations concerned, the line between mono- and polyphyletic is difficult to draw.

As we will never be able to investigate the breeding habits of a population of primates from their scanty and incomplete fossil remains, it follows that the question of mono- or polyphyletic origin of man can never be solved from the available data. The pros and contras in this question are largely of a theoretical nature and depend strongly on the philosophical outlook of the writer.

In theory a polyphyletic origin of man, descending by a number of converging evolutionary lines from various related but distinct populations of primates is possible. Such a convergence of evolutionary lines might be caused by environmental factors, which tend to select a similar type in each line. Such *convergent evolution* is well known in a number of other cases. But, and here lies the bias against a polyphyletic origin of a single species such as man, it has to our knowledge never resulted in so close a resemblance that the populations of the separate evolutionary lines started interbreeding and became one genetically uniform population.

But even when on these and other grounds a monophyletic origin of man is commonly assumed, this does not mean that the higher taxonomic groups of life also must have a monophyletic origin. For this we might turn our attention to Kerkut's seventh assumption, as cited in the preceding section. But, before proceeding any further, we must re-state this assumption more precisely as follows: "Within the vertebrates a well-defined group of fishes gave rise to the amphibia; a group within the amphibia to the reptiles, a group within the reptiles to the birds, and another group within the reptiles to the mammals." The reason for this lies in the fact that it is not the whole "lower" group from which a "higher" group develops, but only from some well-defined forms of the "lower" group.

This stricter version of Kerkut's seventh statement might at first sight seem to favour a monophyletic descent of the various groups mentioned, but this is not really the case. Taking as an example the origin of the mammals, we know that there has been a large number of more or less parallel evolutionary lines in a special but large group of the reptiles, aptly called the "mammal-like reptiles" (cf. ROMER, 1966). During the same period in the Permian representatives of these separate evolutionary lines, which otherwise showed large differences, ranging for instance from herbivorous to carnivorous, all showed comparable evolutionary tendencies from reptile to mammal. The more important of these were the development towards a diversification of the teeth; towards a lighter skull and skeleton; and towards the development of the inner ear. These developments were at different stages in the various evolutionary lines, one group for example having already a diversified set of teeth, but still retaining a heavy skeleton, whilst another had already developed a mammal-like skull and skeleton, but still retained a uniform set of conical reptilian teeth. But in contrast to the descent of man, in which only a single species is involved, the individual populations in the parallel evolutionary lines of the mammal-like reptiles show so many differences in their other characters, that they are classified not only in different species, but even in different genera and families.

For the origin of man we had to postulate one interbreeding population of a higher primate, but for the origin of the mammals we find a large group of different populations of mammal-like reptiles, which did certainly not interbreed. This notwithstanding, they all showed comparable evolutionary tendencies towards "the mammals". So the situation with regard to the origin of man is different from that with regard to the origin of the mammals.

The data from the fossil record are entirely insufficient to decide whether only one evolutionary line of the mammal-like reptiles, or several, or even many, were able to develop into a skeleton which has to be classed as mammalian. A decision on the question of a mono- or a polyphyletic origin of the mammals from the mammal-like reptiles can consequently not be made from the data. We may, however, realize that (*1*) even if we accept a monophyletic origin for the single species of *Homo sapiens* this must not be used as an argument in favour of the monophyletic origin of the whole class of mammals. And (*2*) that the wide variation of the populations taxonomically grouped under the mammals—to cite only the main groups of the mammals, the Monotremata, the marsupials and the placental mammals—seems to point rather towards a polyphyletic origin of this class.

This "feeling", as we may perhaps best describe it, becomes even more evident, when we look at the next earlier step mentioned, i.e. the evolution of the amphibia from a peculiar group of fishes, the *Crossopterygii* of which the coelacanth is the only living representative. Both the *Crossopterygii* and the early amphibians comprise so many widely different groups, that a polyphyletic origin of the amphibia is well within the possibilities.

What this leads up to in relation to the origin of life is that, although a monophyletic descent may seem probable for a single, genetically homogeneous, species, this does not indicate a monophyletic origin of the various organisms grouped together in the larger taxonomic units such as the mammals or the amphibians. The larger the taxonomic unit, the more probable a polyphyletic origin of such a unit becomes. In such taxonomic units a number of separate evolutionary lines of similar morphology, structure or chemistry, which are not related genetically, are grouped together (WHITTAKER, 1969). There is, consequently, from the study of the evolution of later life, no argument to be drawn in favour of Kerkut's second assumption: "Spontaneous generation occurred only once." As stated already, and as will be put forward in more detail in Chapters 16 and 18, the long time during which pre-life and early life have been co-existent on earth, would rather point to a polyphyletic transition from pre-life to early life.

4. THE SPEED OF OCCUPATION

Another point of view which is of importance for our understanding of the history of early life, and which can be drawn from the study of later life, is the speed of occupation of new territories by life. Expressed in the time scale of geologic history, this is instantaneous.

Whenever we find, all during the Phanerozoic, that a variation in the environment has occurred, be it a regression of the sea, leaving dry continental areas, or a transgression of the sea, resulting in the flooding of continental areas, or some other environmental variation, the occupation by the flora and fauna belonging to the new ecology follows immediately. This is in accordance with the studies of volcanic areas recently devastated by paroxysmal eruptions, such as Krakatoa or Surtsey, where in a short number of years a new flora and fauna becomes installed. Of course, during its first colonization, life is still impoverished, but geologically speaking this is mended so quickly, that for our purpose one may speak of instant colonization.

A similar situation is encountered when through an evolutionary change new ecological possibilities are opened up. Such as has been the case, for instance, in the development of the reptiles, or the spreading of the mammals. When during the transition of the Carboniferous to the Permian the reptiles developed, this group was characterized, among other things, by the amniote type of egg. This egg is much larger than that of the fishes or the amphibians, and makes it possible for the embryo to develop fully within the egg. This gave the reptiles a clear advantage over the amphibians, because the juvenile amphibians leave the egg in an incomplete stage, still provided with gills and are dependent upon an aquatic environment. This kept the amphibians confined to marshy areas with enough ponds or lakes to lay their eggs in. The reptiles, in contrast, could lay their eggs anywhere and were no longer confined to the marshes. The result was that already during the Permian the reptiles have swarmed all over the continents and also in coastal and shallow seas, taking full advantage of the new possibility supplied by the possession of the amniote egg. A similar situation is encountered at the end of the Cretaceous, when the "ruling reptiles" were dying out and the mammals at last got their chance. They too in a remarkably short time colonized the territories newly left vacant, and already in the Early Tertiary had taken full possession of the environments left open by the extinction of the ruling reptiles.

For our study of early life we may extrapolate these findings from the history of later life, and, according to the principle of actualism, postulate that newly opened up areas, or other new ecological developments, will have been occupied by early life in a time span which is short in relation to that of geologic history. We will use this postulate in Chapter 15 in tracing the history of life under the two atmospheres.

5. THE DANGERS OF COMPARATIVE BIOCHEMISTRY

As a final point in the enumeration of aspects of the evolution of later life which are of importance for our understanding of the history of early life a word of warning about the dangers of comparative biochemistry seems in order. The basic assumption of comparative biochemistry is that what is more simple, biochemically, in the metabolism of life, is more primitive and consequently older. This assumption is entirely unjustified. It has never been tested, presumably because it will be very difficult to test. Also, quite possibly, it is false.

Nevertheless, this assumption seems to have invaded the field of molecular biology, as evidenced, for instance by the title *Evolving genes and proteins* of a symposium held in 1964 (BRYSON and VOGEL, 1965), in which the evolution of genes and proteins is deducted from their present biochemistry. True, one sometimes finds a qualification, such as: "Experiments ... used to determine molecular, *and by inference*, evolutionary relationships" (TATUM, 1965, p.6, italics by the present author), but on the whole biochemists and molecular biologists tend to equate biochemical sequences with evolutionary successions. It is generally only the taxonomists who speak out against this too facile assumption.

In this case we may do well to remember similar reasonings in the comparative anatomy of the present-day life, which in the past have also been extrapolated without justification to draw up an evolutionary history of later life. Here also "simple" has been cursorily equated with "primitive" and "early", without taking heed of the palaeontological record. Many are the trees of evolution, based only on comparative anatomy, in which members of separate parallel or converging evolutionary lines have been mixed through their anatomical resemblance. Such trees of evolution have often been drawn with a disregard for the stratigraphical position, i.e. the relative age of these forms. Imaginary fore-fathers were supposed to have sired non-related offspring, sometimes tens of millions of years their senior, not on palaeontological proof and stratigraphic reasoning, but only because the anatomy of the supposed fore-father looked "simpler" when compared with that of the postulated offspring. The tenet that "no one can be a fore-father to someone older than himself" has had to be applied rather forcefully from the side of geology in order to cope with this too simple minded trend in comparative anatomy.

To give a single example, in the history of later life the fishes were the earliest known vertebrates, whilst only much later the land was conquered by vertebrates of a more complex anatomy. But this does not mean that we have to regard the whale, the seal or the seacow as being the most primitive amongst the mammals, because of their marine habitat. Instead, they have only later acquired the more "primitive" environment they now share with the fishes. This is, of course, a rather crude example, but it has been chosen expressly, because the type of reasoning found at present in comparative biochemistry is not so very different. "Simple" is no proof, either for "primitive", or for "early". Arranging our present-day microbes in ascending order of remoteness on the basis of the biochemistry of their metabolism is the same as taking the whale, the seal and the seacow and proclaiming them to be the most ancient of mammals. As has been stated before, we do not even know whether our present anaerobic microbes are really that primitive and represent a left-over from early life, or whether they represent a much younger evolutionary off-shoot from later life, just as the sea-going mammals mentioned above also form but a younger off-shoot of the main lines of the continental Mammalia. Arranging our present-day anaerobic bacteria in an ascending order of antiquity only on the basis of comparative biochemistry gives the false impression that we do know much more about the origin and the history of life than we actually do.

REFERENCES

BRYSON, V. and VOGEL, H. J. (Editors), 1965. *Evolving Genes and Proteins*. Academic Press, New York, N. Y., 629 pp.
KERKUT, G. A., 1960. *Implications of Evolution*. Pergamon, London, 174 pp.
ROMER, A. S., 1966. *Vertebrate Paleontology*. Univ. Chicago Press, Chicago, Ill., 468 pp.
TATUM, E. L., 1965. Evolution and molecular biology. In: V. BRYSON and H. J. VOGEL (Editors), *Evolving Genes and Proteins*. Academic Press, New York, N. Y., pp.3–12.
WHITTAKER, R. H., 1969. New concepts of kingdoms of organisms. *Science*, 163:150–160.

Chapter 10

Mechanisms for Concentration, Conservation
and Isolation. The Orogenetic Cycle

1. CONCENTRATION AND CONSERVATION OF PRODUCTS FORMED DURING SUCCEEDING STAGES OF BIOPOESIS

Experimental investigation into the possibility of an origin of life through natural causes has up to now been mainly directed upon the feasibility of inorganic synthesis of "organic" molecules, and upon the type of reactions which may have led from pre-life to life. Or, to use another set of terms, from prebiological to biological systems. But, apart from synthesis, there must moreover have been concentration and conservation of the materials formed during one stage, to provide the basic products for the reactions leading to the next stage.

Although such reactions may occur in dilute solutions, concentration will in most cases be required for a development in quantity of the next stage. On the other hand, conservation mechanisms become necessary when the energy synthesizing the compounds of an earlier stage is harmful to products formed during a later stage. Such as, for instance, the shorter ultraviolet sun rays, which will synthesize the smaller "organic" molecules, but which will at the same time destroy all larger "organic" molecules, from about the size of amino acids onwards.

We have already come across this riddle, and it will be a more or less recurrent theme. It is important to realize at this point, after having reviewed the biological approach, that origin of life on earth through natural causes does imply more than inorganic synthesis of "organic" molecules. Concentration, as a primary condition for the further development of succeeding stages of biopoesis, and conservation of these stages is equally important. The dynamics of the processes involved in all stages of biopoesis, the balance of synthesis and decomposition, will ultimately control the fact whether a continuous evolution from the very simplest "organic" compounds towards life is possible. The half-life, to use a concept borrowed from nuclear physics with which we became familiar in our study of natural radioactive elements in section 7 of Chapter 3, the half-life of the compounds formed during succeeding stages of biopoesis, a factor as yet scarcely known, will play an important part in this development.

2. THE OROGENETIC CYCLE

We must now return to a concept that has been one of the most fruitful in the development of our ideas on the geologic history of the earth, that is the orogenetic cycle. It has been briefly touched upon already in sections 7 and 8 of Chapter 2 as an example of a process which shows variations in rate throughout geologic history. As such it is the main element of UMBGROVE's (1947) "pulse" of the earth. Briefly, the orogenetic cycle consists of three major periods. The initial period is characterized by overall crustal quietness, and occupies variable, but relatively long, time spans. These range from a minimum of 100 m.y., but normally they are much longer, of up to 500 m.y. duration. It is followed by a shorter period of strong crustal movements, of the order of 50 m.y. duration. The cycle is concluded by a third period of transition to the quiet initial period of the next cycle, which is of the order of several 10 m.y.

The initial period is called the *geosynclinal period*, the next following the *orogenetic period*, and the third one the *post-orogenetic period*. We live at present in the post-orogenetic period of the youngest, the so-called Alpine orogenetic cycle.

Orogenetic cycles are known to have succeeded each other since the oldest known crustal rocks (Fig.48, 127). So they form a feature which must be taken into account in our study of the origin of life on earth. Against this background two features of the orogenetic cycle stand out, which, although related, can best be treated separately here. They both stem from the difference in the velocity of crustal movements during a geosynclinal period of an orogenetic cycle on the one hand, and the orogenetic and post-orogenetic periods of a cycle on the other.

During a geosynclinal period crustal movements, including regional uplift, will be more limited than during both the orogenetic and the post-orogenetic periods. Hence, there will be much less weathering and erosion during a geosynclinal period than during the following orogenetic and post-orogenetic periods. And because weathering is based, at least in part, on oxidation, the consumption of free atmospheric oxygen will be less during a geosynclinal period than during orogenetic and post-orogenetic periods. We will return to this point in sections 11 and 12 of Chapter 15.

On the other hand not only crustal movements, but also volcanic activity is higher during the orogenetic and post-orogenetic than during the geosynclinal periods of an orogenetic cycle. Since the major gas produced by volcanic eruptions is carbon dioxide, the CO_2 supply to the atmosphere will be higher during these more active periods of an orogenetic cycle. A higher amount of atmospheric CO_2 will result in a higher percentage of CO_2 in sea water. There is some lag in re-establishing the balance of atmospheric and oceanic CO_2, but this is quite small. It is of the order of 10^3 years, and can thus be neglected.

Another important result of this variation in the rate of crustal movements within successive orogenetic cycles lies in the extreme flatness the surface of the continents must have had during the geosynclinal periods, due to the slow character of the contemporary crustal movements. This applies both to the shallowness of the seas surrounding, and temporarily spreading over the continents—the oceans are different—and to the low altitude, the lowland morphology, of those parts of the continents which were above sea level.

The reason for this lies in the fact that the vertical crustal movements during a geosynclinal period normally are slower than the processes of weathering and sedimentation. Therefore, a rising part of a continent will not develop into a mountain chain, or even into an upland region, because erosion is able to plane it down quicker than it rises. So, though rising, such a part of the crust will still remain a lowland under the conditions prevailing during a geosynclinal period. On the other hand, in a slowly subsiding part of a continent, sedimentation will be able to offset such subsidence, and fill out the basin as fast as it sinks, so that only the shallowest of seas will develop.

It is true that during every geosynclinal period there were belts of several hundred kilometers wide, which showed a somewhat stronger subsidence, and in which sediments at the rate of a thickness of the order of up to 10 km/100 m.y. could accumulate. But even in these belts—the geosynclines proper—sedimentation could normally offset crustal subsidence, so that the subsiding basins were filled with sediments as fast as they sank, and actually the contemporaneous seas were quite shallow also.

3. DATING OF THE OROGENETIC PERIODS OF SUCCESSIVE OROGENETIC CYCLES

Returning now to the first feature mentioned, the stronger crustal movements during the orogenetic period of each orogenetic cycle, and hence the stronger consumption of atmospheric oxygen through the oxidation of rocks brought to the earth's surface by these movements, it is important to know when these orogenetic periods occurred. For we will want to know during which periods in the history of the earth oxygen production by early life was least offset by oxidation of crustal rocks. Or, in other words, during which periods oxygen production was most effective. As we will see, dating of orogenetic periods by way of relative dating is only possible in a roundabout way and is of necessity vague. In absolute dating, on the other hand, it is possible to date the orogenetic periods more directly.

In relative dating the fact is used that in many places we will find an older series of sediments, folded during an orogenetic period, which have become partly eroded as a consequence of uplift during the ensuing post-orogenetic period, and afterwards have become covered by a set of horizontal beds, which may be either of epicontinental or of geosynclinal facies, but which belong in time to the geosynclinal period of a younger orogenetic cycle. Between the older folded beds of the earlier geosynclinal cycle and the horizontal beds of the younger cycle, a so-called angular unconformity has in this way developed, as has been schematically indicated in Fig.8.

During the Phanerozoic the relative age of both the sediments of the older cycle and of the younger cycle can be established, as long as they contain fossils, or as long as they can be correlated to fossil-bearing strata elsewhere. The actual orogeny can, however, only be bracketed as being "younger than the youngest beds of the older cycle", and "older than the oldest beds of the younger cycle". Much depends on the time elapsed in a given fold belt between the folding of the beds of the earlier cycle, and the transgression of the non-conformable beds of the younger cycle.

In absolute dating it has been found that older rocks often have become so hot beneath the geosynclines of younger fold belts that if they did not melt wholesale, they were at least partially remobilized. In those rocks a strong mobilization and migration of part of their material occurs, often including the daughter elements of a naturally radioactive series. The result being that a grain of radioactive material contained in a mineral of a much older rock may lose all or part of its daughter element during a younger orogenetic cycle, even if the rock in question has to all appearances remained solid and did not melt. The radiometric dates will be *rejuvenated*, the rock will be *regenerated*, as are the technical terms.

Careful analysis, using the Rb—Sr method on individual zircons and micas and on whole rocks in the Swiss Alps, has shown that the smaller the mineral grain, the more fully it has been regenerated (JÄGER, 1962; JÄGER and NIGGLI, 1964). This has since been amply corroborated by similar studies in other fold belts, in part using other natural radioactive decay series. The phenomenon of regeneration of older rocks during younger orogenetic cycles is by now quite well understood.

Apparent rejuvenation of at least a part of the older material in the geosynclinal belt of a younger orogenetic cycle explains why, statistically, so many absolute dates have been found to concentrate around rather precise ages (GASTIL, 1960a,b; see Fig.48). Of course, it is but a statistical coincidence. Not every age of an igneous rock indicates an orogeny, as has been pointed out, somewhat superfluously, by GILLULY (1966). But statistically the bunching of radiometric dates around certain rather narrowly defined ages is significant.

This means that for the Precambrian, where we have no fossils to date the sediments older or younger than a given orogeny, we may use a strong concentration of radiometric dates as a direct indication of the age of an orogeny. The ages of major orogenies, as taken from GASTIL (1960a) in Fig.127, are based on this method, whilst a more detailed example, based on the Precambrian of Minnesota, is given in Table XX.

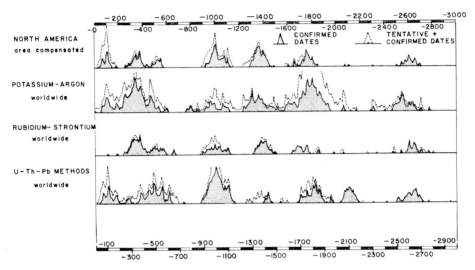

Fig.48. Mineral date abundance plotted against age. The statistical bunching of absolute ages of the Precambrian is very pronounced. In part this is due to the fact that in districts well explored for their mineral wealth a relatively large number of datings of the same rock system will have been taken. But even if an area compensation is applied, such as has been done for North America, this feature remains quite pronounced. It is explained by a rejuvenation of older rocks during a younger orogenetic period. (From GASTIL, 1960a.)

4. THE SEESAW OF LAND AND SEA DURING GEOSYNCLINAL PERIODS

During geosynclinal periods, as we saw, the opposite is true than for orogenetic periods. Crustal movements are at a minimum. The continents were predominantly eroded down to base level, whilst even in the "mobile" geosynclinal belts the subsidence was normally so slow that it could be offset by sedimentation, filling the geosynclinal basins almost up to sea level.

This results in an extreme flatness of both continents and surrounding seas during the geosynclinal periods of any orogenetic cycle. On land there are no mountain chains or uplands, at most an isolated hill here and there; whereas the shallow surrounding seas are built up to wave base. The Amazon Basin, or some of the low lands of northwestern Australia, at present are about the best examples of a topography which must have been all prevalent during geosynclinal periods. During these periods a horizontality controlled the topography, which it is difficult to imagine for us, creatures of a post-orogenetic period, and used to the rugged topographic relief developed as a result of relatively strong crustal movements.

This flatness, this horizontality was most strongly expressed outside the geosynclinal belts. These areas, technically known as *epicontinental*, show sedimentation—and hence subsidence—rates of only several 100 m/100 m.y. The seas were extremely shallow, whilst the adjoining dry land stood only slightly above sea level.

This is attested by the fact that even with the small scale of the crustal movements experienced in these areas, the reactions were enormous. There is a continuous record of almost endless repetitions of flooding and falling dry —technically called *transgressions* and *regressions*—over very large areas. In many cases we find evidence of such repetition of flooding and emergence, sweeping several hundreds of kilometers both ways. The sea swept over an evidently completely flattened continent, only to retreat almost immediately. There were no cliffs, no uplands or mountain chains to bar the way.

There is no evidence of coastal erosion as we know it now. No remnants of a former coastline are found, and hardly ever do we find something like fossil dunes. It all goes to show that during the geosynclinal periods the more stable parts of the continents, the epicontinental areas, were formed by an extremely broad lowland system, bordering upon very shallow, but also very wide seas.

In several of such areas, with ages varying widely within the Phanerozoic, it has been possible to trace individual beds along valley walls, or along a present seashore cliff. There is evidence that a series of such beds were formed at sea level, to become dry land, by emergence from the sea directly after their formation, only to be flooded again before the next bed was deposited. Such sequences are known to be continuous over distances of the order of 50 km (see, for instance PIRLET, 1968). A similar amazing continuity has recently been described by MOORE (1970) from the United States, where the repeated alteration of very shallow seas and low-level land can be followed over hundreds of miles. So here we had a flat area at sea level, or only just above it, at least 50 km wide, which in the next geological event is submerged over its entire width, only to fall dry again immediately afterwards. Unfortunately we have as yet no way of measuring the duration of the individual events which make up such a series. An estimate based on the number of such repetitions within a given geologic system would, however, provisionally lead to a duration of the order of one thousand to ten thousand years for each single event.

The diurnal tides would, at first sight, supply an even much more frequent, and perhaps more important, rhythm of alternation of dry land and sea flood, than the mechanism cited above with time spans running in the thousands of years. I do hesitate, however, to use such a model based on diurnal tides for the earlier history of the earth. For it is not at all sure we had a moon in our earlier history, and without a moon there would be no tides. I have become frustrated personally in my search for an indication of the existence of tides in any of these extremely flat environments described above. Tides have at present in their system of tidal flats and tidal channels a set of clearcut sedimentological features which it would be hardly possible to overlook in older sediments. So, even if the earth had a moon at that time, it may well be possible that the tides did not penetrate into the flat epicontinental seas, being held out by a bordering chain of islands, or by some related mechanism.

The exact time span of these transgressions and regressions is, for our subject, less important than the fact that this phenomenon must have existed through all and every one of the geosynclinal periods of each succeeding orogenetic cycle. We have in this phenomenon a means for an extreme diversification of the environment which could lead to a strong variation in chemical and biochemical development.

Just let us imagine the influence of such extensive flats—50 km by 50 km gives us already a minimum of 2,500 km^2—flooded by the sea, by the "thin soup" of that time, only to fall dry again shortly after. Evaporation on the shallow flats may already have led to a stronger concentration, to a "brine of thin soup", or "thick soup" for short, whereas upon the ensuing regression part of the material formed could have been taken up in the groundwater circulating through porous rocks. Then, after the original material could have had time to develop in this completely new environment, a renewed transgression brought it again into contact with the "thin soup", where newly developed compounds could react with the original material.

Or, if we accept the difficulty of a formation of a "thin soup" in equilibrium with normal ocean water, as discussed in section 8 of Chapter 14, we might, mutatis mutandis, visualize the formation of the "thin soup" on the shallow flats, which then came into contact with ocean water during the next transgression.

A further refinement of this model leads to the assumption that during these seesaw movements parts of the shallow oceans could well become sealed off more or less completely from the open ocean. They would then form large inland basins. The composition of the waters in such inland basins will develop separately from that of the oceans. It would mainly depend on the climate in the basins and in the surrounding lowlands which form the impluvium of such a basin, what type of waters they would contain. Schematically, one can make a distinction between wet climates, in which precipitation is higher than evaporation, and dry climates in which evaporation predominates. In wet climates fresh water basins would normally develop, whereas dry climates would lead to the formation of soda lakes.

We will return to the existence of vast inland basins during geosynclinal periods in sections 11 to 14 of Chapter 13, where we will discuss the Precambrian iron formations. At this point we have only to realize, how vast areas of shallow seas, of inland basins and of adjoining lowlands formed the morphological characteristic of the earth's surface during the geosynclinal periods of successive orogenetic cycles, and how this morphology was again and again influenced by the widely sweeping recurrent transgressions and regressions of the sea.

BERNAL has in 1961 proposed the theory that life had originated on the shore of the ocean. The reason being that on the sea shore the "organic" materials of the dilute solutions forming the "thin soup" could become concentrated by adsorption on clays. Only in this concentrated form could they, in the ideas of Bernal, serve as a basis for the further evolution of life. This certainly is a possible model for the required concentration of primary "organic" materials. But during all of the geosynclinal periods in the succession of orogenetic cycles this concentration needs not to have taken place only along the shores of the oceans. During these periods, and they form by far the larger part of geologic history, such concentration could also be achieved in the wide areas affected by the seesaw of land and sea. Instead of only along the shore lines, which indeed are but a very minor part of the surface of the earth, such concentration could well take place over most of the area of the flattened continents.

Taking Bernal's model out of the background of our own present-day post-orogenetic topography, with its emphasis on the narrow shore lines as the meeting point of sea and land, and into the setting of the wide areas of alternating transgressions and regressions during a geosynclinal period, consequently strengthens its practical value.

5. WATER-CIRCULATION IN THE SEAS AND ON LAND

In view of the importance of this seesaw of land and sea during larger parts of the geologic history, it is well to realize the differences existing in the circulation of water in the sea and on land. If we temporarily neglect the water drawn from the sea into the atmosphere in the form of spray or of water vapour, we find that in the seas, and also in lakes, water circulates only in the water body itself. On land, on the other hand, water will penetrate into the solid earth, and circulates as groundwater in the upper part of the lithosphere.

In seas and in lakes the water body has no more contact with the material sedimented on its bottom after this is sealed off by the next layer of sediment. On land, there is ample contact between rain water and ground water and at least the upper part of the lithosphere.

On land this contact leads to the formation of soils. Because of the importance soils may well have had both in the concentration and in the conservation of "organic" materials during regressive periods of the sea, the two major types of soils must be mentioned here. The distinction between these two major soil types rests on the climate, and more in particular on the balance between rain fall and evaporation, under which they have been formed. In pedology a great number of further distinctions can, of course, be made in order to arrive at a practical classification of the soils of the world. The question, whether the rain fall is seasonal, or distributed equally over the year, is, for instance, of prime importance. But for our problem it is not necessary to go into such details. It suffices to note that in climates where rain fall predominates over evaporation leached soils of the *podsol* or *lateritic* type will be formed, whereas in the other case enriched soils of the black earth or *chernosem* type develop.

6. MECHANISMS FOR CONCENTRATION AND FOR CONSERVATION

To summarize what has been said in the preceding sections Fig.49–52 give a schematized illustration of what might have been processes of concentration and of conservation during the time when primitive life developed from the simple "organic" compounds of the "thin soup".

As indicated in Fig.49 we must distinguish between the deeper seas, the shallow seas, the littoral zone and the land. The deeper seas, which can be further subdivided in the ocean troughs, the average ocean and the continental slope, in this view comprise all seas beneath the wave base. That is all seas deeper than about 30 m. The phototrophic zone, in which enough sunlight filters through to permit ample organic photosynthesis, reaches to about 50 m depth, and consequently extends beyond the wave base.

For both the deeper and the shallow seas the concentration mechanisms of formation of membranes at the water–air interface and that of formation of coacervate droplets through wave action are operative (*a* and *b* in Fig.50 and 51). There is, on the other hand, no conservation mechanism at the sea's surface, all "organic", and for that case all organic materials too, being exposed to the destructive forces of the surroundings. That is to ultraviolet radiation under an anoxygenic, and to oxidation under an oxygenic atmosphere.

There is a certain measure of conservation, but no concentration in the waters of the deeper sea, from a depth of 30 m to the ultimate depth of over 10 km. There is concentration of material when it settles on the sea floor.

In the shallow seas there is little concentration, except that of the formation of membranes and coacervate droplets at its surface, already described. In this environment the water is agitated down to the bottom at every storm, resulting in seaward transport of bottom materials.

In the littoral zone, on the other hand, considerable concentration may be found. This can take the form of a concentration of heavier material through wave action, leading to the formation of placer deposits. But, depending upon the local topography and the climate of a certain littoral zone, it can also take the form of a concentration of soluble material of the sea water, and the formation of evaporites. Just as is the case at the sea's surface, there will, however, be no important conservation mechanism operative in the littoral zone, as both placers and evaporite deposits will be equally exposed to atmospheric influence.

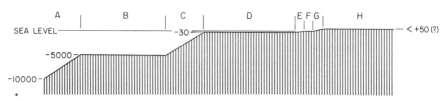

Fig.49. Schematic model of the various environments above and below sea level, that in their different ways are important in view of mechanisms of concentration and conservation. (Not to scale!)

Key to Fig.49–52:

A = ocean troughs; B = average ocean; C = continental rise and slope; D = shallow seas; E = coastal sea slope; F = littoral zone; G = coastal land slope; H = base-levelled continent.

$a–c$ = processes at or near sea level. a = formation of membranes at the water–air interface (see Fig.43); b = formation of coacervate droplets through wave action (see Fig.41); c = transportation along the surface of the sea through wind and wave action.

d, e = processes at the bottom of shallow seas. d = seaward transport of bottom material; e = this material will eventually reach the edge of the continental slope and will help in building out the shallow sea.

f, g = processes in the ocean. f = sinking and eventual settling of material heavier than water; g = ocean currents effective in mixing water, but not in concentration of suspended matter, except on the ocean floor. Here the sinking matter is swept away in some areas, concentrated in others. Water in the ocean troughs is practically stagnant.

$h–i$ = processes in the littoral zone. h = possible concentration of heavier material in coastal placers; h' = possible concentration of soluble materials by evaporation; i = possible seaward transportation of concentrated lighter materials.

$j–o$ = processes on land; $j–l$ = podzolic or lateritic types of soils; $m–o$ = chernosem types of soils. j = groundwater leaching the topsoil down to depths of several meters or even several tens of meters; k = part of the leached material falls out at groundwater level, forming the cementation zone; l = the remainder is transported seaward by groundwater movement; m, o = groundwater leaching the top layers without a clear break, both above and below groundwater level; n = leached material falls out at or near the surface of the topsoil through evaporation.

It is only on land that important concentration and conservation of "organic" materials seems possible. At the onset of a regression, when the land rises out of the sea, or when the sea retreats from the land, this has of course the same effect, the upper layers of the lithosphere will be drenched by the "thin soup". The amount of "thin soup" included in the upper layers of the lithosphere depends mainly on the nature of these layers. If the sea bottom is formed by crystalline basement, it is minimal. But if it is formed by unconsolidated sediment, such as sand, it may take up about half the volume of these sediments, and reach down to several tens of meters. When the land rises further, circulation of groundwater starts, because there is now a gravitational force acting upon the water situated above sea level. The groundwater will tend to flow towards the sea, whilst rain fall will in part be absorbed into the earth, sustaining the groundwater flow.

The difference between the sediments below the sea level and those above it therefore lies in the fact that in the first place none or hardly any circulation takes place. The former sea water, included in the bottom sediments, is fossilized in place. Small scale reactions of adsorption type may take place, but there is no wholesale movement of the water body, together with its included materials. This only starts when the land rises above the sea.

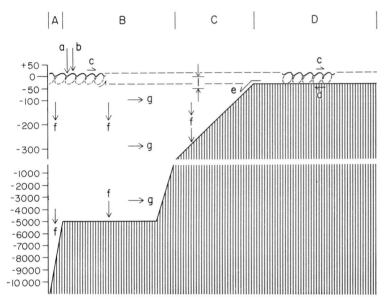

Fig.50. Schematic model of water circulation and of possible mechanisms of concentration and of conservation in the sea. (Not to scale! For key see Fig.49.)

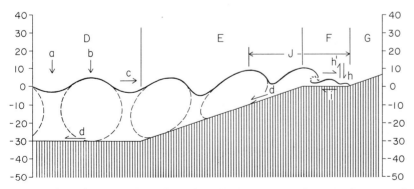

Fig.51. Schematic model of the circulation of sea water near the coast. (Not to scale! For key see Fig.49.)

Through the influence of groundwater movement soil formation will start. In Fig.52 the two most distinct types of soil, the podsol and the chernosem soils and their influence in concentration have been indicated. Not only do the various forms of soil formation and the groundwater movement offer concentration mechanisms, but in this case we also have conservation mechanisms at the same time. A thin layer of soil will already provide shelter against ultraviolet radiation. Whereas, dependent upon the speed of the groundwater stream, all free oxygen will be used up relatively quickly in more or less stagnant areas, so that anoxygenic, in this case truly anaerobic, conditions, may readily develop.

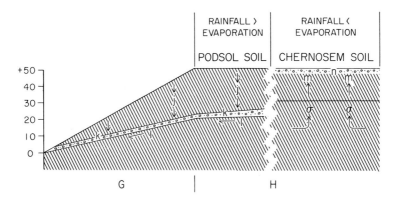

Fig.52. Schematic model of the circulation of fresh water on land. In the lowland area H the influence of two different types of climates are indicated. At left rainfall is higher than evaporation, at right the reverse is true. (Not to scale! For key see Fig.49.)

But even so, a state of equilibrium will soon be reached, leading to a situation which overall is static. The original "organic" material may have evolved a single step, or perhaps a couple of steps, toward the next stage in biopoesis, without being able to proceed any further. An incentive for further evolution may then well be supplied by the next transgression, when the soils formed on the temporary land surface are again doused by the waters of the "thin soup", and may react with more of the primary "organic" materials.

A further sophistication of this model is arrived at, when the next transgression is so unimportant that a lagoonal environment is formed, in which evaporation occurs. The soils formed on land will then be flooded, not by a "thin soup", but by a "thick soup", and the reactions might be considerably speeded up. In the model of dominant horizontality for the geosynclinal periods, this sophistication seems not to be far fetched at all. For the smaller the crustal movements, the greater the probability that lagoonal conditions with their corollary evaporation do develop.

REFERENCES

BERNAL, J. D., 1961. Origin of life on the shores of the ocean. In: M. SEARS (Editor), *Oceanography*. Am. Assoc. Advan. Sci., Washington, D. C., pp.95–118.

GASTIL, G., 1960a. The distribution of mineral dates in time and space. *Am. J. Sci.*, 258:1–35.

GASTIL, G., 1960b. Continents and mobile belts in the light of mineral dating. *Intern. Geol. Congr., 21st, Copenhagen, 1960, Rept. Session, Norden*, 9:162–169.

GILLULY, J., 1966. Orogeny and geochronology. *Am. J. Sci.*, 264:97–111.

JÄGER, E., 1962. Rb-Sr age determinations on micas and total rocks from the Alps. *J. Geophys. Res.*, 67:5293–5306.

JÄGER, E. and NIGGLI, E., 1964, Rubidium–Strontium Isotopenanalysen an Mineralien und Gesteinen des Rotondogranites und ihre geologische Interpretation. *Schweiz. Mineral. Petrog. Mitt.*, 44:61–81.

MOORE, R. C., 1970. Stability of the earth's crust. *Geol. Soc. Am. Bull.*, 81:1285–1323.

PIRLET, H., 1968. La sédimentation rythmique et la stratigraphie du Viséen supérieur V3b, C3c inférieur dans les synclinoriums de Namur et de Dinant. *Mém. Acad. Roy. Belg.*, 2, 17, 4, 98 pp.

UMBGROVE, J. H. F., 1947. *The Pulse of the Earth*, 2nd ed. Nijhoff, The Hague, 358 pp.

Chapter 11 | Where to Look for Remains of Early Life. The Old Shields

1. PAUCITY OF EARLY RECORDS

Examining geology further in our search for facts about the origin of life, we find them to be deplorably scarce. There are few fossil finds from the early history of the earth. This stems from two main reasons, one being the small size and the softness of early life forms, the other the great age of the rocks which contain the fossils. We have already noted the importance of microbes for early life in section 4 of Chapter 8, and we will go more fully into the matter of the smallness and the softness of early life in the next chapter. In this chapter we will present a summary of the areas in which the rocks dating from the early history of the earth can be found.

Most of the rocks formed in those early days are not visible for inspection now. They are buried below younger rocks, mainly sediments, in varying thickness. These may range from several hundred meters to several tens of kilometers, dependent upon the rate of subsidence of the crust since the beginning of the Phanerozoic.

And although in those areas where the sedimentary cover is less than ten kilometers thick, a steadily growing number of drill holes reach the older rocks, or the "basement", as it is called in oil geology, and thus the exploration for oil has consequently given us a much greater insight into the structure of those basement rocks, even when covered under thick series of sediments, this has not helped us much in our search for facts about early life. The chance that in a drill core, several centimeters long and a couple of centimeters across, one of those extremely rare fossils of the Precambrian could be found, is negligible. We have to turn our attention to those areas where Precambrian rocks at present crop out at the earth's surfaces, regions technically known as the *old shields*.

Even here, most of the rocks exposed have suffered so much during their later history that every vestige of early life they might have contained has been destroyed. For one thing, although exposed at the surface now, they have often

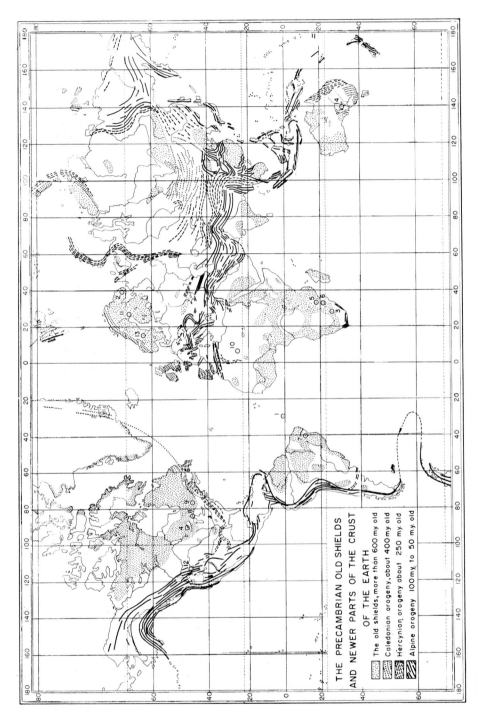

THE PRECAMBRIAN OLD SHIELDS
AND NEWER PARTS OF THE CRUST
OF THE EARTH

The old shields, more than 600 my old
Caledonian orogeny, about 400 my old
Hercynian orogeny about 250 my old
Alpine orogeny 100 my to 50 my old

been buried underneath younger rocks for some time during the Phanerozoic. On the other hand, large areas have become folded, or intruded by extensive bodies of magma, forming igneous rocks such as granites when cooling. All such mishaps expose the rocks to a temporarily elevated temperature and pressure. This has the effect of mobilizing the mineral constituents of the rocks, because under higher temperature and pressure the melting point of certain minerals is approached. These will tend to recombine and to recrystallize in large crystals, either of a similar composition, or even of a composition different from the original minerals. But in both cases primary rock structures, such as fossils, will be destroyed.

2. THE OLD SHIELDS

As we saw, the old shields are formed by those areas on earth where Precambrian rocks crop out over large regions, without a cover of younger rocks. From Fig.53 we learn that, in a certain sense, the old shields form the cores of the continents. There are a Canadian, a Fennoskandian and a complex Asian Shield on the northern continents, whereas we find more or less corresponding shields

Fig.53. Geologic sketchmap of the world, showing the extent of the Precambrian old shields and of the three main orogenies of the Phanerozoic.

1 = St. Peter and Paul Rocks. Mantle material thrust upwards in a major fault system. 4.5 billion years old. (See section 14 of Chapter 3.)

2 = Kola Peninsula granites. Oldest known crustal rocks. 3.3 billion years old. (See section 14 of Chapter 3.)

3 = Swaziland System (Onverwacht and Fig Tree Series). Oldest known slightly altered sediments. Over 3.2 billion years old. (See section 18 of Chapter 12.)

4 = Soudan Iron Formation. Sediments with molecular fossils, over 2.7 billion years old. (See section 9 of Chapter 12.)

5 = macroscopic biogenic deposits. Algal limestones described by MacGregor (1940). Over 2.7 billion years old. (See section 8 of Chapter 12.)

6 = Witwatersrand complex. Gold–uranium reefs with non-oxidized pyrite sands. Over 2 billion years old. (See section 7 of Chapter 13.)

7 = Serra de Jacobina gold–uranium ores, comparable to the Witwatersrand complex.

8 = Gunflint Series. Oldest known well-preserved microflora, about 2 billion years old. (See section 13 of Chapter 12.)

9 = Blind River gold–uranium ores, comparable to the Witwatersrand complex. 1.8 billion years old. (See section 10 of Chapter 13.)

10 = algal reefs of Conophyton. (See section 5 of Chapter 12.)

11 = algal limestones. (See section 5 of Chapter 12.)

12 = well-preserved microflora in the Late Precambrian Belt Series. Over 1.1 billion years old. (See section 6 of Chapter 12.)

13 = Dala Sandstones earliest red beds (?). 1.4 billion years old. (See section 1b of Chapter 13.)

14 = Ediacara fauna. Oldest known metazoan fauna. Between 0.6 and 1 billion years old. (See section 4 of Chapter 12.)

(From UMBGROVE, 1947.)

in Brazil and surroundings, in South Africa and in Australia on the Southern Hemisphere. Moreover, there are smaller, detached areas, such as Madagascar and the subcontinent of India. However, as we see best illustrated in South Africa, old shields do not always form the core of continents. They are not always surrounded by younger rocks, but may also be cut off directly by the ocean shores.

As anyone familiar with geology knows, there has recently been much theorizing about the former positions of the continents on our globe, and continental drift of larger dimensions is now almost generally accepted. Up till the present, this applies, however, mostly to the period beginning with the Permian. We have much less data about the position of the continents in the earlier history of the earth, and continental drift has, as far as we know, not had a particular influence on the early history of life.

In most textbooks on geology the old shields are indicated as quite different from the newer regions of the earth's surface. In a well-known illustration in UMBGROVE's *Pulse of the Earth* (1947), here reproduced in Fig.53, we see how the old shields are indicated by stippling, without any indication of pronounced structural directions. This is in clear contrast to the areas in which younger rocks crop out, and where the main directions of the fold belts of the three major orogenies of the Phanerozoic are indicated.

It must be stressed that this distinction, as suggested in so many textbooks of geology, is but an apparent one. It follows from the break in classical geology between all of the Precambrian and the younger history of the earth. As we have seen in Chapter 3, the only difference is that from this "break" onwards there are sufficient fossils to permit relative dating and the recognition of detailed successive stratigraphic units for the younger history of the earth. In contrast, as we saw in section 4 of Chapter 3, the divisions of the Precambrian into Archean and Algonkian, or into Azoic and Proterozoic, such as used formerly, are devoid of any real basis.

Only now, with the help of our modern absolute dating techniques, has the unravelling of the old shields become a practical possibility. Before that time conclusions about age relationships, not only between rocks from two different old shields, but even for separate areas within the same old shield, were based on assumptions rather than facts.

As an example of a more modern division of the Precambrian, based on absolute dating, Table XV gives the major subdivisions of the Precambrian of Minnesota. So the apparent difference between the old shields and the younger areas, as indicated in maps such as Fig.53 does not indicate that the old shields are different from those regions where younger rocks crop out, but only that we know next to nothing about the old shields. So one has to take recourse to an indication like stippling, in which no preferred orientation can be given.

TABLE XV

SUBDIVISION OF THE PRECAMBRIAN OF MINNESOTA[1]
(From GOLDICH et al., 1961; cf. Table XXI)

Canadian usage	Era	System	Orogeny	Years ago ($\times 10^9$)
Paleozoic	Paleozoic	Cambrian		
				– –0.6– –
			Grenville	—1.0—
Proterozoic	Late Precambrian	Keweenawan		
			Penokean	—1.7—
	Middle Precambrian	Huronian		
			Algoman	—2.5—
		Timiskamian		
	Early Precambrian		Laurentian	– –?– –
Archean		Ontarian		

[1] In this region of the Canadian Shield the Algoman orogeny has resulted in widespread strong metamorphosis and the emplacement of many intrusions of igneous rocks. This has, in the old days of relative dating, been taken as the border between Archean and Proterozoic. Elsewhere, however, either a younger or an older orogeny may have had the same effect, and the border between Archean and Proterozoic was shifted to that orogeny. In fact, all strongly metamorphosed rock series have formerly been assigned to the Archean, and in a given area the border with the Proterozoic only indicates the last strong Precambrian orogeny *in that area*. It was not, and could not, be applied to a constant age.

3. STABILITY OF OLD SHIELDS DURING THE PHANEROZOIC

The indiscriminate stippling of the old shields in Fig.53 does, however, indicate a difference between the old shields and the rest of the continents, i.e. a difference in crustal stability. All during the Phanerozoic the old shields have only undergone relatively small vertical movements, either up or down, and often alternating in sense with time. Outside of the old shields, on the other hand, there has been the activity of the three major orogenies of the Phanerozoic, and of some minor periods of unstability, in which not only vertical movements, but also folding of the crustal rocks has taken place.

This stability is not limited to the actual area of the outcrop of the Precambrian rocks, but extends over a narrower or a wider margin into regions where the Precambrian is covered by horizontal rocks of the Paleozoic. These regions, which also have undergone no folding during the Phanerozoic, structurally belong to the same province as the old shields. The difference is a matter of erosion of the Paleozoic rocks, which formerly have covered far wider areas of what now is the old shields. These regions, which are left blank in Fig.53 are called the "Stable

Interior" on the North American continent. It follows that the old shields, together with the surrounding stable regions, remained relatively stable during the younger history of the earth. Newer orogenetic cycles led to the formation of fold belts outside the earlier areas, or even more or less concentrically around them. This is, however, no absolute law, for many times also a newer fold belt cuts one or more earlier fold belts. In short, parts of all the continents became stabilized at the beginning of the Phanerozoic, and no newer fold belts developed in these areas. Why this is so is still one of the enigmas of geology. Formerly a rather naive comparison with corrugated iron was alluded to, corrugated iron being stronger than sheet iron. This could, however, not be extrapolated to the crust of the earth, which has quite different material constants, whereas it also does not explain why in so many cases newer fold belts did develop, not outward of older fold belts, but cross-cutting them.

The apparent stability of the old shields does, however, only apply to the Phanerozoic. It is only during the Phanerozoic that the old shields remained relatively stable, whilst younger fold belts developed elsewhere. It does not hold for the earlier history of the earth, and the internal structure of the old shields is just as complex as that of the parts of the crust affected by the younger orogenetic cycles.

4. COMPLEX STRUCTURE OF OLD SHIELDS

Just as in the areas outside the old shields, a succession of older and younger orogenetic fold belts are found to make up the structure of each old shield. And, just as is the case in the younger history of the earth, part of the areas of the older orogenetic cycles have no more been affected during younger cycles. The old shields consequently are not a simple and uniform area, forming the core of the continents. They themselves are formed of older and still older cores, in part surrounded, in part cut up by younger fold belts. This situation is clearly expressed in Fig.54 and 55 for the North American continent.

Realization of this complex structure of the old shields is important in our search for areas where vestiges of early life may be found. For it is only in the areas that remained stable since their formation that we may hope to find such remains. In those areas relatively thin series of rocks may have become deposited which were not too much affected by later orogenies. Consequently, in these areas the rise of temperature due to the cover by younger beds is not excessive, whereas there also is not too much magmatic activity. Only in such areas, technically called *cratonic*, is there any chance that original rock structures, such as fossils, may be preserved. In the major Precambrian foldbelts, on the other hand, destruction of earlier rock structures is almost certain through the accompanying metamorphism and magmatization. An example of the difference in the degree of metamorphism of rocks of similar age, in a major fold belt and in a relatively little disturbed epicontinental area, can be found in the Cambrian of Europe.

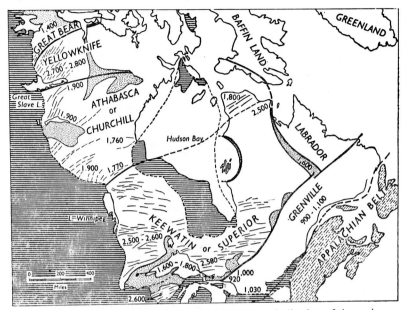

Fig.54. Schematic map of the Canadian Shield, with an indication of the various provinces and the ages of the latest granites belonging to each province. The Appalachian fold belt, bordering the shield to the southeast, is indicated, as are the horizontally hatched areas of Paleozoic and younger rocks of the Stable Interior, which border the shield to the south and west.

The dotted areas on the shield represent those areas where Precambrian rocks have suffered little or no later folding. When the Canadian Shield is compared with the African Shield (see Fig.56) it will be seen that such cratonic areas are much larger and more widely spread on the African Shield.

The structure of the Precambrian, characterized by the existence of well-defined fold belts, is not only found on the shield itself, but extends also under the cover of younger rocks, as can be seen from Fig.55. (From HOLMES, 1966.)

During the Cambrian sediments were laid down all around, and possibly all over, the Fennoskandian Shield. In Norway and Scotland a geosynclinal basin developed, which was filled with a thick pile of sediments, and subsequently folded and magmatized during the Caledonian orogeny. Farther east, on the other hand, around the Baltic and up to the Ural Mountains, thinner series of sediments developed in an epicontinental facies, which were not disturbed by folding. The Cambrian rocks of the Caledonian fold belt are now all more or less strongly metamorphosed, forming slates, schists and gneisses. But in the epicontinental deposits around Moscow the Cambrian clays are still so soft that they can be used in brick making.

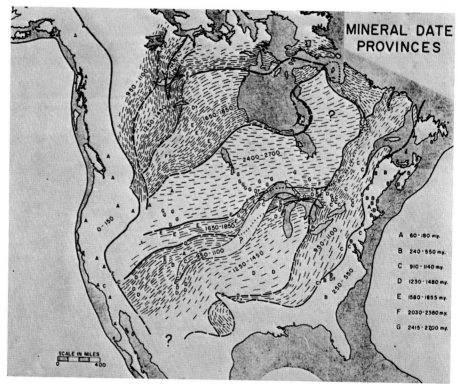

Fig.55. Schematic map of the complex structure of the Precambrian of North America.

Due to the large number of deep exploratory oil drillings, which have penetrated into the Precambrian basement, the structure of the Precambrian, in its general aspects, is now known far beyond the area where it actually crops out in the Canadian Shield (see Fig.53 and Fig.54).

The crust is formed by a complex system of Precambrian fold belts (G–C), that are quite similar to the younger fold belts, such as the Appalachian (B) and the Cordilleran (A).

Schematically, the age of a fold belt is the same as that of the youngest mineral dates found in it. Older mineral dates are recorded from minerals which have not become completely regenerated during the younger orogeny. In detail two consecutive orogenies may have been lumped together, such as the Caledonian and the Hercynian orogenies in the Appalachian belt of B.

But such details apart, it is seen that there are quite distinct fold belts of various ages. Moreover, younger fold belts may cross freely over older fold belts. (From GASTIL, 1960.)

Of course, the farther we go back into the early history of the earth, the smaller are these oldest cores, which were already stabilized during an even older orogenetic cycle, and which have escaped mobilization during one of the younger orogenetic cycles. So, the older the rock, the greater the chance that some unfortunate mishap has occurred which destroyed all vestiges of early life it might have carried. The abundant fossils from the Cambrian onward are "only" one half billion years old at most. But as we will see in the next chapter, research into early life will refer us to the 3 billion year age bracket.

The devastating influences of each succeeding orogenetic cycle have seriously limited the extent to which we may be able to trace back the existence of life on earth. I am alluding to the oldest crustal rocks found so far, the 3,3 billion years old granites of the Kola Peninsula in northern Europe. There is a chance of finding in these rocks minerals which still retain some of the radioactive decay history of the earlier sediments. But the metamorphism of these older sediments has been so intense that all traces of life have disappeared. The only chance we have to come across still older vestiges of life lies in other, less well studied old shields, where a less disturbed core formed by those older sediments might some time be

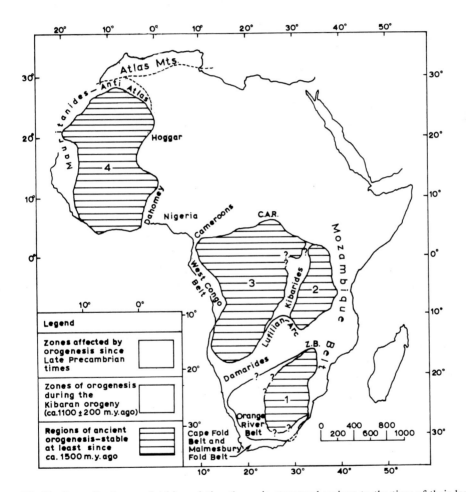

Fig.56. Generalized map of Africa relating the main structural regions to the time of their last orogenetic deformation. The four older cratons are: (*1*) the Rhodesia–Transvaal craton; (*2*) the Tanzania craton; (*3*) the Angola–Kasai craton; (*4*) the West African craton. The Swaziland System is found in the Rhodesia–Transvaal craton. (From CLIFFORD, 1968.)

found. Until that time the complex history and structure of the old shields seems to have limited the period from which we are able to find indications of life on earth. This tectonic history of the old shields has of course nothing to do with the history of life on earth, and the later destruction of early vestiges of life is entirely gratuitous.

In section 18 of the next chapter we will see, how the oldest known sediments which have escaped strong metamorphism, and which are thought to contain vestiges of former life, are found in South Africa, in the Fig Tree and Onverwacht Series of the Swaziland System (see number 3 in Fig.53). The reason why Africa as a whole is more propitious for the preservation of ancient sediments than are the other old shields, lies in the fact that Africa contains a relatively large number of cratonic areas of very old age, which have not suffered the effects of a younger orogeny (Fig.56). To state, however, that the sediments of the Swaziland System are the oldest rocks in which it will ever be possible to find remnants of early life on earth, as has been done by ENGEL et al. (1968) seems to be premature. Although it is very probable that no older sediments are preserved in the Canadian and the Baltic Shields, it still remains possible that the shields of Brazil and Asia harbour still older cores.

REFERENCES

CLIFFORD, T. N., 1968. Radiometric dating and the pre-Silurian geology of Africa. In: E. I. HAMILTON and R. M. FARQUHAR (Editors), *Radiometric Dating for Geologists.* Interscience, New York, N. Y., pp.299–416.
ENGEL, A. E. J., NAGY, B., NAGY, L. A., ENGEL, C. G., KREMP, G. O. W. and DREW, C. M., 1968. Alga-like forms in Onverwacht Series, South Africa: Oldest recognized lifelike forms on earth. *Science,* 161:1005–1008.
GASTIL, G., 1960. The distribution of mineral dates in time and space. *Am. J. Sci.,* 258:1–35.
GOLDICH, S. S., NIER, A. O., BAADSGAARD, H., HOFFMAN, J. H. and KRUEGER, H., 1961. The Precambrian geology and geochronology of Minnesota. *Minnesota Geol. Surv., Bull.,* 41:1–193.
HOLMES, A., 1966. *Principles of Physical Geology,* 2nd ed. Nelson, London, 1288 pp.
McGREGOR, A. M., 1940. A Precambrian algal limestone in Southern Rhodesia. *Trans. Geol. Soc. S. Africa,* 43: 9–16.
UMBGROVE, J. H. F., 1947. *The Pulse of the Earth,* 2nd ed. Nijhoff, The Hague, 358 pp.

Chapter 12 | The Fossils

1. THE VARIOUS KINDS OF FOSSILS

Normally we think of fossils as the petrified remnants of former organisms. In fact it is not absolutely necessary that the organic material has been really petrified, that is, replaced by inorganic molecules. But although a number of fossils has indeed still retained at least a part of their original organic material, this is definitely an exception, whilst it occurs only with younger fossils. The fossils we normally visualize are only the morphologically preserved parts of former life, such as shells, teeth, leaves, etc. There are, however, still other kinds of remnants of former life, which I should like to include in this chapter. These are, first, the macroscopic biogenic deposits, and second, the molecular biogenic deposits, the molecular or chemico-fossils[1].

The macroscopic biogenic deposits consist of material deposited by organisms during their metabolism[2]. At present the best known and the most voluminous of such deposits are limestones formed by algae. These are normally found in limestone areas, in which the well waters are saturated with carbonate. Because of their non-specific nature, macroscopic biogenic deposits at best tell us in a vague way only by what sort of organisms they were deposited. They do, however, prove that life was present at the time when their deposition took place.

---

[1] Professor Eglinton used the word "chemicofossil", which I subsequently borrowed (RUTTEN, 1966), in a paper read before the "Geologische Vereinigung" in 1965 (see also EGLINTON and CALVIN, 1967). In the publications on this subject, the term "molecular fossil" is, however, more often used (CALVIN, 1965, 1969). It therefore seems best to use the second term.

[2] This distinction between shells and biogenic deposits is in a sense not realistic, because the material from which shells are built is also deposited during metabolism. But shells, such as of mollusks and corals, are secreted by definite areas of an organism. Therefore they are intimately related to the morphology of the organism that secretes them. Shells, and also teeth and bones, are therefore different, morphologically distinct, from species to species. A more exact definition would have us separate the morphologically specific biogenic deposits (or fossils sensu stricto) from the morphologically non-specific deposits.

Molecular fossils are biogenic molecules. They form a new group of fossils, which has only been detected in the sixties. These biogenic molecules consist of the stable parts of larger organic molecules, which formed the building parts of former organisms, but which themselves have become degraded during fossilization. Of the original biological materials which once built up these organisms, nothing is left. So molecular fossils do not give us an indication on the specific nature of the organisms they once belonged to. They only tell us that life was present on earth when the organisms from which they are derived lived. But, because it so happens that some of these molecular fossils can be interpreted as the degradation products of very special molecules such as chlorophyll, they may tell us something of the metabolism of the organisms from which they stem.

To sum up, we will in this chapter not only study what is normally called fossils, but all remnants of life dating from the early history of the earth. These are:

(*1*) Macroscopic biogenic deposits.

(*2*) Molecular biogenic deposits, the molecular fossils.

(*3*) Fossils sensu stricto.

2. NON-LIVING AND LIVING IN GEOLOGY

"Returning now to our speculations upon the distinction between non-living and living in the present-day world (see section 2 of Chapter 4), it is well to realize at this point that in geology this distinction is quite easy. In geology we meet none of the difficulties the biologist encounters when he tries to distinguish between non-living and living.

"To be at all recognizable as remnants of former living organisms, fossils must have preserved some clearly organized shape or structure, distinguishable as such by the eye or by the microscope. Only rarely do we have some idea of how these organisms died, for instance when they have been buried by volcanic ash, or when they slid into an asphalt pool. Moreover, we have only the vaguest notions as to why, when and how our fossils-to-be were preserved, how and when the fossilization process, the degradation of the original organic matter and its replacement by mineral substances took place. Our fossils can no more yield the fresh preparations biologists study under the electron microscope, nor the extracts they prepare in an ultracentrifuge. Fossil remains, it must be stated most strongly, are not only dead but also buried and fossilized.

"It follows that to be recognizable as a fossil the organisms from the early history of life must have already had some organization, some kind of structure. Organisms capable of being detected as such in fossilized state after billions of years may well have belonged to the "lower" organisms, to microbes or algae-like forms, but nevertheless the organization of such forms has progressed already a long way from that borderland between non-living and living the biologist is interested in. We can hardly doubt that organisms which could turn into recognizable fossils were already fully alive and did not belong to the borderland between pre-life and life."

Such was the gist of a statement I could still make in 1962. Transported on a molecular scale this also holds good. We all know the difference in nature between inorganic and organic compounds. But since then we have learned that this only holds good for the later part of the earth's history, for the Late Precambrian and the Phanerozoic. It is no more than the story pertaining to the "normal" geology of the Late Precambrian and the Phanerozoic, when, under the oxygenic atmosphere, inorganic synthesis of organic compounds was impossible, and when it is fairly easy to distinguish fossilized life from inorganic growths. The statement cited above does not, however, apply to the Early and Middle Precambrian, when the earth still possessed its primeval anoxygenic atmosphere. Under the primeval atmosphere, during Early and Middle Precambrian, the situation has been quite different. For one thing, inorganic synthesis of "organic" compounds was, as we have seen earlier, possible at that time. And for another, life was in these early days still so primitive that it is often impossible, from the morphology alone, to tell whether we have to do with a fossilized life form or with a fossilized "organized element" which has been synthesized inorganically.

In this chapter we will see in sections 4–6 that we find no difficulty in recognizing a rather varied spectrum of fossilized life from the Late Precambrian. For the Middle Precambrian too, we find fossilized life well represented, as described in section 13. But already in the Middle Precambrian many of the so-called fossils are so simply built that, as we will see in section 14, no distinctive features can be found to separate fossils from "organized elements". This applies a fortiori to the Early Precambrian, where no structures have as yet been found which can without doubt be interpreted as fossils (see section 18).

A similar situation is encountered on the molecular scale. During the Early and Middle Precambrian inorganic synthesis of "organic" compounds has not only been possible, but must have been a common, highly varied and widely disseminated process. The successes attained in synthesizing "organic" methods, as related in Chapter 6, indicate that all compounds necessary for life, nucleotides and proteins included, could at that time be formed inorganically. So for this early period of the earth's history, a compound must have very special qualities before we can be satisfied that it really once did belong to a biogenic, a really organic, molecule, which, in turn has formed part of a living organism. We will go into this matter in sections 9–11, when discussing the molecular fossils. As a parallel to the situation encountered in the study of morphologically preserved fossils, cited above, we will find that trustworthy indications for the existence of life during the Early Precambrian have not yet been supplied by the molecular fossils either.

This attitude, so infinitely more cautious than that prevailing till fairly recently, and cited in the first part of this section, rests upon the results of studies in Precambrian geology dating of the last decade. Even now some authors indiscriminately describe "fossils"—both morphological and molecular—from very old sediments and of a very primitive morphology and molecular structure, which cannot stand up to the more rigorous tests which have to be applied at present.

The attitude still prevails that materials from older sediments prove the existence of contemporary life if and when they resemble present-day life. Whereas with our knowledge of the molecular structures and the morphological forms pre-life is able to develop, it behoves us to look for criteria which can really distinguish between pre-life and life. As an example, we might cite how in 1965 and 1966 Professor H. C. Urey, whilst critically reviewing the pros and contras of the supposedly biological origin of material in the carbonaceous meteorites (see section 6 of Chapter 17) could still write: "It seems safe to say that if similar materials were found on earth, no one would question its biological origin." Thereby disregarding that such carbonaceous material could have been formed a-biogenically, that is by inorganic processes, on the earth just as well as on the parent body of the meteorites. And although the statement was repeated in 1969 (NAGY and UREY, 1969), it is my contention that the same discretion should apply to our study of terrestrial life-like remains of the Early and Middle Precambrian, as in the study of the carbonaceous meteorites. On our study of terrestrial samples of the earliest history of the earth we are now also approaching the border line between non-living and living.

Summarizing the contents of this section, we may stress again that for the later history of the earth, for the Late Precambrian and the Phanerozoic, there is generally no difficulty to distinguish between non-living and living in geology. But for the earlier history, under the primeval atmosphere, we find it very hard to draw the exact boundary between non-living and living, in geology too.

3. LATE PRECAMBRIAN FOSSILS

The paucity of the early fossil record of life on earth, to which attention has been drawn already several times, stands, perhaps, in need of qualification. For there exists a rather extensive literature on fossils from the Precambrian. However, the majority of these fossils were found in sediments of Late Precambrian age. They are not so very much older than, and form the precurors of, life during the Cambrian. Although they have no direct relation to the problem of the origin of life, three examples will be cited here, in order to stress the difference between life during the Late Precambrian and that during the earlier history of life, to which the remainder of this chapter will be devoted. These are: (*1*) a Late Precambrian fauna from Australia; (*2*) Late Precambrian reefs of algal limestones, and (*3*) a rich microflora of the Late Precambrian of North America. They show that life during the Late Precambrian was already well developed and diversified into members of both the plant and animal kingdoms. Basically the only difference between life during the Late Precambrian and that during the Early Phanerozoic is that during the latter period some groups of animals developed shells, skeletons and other hard parts, which are more suitable for fossilization.

Another practical reason which for so long a time has strengthened the apparent break between Late Precambrian and Phanerozoic, is that, until a short time ago, all geologists studying the Precambrian were so-called "hard rock" geologists, mainly interested in igneous and metamorphic rocks. That is where most of the mineral wealth of the Precambrian is found, and geologists working for ore companies could, without risk of dismissal, only have a superficial look at sediments. Since the last few decades this attitude has, however, changed and the scientific interest in life during the Precambrian has drawn a much wider attention for this early period of the earth's history; an attention as yet mainly drawn from members of universities. This has already resulted in the new discoveries reviewed in this chapter, discoveries which will doubtless be followed by many more during the years to come.

4. A LATE PRECAMBRIAN FAUNA

Late Precambrian animal life is represented by fossil remnants of Radiolaria, Coelenterata, annelid worms and even Crustacea, indicating the existence of a rich and varied fauna during this period (GLAESSNER, 1961, 1962, 1966; CLOUD, 1968; WADE, 1969; see Fig.57). Life of the Late Precambrian was not restricted any more to microbes and plants. The organisms which lived during the Late Precambrian belonged to a much higher level of organization than those we must look for in our study of the origin of life. Segmented animals like annelid worms and crustaceans are as far removed from the origin of life, as a jet aircraft is from a wheelbarrow.

Consequently, these faunal assemblages of Late Precambrian times, in spite of the publicity they have received, add but little to the knowledge of early life. They are all found in sediments laid down a short time before the beginning of the Phanerozoic, whilst they form the ancestry of the animal world of the Phanerozoic. Their absolute age presumably will not be much higher than the 0.6 billion years which is now accepted for the base of the Cambrian, and probably will be lower than one billion years.

Fig.57. Reconstruction by M. F. Glaessner and M. Wade of the Ediacara fauna of the Late Precambrian in South Australia. (For location see Fig.110.)

The fossils can be grouped into three classes, the Coelenterata, the Annelida and the Arthropoda. Although several forms have died out already before the beginning of the Cambrian, there is in principle only one difference between the fauna from the Precambrian and the later ones, i.e., that non of the Precambrian animals have worn shells (GLAESSNER and WADE, 1966; GOLDRING and CURNOW, 1967). The individual animals portrayed are: 1–10 = Coelenterata. 1 = *Ediacara flindersi*; 2 = *Beltanella gilesi*; 3 = *Medusinites asteroides*; 4 = *Mawsonites spriggi*; 5 = *Cyclomedusa davidi*; 6 = *C. plana*; 7 = *Conomedusites lobatus*; 8 = *Rangea longa*; 9 = *Arborea arborea*; 10 = *Pteridinium simplex*; 11–14 = Annelida; 11 = *Spriggina flounderi*; 12 = *Dickinsonia costata*; 13 = *D. elongata*; 14 = *D. tenuis*; 15, 16 = Arthropoda; 15 = *Parvancorina minchami*; 16 = *Praecambridium sigillum*; 17, 18 = other organisms. 17 = *Tribrachidium heraldicum*; 18 = subsphaerical gelatinous bodies. (Courtesy Professor M. F. Glaessner.)

The Ediacara fauna, for instance, given here as an example of the faunas described from the Late Precambrian, does not give us a clue to the "age of the animals". Or as Professor CLOUD (1968) prefers to call this point in the history of the earth, to the "age of the earliest metazoans". The diversification of the Ediacara fauna points to a rather longish preceding evolution. But in how little time, as measured in absolute age, this evolution might have been condensed, or over how long a period it might have been drawn out, we have, as yet no idea. In Fig.135 the separation of the animals from the main stream of the organisms has therefore been indicated by a question mark.

5. ALGAL LIMESTONE REEFS

Similar considerations apply to the reefs of algal limestone, which abound in certain layers of the Late Precambrian of every old shield. The similarity is, however, less strict, because similar reefs of algal limestone are also found in the Early and Middle Precambrian. Such remains are normally described as belonging to stromatolites, which is, however, but an artificial taxonomic group (see section 12 of Chapter 14). They are formed by masses of limestone, sometimes globose, sometimes tubular. They may have a distinctive microstructure because they are formed by limestone deposition around algal bodies, but they have no relation to the specific characters, nor to the morphology, of the organisms by which they were deposited.

Algal limestones may occur in stratiform layers, but also in colonies of considerable size. The latter may form reefs, or, to use the technical term, *bioherms*. Such reefs grew in situ as mounds constructed by the lime-depositing organisms and stood out above the surrounding floor of lakes or oceans, just as do the present-day coral reefs. Algal reefs of the group of stromatolites are the earliest known bioherms in geologic history.

We have no direct ways of dating these algal limestones within the Precambrian. As a case in point, we might cite the common reefs of *Conophyton* in the vicinity of the Hoggar in the central Sahara, as described by GRAVELLE and LELUBRE (1957; see Fig.58, 59). The beds in which these reefs are found are

Fig.58. Reef of *Conophyton* in the Pharusian. Precambrian of the Hoggar, Sahara. (After GRAVELLE and LELUBRE, 1957.)

definitely older than the so-called Infra-, or Eocambrian, which is latest Precambrian in age and forms a conformable series with the overlying Phanerozoic. On the other hand, these beds, indicated by the local stage name of the *Pharusian*, are definitely younger than the metamorphic Precambrian basement of this part of Africa. In the local Sahara stratigraphy, the Pharusian, together with the reefs

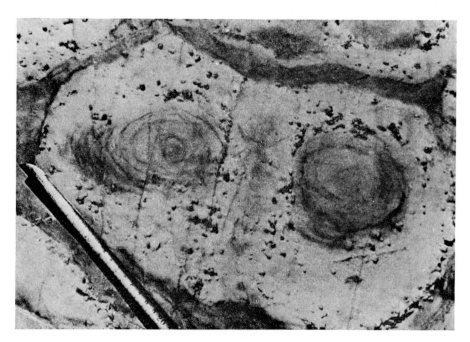

Fig.59. Detailed view of two colonies of *Conophyton* in transverse section. Pharusian, Precambrian of the Hoggar, Sahara. (After GRAVELLE and LELUBRE, 1957.)

of *Conophyton*, are classed as "Middle Precambrian". But although a local stratigraphy can be established on various different forms of *Conophyton* (BERTRAND, 1969), it remains to be seen what is meant by that term. It has, as yet, no definite meaning in absolute dating and as such it is not comparable of, say, the Middle Precambrian of Minnesota (see Table XX).

    Apart from some of the old shields, where, according to some geologists, a broad division of the Late Precambrian is possible by the use of these algal limestones as index fossils (GLAESSNER, 1968; SALOP, 1968, for Russia; and GLAESSNER et al., 1969, for Australia), no general stratigraphic subdivision has as yet been based on the algal limestones. In most cases it is even not possible to tell apart algal limestones from the Late Precambrian from those of the Early and Middle Precambrian, such as will be described in section 8 of this chapter.

A piece of information which at first sight would follow readily from these algal limestones is the constitution of the contemporary atmosphere. For apart from the *Conophyton* illustrated in Fig.58 and 59, other genera of stromatolites are also found, of which the most common form is *Collenia*. Now the latter genus not only forms bioherms during the Early Phanerozoic, but it also occurs on present-day reefs. So, because it is found to grow in an actualistic, oxygenic atmosphere, we might postulate the atmosphere of the Precambrian at the time of the formation of algal reefs of *Collenia* has already been oxygenic, too. This is, however, no foregone conclusion, because, as we will see in section 12 of Chapter 14, the algae are not a genetic, but only a morphological group. Stromatolites, as we saw, are defined as comprising masses of limestone, globose or tubular, which have no clear microstructure. Various species of *Collenia*, the most widely distributed genus of stromatolites, consequently may have not only quite different genetic ties, but also a quite different sort of metabolism. So, although species of stromatolites which have occurred during the Phanerozoic have lived under an oxygenic atmosphere, there may well have been Precambrian organisms, similar in outer form and hence classed in the same artificial taxonomic group, which have lived under the anoxygenic atmosphere (see section 8 of this chapter).

6. A LATE PRECAMBRIAN MICROFLORA

An example of fossils found in Late Precambrian rocks hitherto regarded as non-fossiliferous is the microflora of the Belt Series in the Rocky Mountains of North America (PFLUG, 1965, 1966a). This thick sequence of clastic rocks contains a large number of siliceous members, such as layers of flint and of siliceous limestone. Its age is considered to be around 1.1 billion years. The flora consists of very small organisms, which, in a general way, can be classed with the Algae, but a more specific relation could not be established. They are therefore classed as *Algae incertae sedis*.

For an evaluation of the wealth of forms exhibited by these fossils, as well as for an impression of their excellent state of preservation, the reader is referred to the beautiful illustrations of the original publication (PFLUG, 1966a). Here one plate is reproduced in Plate I because this shows how one has succeeded in reconstructing the various forms of the life cycle for the most commonly occurring species of these fossil algae.

Methodologically Pflug's remarks on the difficulties encountered in searching for fossil remains in the Precambrian are well worth noting. First comes the delicate nature of these microscopic organic remains. All "maceration" techniques, such as used to isolate plant remains from younger rocks, have proved useless. In these techniques the rocks are attacked with strong acids. But in the case of the minute Precambrian remains, the latter are destroyed too. A second factor is formed by the extremely small dimensions of these fossils. This can, for instance, be seen in Plate I, which is magnified 2,350 times. Normal magnifications of microfossils, even of the "nannofossils" of the Phanerozoic, does not exceed 500 times. It is therefore completely useless to search for this type of Precambrian fossils with a hand lens, or even with a stereomicroscope, which also does not magnify more than a couple of hundred times.

PLATE I

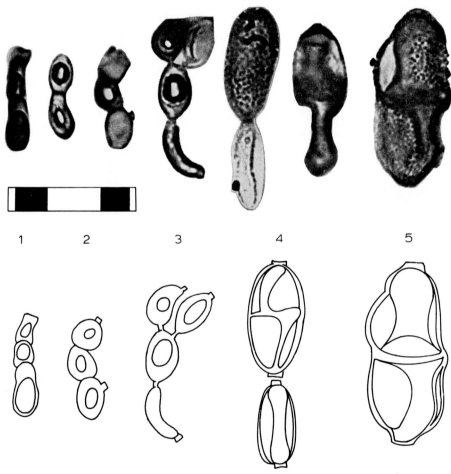

Various forms of the life cycle of the new *Fibularix* from the Late Precambrian Belt Series of North America. 3 × 2,350.
1, 2. Cellular filaments with gas vacuoles.
3.   Branched-cellular filaments
4, 5. Cysts.
(From PFLUG, 1966.)

A new technique had to be developed, consisting of a first step in which surfaces of likely rocks are highly polished and studied under a microscope with high magnification, using an oil-immersion objective, and under strong illumination. By these means the surface of the rock samples is made slightly transparent, and traces of microfossils can be detected. Those rock samples which are found to be fossiliferous are then crushed and finely ground. The rock powder is thereupon subjected to a strong electrostatic field, which separates mineral grains from grains containing a large amount of organic material and from those consisting entirely of organic material. In this way organic remains are not destroyed, as was the case in the maceration techniques. Once concentrated in the electrostatic field, they can be mounted on glass slides for microscopic study.

One must admit that searching for fossils in the Precambrian is quite distinct from a pleasurable fossil hunt in a limestone quarry on a sunny afternoon.

7. EARLY PRECAMBRIAN REMAINS

Retracing the earth's history beyond these Late Precambrian remains we will only have a small number of finds. Historically, the first find was that of biogenic limestones from Southern Rhodesia, described in 1940 by MacGregor (1940, see the next section). They still are amongst the oldest known proofs for the existence of life, their age being higher than 2.7 billion years. Then came the description of the first real fossils from the Canadian Shield by Tyler and Barghoorn in 1954 (see section 12 of this chapter). Their age is only imperfectly known, but is assumed to be around 2 billion years. They are, consequently, around one billion years younger than the biogenic limestones from Southern Rhodesia. The epochal find of Tyler and Barghoorn of fossils far older than anyone had expected to be possible was then followed in 1964 by the detection of the molecular fossils by Eglinton et al. (1964a, see section 9 of this chapter). The oldest of these organic molecules, which, for various reasons detailed later in this chapter, are thought to be of biological origin, are also found on the Canadian Shield and their age too is higher than 2.7 billion years. Consequently the oldest known macroscopic biogenic deposits—the limestones of MacGregor—and the oldest reliably known molecular biogenic deposits fall in the same age group.

The successes reported above have led to an intensive hunt for still older sediments in which biogenic deposits or real fossils could be found. The areas which are propitious for such a search are, however, limited. As we saw in Chapter 11, they are to be found only in those parts of the old shields which have not undergone strong folding or metamorphism since those very early times. The minute fossil remains of the ancient history of life are destroyed by such crustal disturbances, whereas the molecular fossils too stand no chance of surviving a real "revolution" of the earth's crust.

There is at present only one area known, which, more or less miraculously, has not become involved in any later major disturbance since the earliest Precambrian. That is the area occupied by the Swaziland System, comprising the Onverwacht and Fig Tree Series in South Africa, which is over 3 billion years old. Both fossil remains and molecular fossils have been described from beds of this formation, which, despite their venerable age, have only been slightly metamorphosed. However, as we will see in sections 11 and 17 of this chapter, neither the so-called organic molecules, nor the structures interpreted as fossils are as convincingly of a biological origin as are the remnants found in the later part of the Early and in the Middle Precambrian.

We will describe the various finds of remnants of early life in the following sequence: macroscopic biogenic deposits; molecular biogenic deposits or molecular fossils; and lastly the fossils sensu stricto.

8. MACROSCOPIC BIOGENIC DEPOSITS

The setting of the oldest macroscopic biogenic deposits known up to now is South Africa, and more particularly the Bulawayo area in Southern Rhodesia. The Bulawayan System forms the great bulk of the Rhodesian Precambrian belts and consists of basaltic and andesitic lavas alternating with volcanic breccias and with thin interbedded sediments, such as banded iron formations, graywackes, conglomerates and scattered limestone intercalations. Of the latter the Dolomite Series, which is quarried for lime in the Bulawayo area is of particular interest.

The Dolomite Series consists of limestone and of dolomite (limestone containing magnesium). In the limestone peculiarly layered structures occur in definite beds. On the upper surface these beds are characterized by a series of domes, or of finer dentate structures. In cross-section they are found to be built up by sequences of fine layers which more or less follow the dome-like or dentate forms of the surface. These layers, which may be as much as 3 mm apart in the central part of the domes, do narrow considerably outwards. Better than by words, the form of these structures can be conveyed by pictures taken of the quarry wall, which indicate their general form, and from polished and etched surfaces, in which their detailed structure can be studied (Fig.60–64).

The distinctive characters of these deposits are their dome-like or dentate form; the fact that they occur in definite beds, not scattered throughout the limestones; and their finely layered microstructure. The latter shows a certain regularity, which is not as well pronounced as in skeletal remains, but which is well known from limestone deposits secreted by algae during the later periods of the earth's history. The structure, on the other hand, is quite distinct from that of inorganic rhythmic deposits, which never shows the alternation of more regularly layered

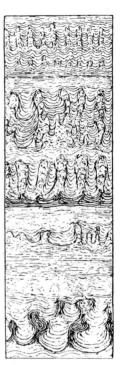

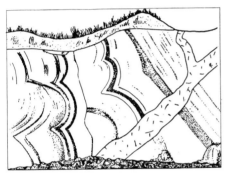

Fig.60. Face of a limestone quarry near Bulawayo, Southern Rhodesia. Length about 5 m. Precambrian Dolomite Series with biogenic deposits formed at the time of sedimentation of the limestone. The limestone is cut by a younger intrusive granitic dyke (actually a pegmatite). The absolute age of these dykes has been established at 2,650 m.y., giving a minimum age for the biogenic deposits which forcibly must be still older. (From MacGregor, 1940.)

Fig.61. Sequence of biogenic deposits in parallel beds in Precambrian limestones of the Dolomite Series, Southern Rhodesia; × 1/40. (From Young, 1940a.)

parts with the quite differently built "columns", which is so typical for the structure of the deposits of the Dolomite Series. It follows that in 1940 MacGregor justly concluded that these are biogenic deposits, formed at the time of deposition of the limestone series.

In view of the importance of the biogenic limestone deposits of the Bulawayan System for establishing a minimum age for the existence of life on earth, the question of the biological origin of these structures has been carefully studied by Young (1940a,b). The peculiar structures of inorganic genesis which are found in limestones also show quite a variety of forms, and may be banded or laminated, wavy or dentate. One particular structure, the so-called "cone-in-cone", would, perhaps, most closely resemble the "dentate bands" of the Dolomite Series. It is, however, quite different in its microstructure, and it has never been held for an algal limestone deposit. So, although not every abnormal limestone is biogenic,

Fig.62. Polished and etched surface of biogenic deposits in Precambrian limestone. Dolomite Series, near Bulawayo, Southern Rhodesia. Fine lamellar structure in the so-called dentate band indicates biogenic deposition, rather than inorganic crystallization of limestone; ×2/3; scale shows inches. (From MacGregor, 1940.)

Fig.63. Thin section of biogenic Precambrian limestone deposit. Dolomite Series, near Bulawayo, Southern Rhodesia. Slightly reduced. Note the intricate structure of the central "column", which is typical for biogenic deposits and not known from inorganic limestone deposits. (From MacGregor, 1940.)

Fig.64. Weathered face of stratified Precambrian dolomite, more than 1,780 m.y. old, near Ollitervo, Tervola (Finland). The rock has been formed by stromatolites and biogenic deposits were deposited layer after layer. (From HÄRME and PERTTUNEN, 1963. Courtesy Dr. M. Härme.)

there is enough difference between inorganically and organically deposited structures. For the Dolomite Series Young has carefully distinguished between undoubtedly biogenic structures (YOUNG, 1940a) and structures which cannot be distinguished undoubtedly from inorganic rhythmic structures, and which therefore might also have been formed inorganically (YOUNG, 1940b). The structures first described by MACGREGOR thus underwent a most thorough critical appraisal, before they were accepted as proof for the existence of life on earth at the time of their formation.

The areas where these ancient limestones are still preserved are widely scattered throughout the outcrops of the Bulawayan System, so it is difficult to establish a detailed stratigraphy for the Dolomite Series. All members of the Bulawayan System have, however, been intruded by younger dykes of granitic rocks (actually by *pegmatites*, a very coarse-grained variety of granite), for which an absolute age of 2,640–2,650 m.y. has been established (NICOLAYSEN, 1962). The Bulawayan Series must therefore be still older, and this forms the basis for the commonly accepted age of the Dolomite Series and its algal limestones of "over 2.7 billion years". Several authors have, however, correlated the Bulawayan System in Southern Rhodesia with the Swaziland System in South Africa, which would bring the age of the algal limestones of the Dolomite Series at "over 3.2 billion years". Since this correlation is not yet commonly accepted, I have in this book used the more conservative lower age of "over 2.7 billion years".

Although at present still the oldest known, the biogenic deposits described by MacGregor are by no means the only examples of such formations occurring in the older Precambrian. In Fig.64–67 we show, for instance, a prolific development of comparable structures from the Precambrian of Finland, as described by HÄRME and PERTTUNEN (1963), whereas we will meet with similar deposits from the Canadian Shield in section 14 of this chapter. The Finnish remains have been dated as "older than 1,780 m.y.", and they probably belong to the Middle Precambrian. This deposit is mainly cited because of the close resemblance of its structures, both with the Early Precambrian remains of Southern Rhodesia, as with those described in section 5 of this chapter from the Sahara. A comparison of these structures will make it plain to the reader why we, geologists, do not hesitate to admit the biogenic nature of such deposits.

Fig.65. Detail of the stromatolite structures of the Precambrian of Finland of Fig.64. (Courtesy Dr. M. Härme.)

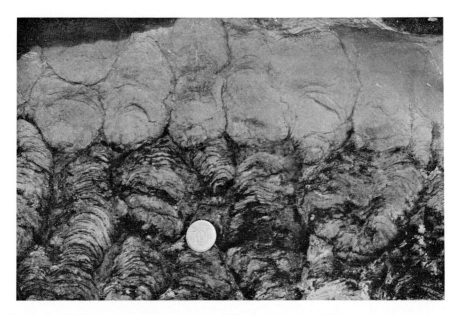

Fig.66. Detail of the stromatolite structures of the Precambrian of Finland of Fig.64 showing the succession of two series of layers with a different structure, probably the result of a deposition by different kinds of organisms. (Courtesy Dr. M. Härme.)

Fig.67. A bedding plane in the stromatolite structures of the Precambrian of Finland of Fig.64 showing the horizontal outline of individual colonies of stromatolites. (Courtesy Dr. M. Härme.)

A question can be raised about the possibility of algal organisms depositing limestone under the anoxygenic primeval atmosphere. Limestone deposition is, however, entirely compatible with several forms of the present-day anaerobic life as will be set forth more in detail in section 10 of Chapter 14, so there seems to be no difficulty in this aspect.

9. MOLECULAR BIOGENIC DEPOSITS: THE MOLECULAR FOSSILS

As we saw in the first section of this chapter, molecular fossils are chemically recognizable remnants of biological substances. They normally consist of stable parts of much larger molecules, which have become degraded during the many years since their deposition. The detection of molecular fossils in ancient sediments is made possible by the improvement of the techniques of chemical analysis, such as reviewed in section 9 of Chapter 6.

The bulk of the hydrocarbons found in recent and in not-so-very-old sediments is biogenic. They form natural gas, oil, asphalt, coal and related substances. The presence of complex hydrocarbons in ancient sediments consequently suggests that life was already present during the time of their deposition. And because such compounds have been found in the oldest sediments known at present, whose age is over 3.2 billion years, this further suggests that life was already present on earth at that early date.

There are, however, several pitfalls possible. The first is that natural gas and oil are highly mobile through the crust of the earth, and may easily have migrated into rocks quite unconnected with their origin. Most of the oil and gas industry rests in fact on the mobility of these substances which has permitted them to percolate and move into reservoir rocks from which they can be economically extracted. Secondly, the experimental work reported in Chapter 6 shows that inorganic synthesis of "organic" compounds is possible under the conditions of the primeval anoxygenic atmosphere. And third, the amount of hydrocarbons in the very old sediments of Early Precambrian age is very small indeed. The possibility of contamination during sampling and subsequent analytical procedures must consequently be checked for very carefully.

The first pitfall is circumvented by excluding from this type of research all oil which has accumulated in large enough quantities to be of commercial value. Instead, attention is focussed on fine-grained rocks, the clayey *shales* and the siliceous *cherts*. Although these contain biogenic compounds in far lesser quantities, these offer some evidence of having stayed in place during the life time of the sediment. The first analyses were made on oil shales, but more refined techniques now permit the analysis of normal shales and cherts which have only a very low content of carbon compounds.

The second difficulty can be solved, because during synthesis of organic compounds, be it through inorganic or organic reactions, a molecule is always built up step by step. During degradation, on the other hand, a complex molecule may lose a set of building blocks in a single step, when the chemical bonds within such a subunit are stronger than those at some locality elsewhere in the parent molecule. It is consequently possible to prove the biogenic nature of some carbon compounds from their relative abundance. An example will be given in the next section.

As to the third difficulty, HARE (1965) has shown that the modern analytical techniques are so sensitive that even a single thumbprint put on the inside of his apparatus did completely falsify the record. Extreme caution in the handling of the samples is therefore called for. It is, for instance, not permitted to wrap rock samples in newspapers, for in that case an analysis of the printer's ink will be found, not of the molecular fossils of early life.

Amongst the various biogenic compounds found in nature the carbohydrates have up to now yielded the most interesting results. Some of these will be reviewed in the next sections. Other compounds have, however, also been found in old sediments. SCHOPF et al. (1968) supplied an overview of the amino acids found in ancient sediments. Abelson and Hoering (cf. ABELSON and PARKER, 1962; HOERING, 1962) and BURLINGAME et al. (1969) have demonstrated that the amorphous carbonaceous matter called *kerogen*, which is found in sediments including those of Precambrian age, could be attacked and analyzed by modern chemical methods. They found that kerogen contains residues of fatty acids in which the stable isotopes of carbon, $^{13}C/^{12}C$ would strongly point to a biological origin. However, in section 9 of Chapter 14 we will see that this cannot longer be accepted as such. Moreover, it is not sure whether the molecules which can at present be extracted from kerogen really formed part of the original biological material, and hence whether we have to do with real molecular fossils. Further research on this promising question will, however, no doubt yield more results in the near future.

The search for molecular fossils has been started by professors G. Eglinton and M. Calvin, and the hunt has been on only since the early sixties (EGLINTON et al., 1964a, 1966; BELSKY et al., 1965; JOHNS et al., 1966). Their findings have since been corroborated by other investigators (cf. W. G. Meinschein in BARGHOORN et al., 1965; ORÓ et al., 1965). The subject is fully treated in Professor CALVIN's new book (1969) and in EGLINTON (1969). A detailed account is given in the two roneotyped theses of MCCARTHY (1967) and VAN HOEVEN (1969), available from the Lawrence Radiation Laboratory at Berkeley.

Starting with the Green River Shale of Colorado, which is no more than 60 m.y. old, and which was used as a test to see what the analytical techniques could accomplish, it was found that natural hydrocarbons could be detected in older and yet older sediments. And, more valuable still, it is nowadays possible not only to detect hydrocarbons as a somewhat vague group, but to find out exactly what compounds are present in many of the hydrocarbon fractions. The nature of some of these molecular fossils will be detailed in the next section. Moving downward in the stratigraphical column, the method tested in the Green River Shale was then applied to Precambrian cherts and shales. The results are given in Table XVI.

It follows from Table XVI that the oldest rocks in which so far molecular fossils have been definitely reported are at least 2.7 billion years old. They belong to the same age bracket as the Precambrian rocks of Southern Rhodesia in which the biogenic deposits described by MacGregor, as related in the preceding section,

TABLE XVI

THE MAJOR LOCALITIES OF MOLECULAR FOSSILS[1]

Formation	Location	Age in billion years	Bibliography
Green River	Colorado, U.S.A.	0.06	EGLINTON et al. (1964, 1966)
Antrim Shale	Michigan, U.S.A.	0.265	BELSKY et al. (1964), EGLINTON et al. (1964, 1966)
Nonesuch Shale	Michigan, U.S.A.	1	EGLINTON et al. (1964, 1966), BARGHOORN et al. (1965)
McMinn Shale	Northern Territory Australia	1.6	HOERING (1965), ABELSON and HOERING (1965)
Gunflint Chert	Ontario, Canada	2	ORÓ et al. (1965)
Witwatersrand	South Africa	~2.5	PRASHNOWSKY and SCHIDLOWSKI (1967), PFLUG et al. (1969), SCHIDLOWSKI (1970)
Soudan Iron Formation	Minnesota, U.S.A.	over 2.7	BELSKY et al. (1965), JOHNS et al. (1966)
Onverwacht and Fig Tree Series ?	eastern Transvaal, South Africa	over 3.2	HOERING (1965), ORÓ and NOONER (1967), KVENVOLDEN et al. (1969)

[1] From the Gunflint Formation comes the famous microflora of TYLER and BARGHOORN (1954), described in section 13 of this chapter. CLOUD et al. (1965) have described carbonaceous microstructures from the Soudan Iron Formation, which, on insufficient grounds, are thought to be biogenic. Carbonaceous microstructures from the Fig Tree Series, described by BARGHOORN and SCHOPF (1966), SCHOPF and BARGHOORN (1967), SCHOPF (1968), and PFLUG (1966b, 1967), are not as convincing, and could perhaps better be called "organized elements" pending further research. A similar attitude must prevail in regard to the carbonaceous structures found in the underlying Onverwacht Series, described by ENGEL et al. (1968) and NAGY and UREY (1969).

are found. Both dates represent minimum ages, for in both cases the age of the sediments has not been dated directly. Instead, the age given is that of a younger granite, which cross-cuts, and is intrusive in, these sediments. We may assume from the regional geology of the Canadian Shield that the age of the Soudan Iron Formation is not very much higher than the 2.7 billion years given, because it is found in the middle of the Early Precambrian (see Table XVI). The age of the biogenic deposits of Southern Rhodesia is, on the other hand, not so limited. As we saw in the preceding section it is even correlated with the Swaziland System of the Transvaal, in which case it would indeed be much older. We have, however, adopted here a conservative age estimate for both sedimentary series.

Similar molecules as those found in the rocks described above have also been found in the Onverwacht and Fig Tree Series of the Swaziland System just mentioned, which are over 3.2 billion years old. After a preliminary study by HOERING (1965), this has been affirmed by ORÓ and NOONER (1967). The actual compounds found in the Swaziland System differ, however, considerably from those found in the Soudan Iron Formation and younger beds. This led Oró and Nooner to leave as undecided the genesis of the Fig Tree and Onverwacht hydrocarbons. They could not state definitely whether these compounds were formed biologically or through the early inorganic processes of photosynthesis. We will assess the reasons for this caution at the end of the next section, whilst a discussion of the importance of the optical activity of amino acids from the Fig Tree Series, as reported by KVENVOLDEN et al. (1969) will be discussed in section 2 of Chapter 14. This is the reason why a question mark accompanies the Onverwacht and Fig Tree Series in Table XVI.

It is not so very important, I think, whether the oldest molecular fossils of known date stem from the Soudan Iron Formation, which is older than 2.7 billion years, or from the Onverwacht and Fig Tree Series of the Swaziland System, which are older than 3.2 billion years, for they both belong to the very early history of life, to the *Early Precambrian* (see Table XXII). Further results of the very active research going on in this field will certainly improve our knowledge on this count in the near future. Personally I believe that evidence for life even in these oldest sediments known up to know will be forthcoming, but at present it seems not safe to assume this. As a matter of caution we will therefore limit our considerations to the molecular fossils found in the Soudan Iron Formations and in younger rocks. As stated already in the preceding section, it is pleasing that the age of the oldest known macroscopic biogenic deposits is similar to that of the oldest known molecular biogenic deposits, both being over 2.7 billion years old, proving that life existed on earth already 2,700,000,000 years ago.

An additional important fact is that from a couple of the isoprenoid alkanes found, from *phytane and pristane*, the inference can be drawn that molecules similar to chlorophyll were already present at that early date. This would prove, not only that life was in existence 2.7 billion years ago, but that it had developed the capacity of organic photosynthesis. As we will see in Chapter 15, this does not mean that the earth had at that time acquired an oxygenic atmosphere. It only indicates that the biogenic formation of free oxygen had started.

Amongst the molecular fossils not only isoprenoid alkanes are found, but other substances too. Their nature will be discussed in the next section. The complexity of these hydrocarbons indicates that life had developed a marked diversity as far back as 2.7 billion years ago. And, if we take into consideration its possible existence during the deposition of the Swaziland System, over 3.2 billion years ago. Since the oldest known rocks are 4.5 billion years of age, this means that the early evolution, from the "organic" materials of the "thin soup" towards early life must really have taken place in the earliest geologic history of the earth.

10. CHEMICAL STRUCTURES OF THE MOLECULAR FOSSILS

The best known of the molecular fossils belong to the saturated hydrocarbons or alkanes, the amino acids and the fatty acids. Normal, branched and cyclic alkanes of widely varying C number have been studied most intensively.

The simplest alkanes are those in which the carbon atoms are arranged in a straight chain, the so-called *normal alkanes*. The simplest form is methane, CH_4, and they form a series in which each new compound has one more atom of carbon and two of hydrogen. The general formula of this series consequently is C_nH_{2n+2}. But as we have seen in section 9 of Chapter 6, the more complex molecules, from butane, C_4H_{10}, onwards, can be built in more than one way, always with the same number of atoms of carbon and of hydrogen. They form isomers of the straight alkanes of the same molecular weight and may be either branched or cyclic. They include a group of biologically important materials, the *isoprenoids*, built up by one or more isoprene units. In the literature the terminology concerning the straight and branched alkanes is given as follows.

Normal alkanes for the straight, unbranched series with the general formula of $CH_3-(CH_2)_{n-2}CH_3$.

Iso-alkanes for the branched series showing so-called terminal branching, with the general formula of:

$$CH_3 - \underset{\underset{\displaystyle CH_3}{|}}{CH} - (CH_2)_{n-4} - CH_3$$

Anteiso-alkanes for those compounds in which the branching takes place in the next step in the chain. Their general formula is:

$$CH_3—CH_2—CH—(CH_2)_{n-5}—CH_3$$
$$|$$
$$CH_3$$

The *isoprenoid unit* consists of an alkane with four carbon atoms in a row and one carbon atom at a branch.

In the analysis of the various kinds of alkanes found in ancient sediments it is necessary to separate the normal, branched and cyclic alkanes. A first step is the separation of the normal from the branched and the cyclic. For this a molecular sieve with holes 5 Å in diameter is used, which effects a partial separation, as the straight alkanes are retained, whereas most of the branched and cyclic alkanes pass through. Dissolving the sieve in hydrofluoric acid makes it possible to collect the group of the branched and the cyclic alkanes. These can be further distinguished by the fact that the branched isoprenoids found in nature have a lower number of carbon atoms, and hence a lower molecular weight, than the cyclic forms. For the first group the important compounds show a distribution around 20 C atoms, whereas the second group lies around 30 or even more C atoms. Both groups, that of the normal and that of the branched and cyclic alkanes is then further analysed in a gaschromatograph, in which the individual compounds of each group are separated according to their C number. These may then be collected and used for a structural analysis in the mass spectrometer, such as is described later in this section, or part of the sample may be fed directly into the mass-spectrometer, such as indicated in Fig.68.

In the relatively young Green River Shales, already mentioned in the preceding section, which are only 0.06 billion years old, a full spectrum of straight, branched and cyclic alkanes could be extracted from the bituminous material (Fig.69). But, as Fig.70 and 71 show, molecules of higher molecular weight are present in smaller amounts in older sediments, such as the Nonesuch Shale and the Soudan Iron Formation. The reason for this relative absence of the larger molecules in older sediments might either lie in the fact that such compounds did not yet exist at that date, or that they have been subsequently destroyed during the diagenesis and/or the metamorphism of the sediments in question. As this relative absence of molecules of higher molecular weight is already apparent in the Antrim Shale, 0.265 billion years old, that is in the middle of the Phanerozoic, when, as is known from the fossil record, life was already fully and abundantly developed, the second explanation seems to be the more reasonable.

It must further be noted that the type of the environment during deposition might also have had a profound influence on the composition of the alkanes. According to CALVIN (1965, 1969) the organisms of our present-day life have a different alkane spectrum, whether they are of marine or terrestrial (limnic) origin. It is possible that variation in the abundance in the normal alkanes, which show

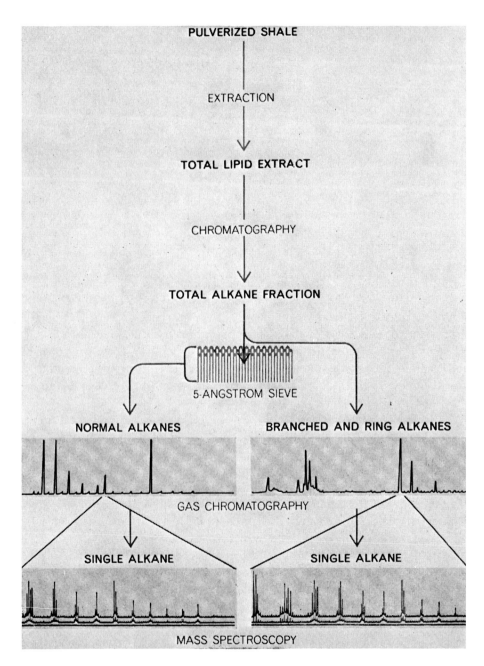

Fig.68. Schematic chart of the analytical procedure for identifying chemical fossils. The last two steps, the gas chromatography and the mass-spectroscopic analysis are nowadays performed in a combined instrument, which records both the chromatographic and the mass-spectrometric data of one sample (see Fig.40). (From EGLINTON and CALVIN, 1967.)

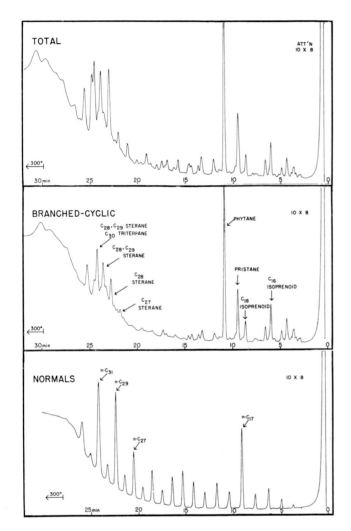

GREEN RIVER SHALE (COLORADO),~60×10⁶ YRS. ALKANE FRACTIONS

Fig.69. Gas chromatogram of Colorado Green River Shale. Upper graph: total extract; middle and lower graphs: branched and cyclic, and straight-chain alkanes respectively, as separated on a molecular sieve with holes 5 Å in diameter.

The horizontal axis is the time axis for the chromatogram. The straight-chain alkanes migrate slower during analysis than the branched and cyclic. C_{17} straight-chain (lower graph) therefore corresponds in position, that is in time-since-the-start-of-the-analysis to pristane, the branched isoprenoid C_{19}. The structure of the molecules forming each peak is not directly evident from the chromatogram. It can be ascertained by collecting the material from each peak and further analyzing it in a mass-spectrometer.

Note that the branched isoprenoid C_{17} is lacking. (From JOHNS et al., 1966.)

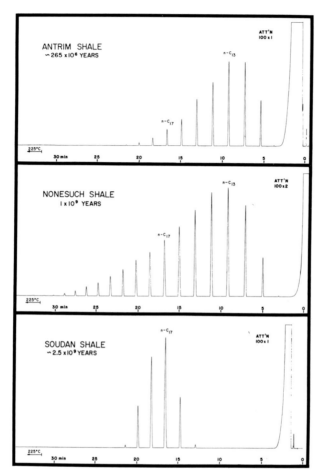

Fig.70. Gas chromatogram of the straight-chain alkanes from the Antrim, Nonesuch and Soudan Formations.

Compared with the much younger Green River Shale the spectrum of the straight-chain alkanes is much reduced in the older sediments. There is, however, no clear correlation with age. The Soudan Formation, incidentally, is not a shale but a chert. (From CALVIN, 1965.)

a maximum for the compound C_{17} in the Green River Shales, of C_{13} for the Antrim and Nonesuch Shales, and again of C_{17} for the Soudan Iron Formation (Fig.69–70) can be explained in this way. However, in the absence of reliable fossil evidence, we do not even know, in the case of Precambrian sediments, whether they have been laid down in the oceans or in fresh water lakes (see section 14 of Chapter 13). Much more research is necessary along this promising line of enquiry, if we are to know more about these features of early life.

Turning now to the interpretation of the alkanes found in ancient sediments, it has already been stated that their most important aspect lies in the fact that

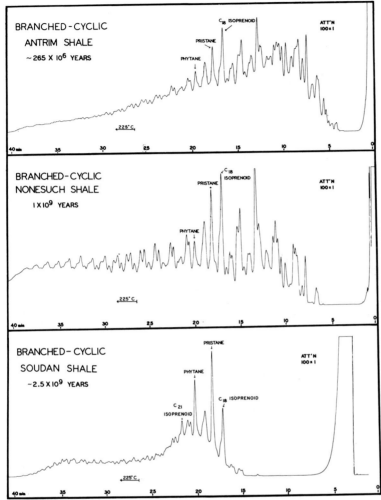

Fig.71. Comparison of the gas chromatograms of the branched and the cyclic alkane fractions of the Antrim, Nonesuch and Soudan Formations.

Compared with the much younger Green River Shales the higher molecular weight branched and cyclic alkanes are less represented in the older sediments. This is thought to be the result of post-depositional destruction of these compounds during the later history of the sediments. (From JOHNS et al., 1966.)

they may represent fragmentation products of chlorophyll or of chlorophyll-like products. Present-day chlorophyll consists of a central part—formed by a single metal atom surrounded by a chlorin ring of carbon, hydrogen and nitrogen—to which, amongst other groups, a long tail, the so-called *phytyl chain* is attached, which is a branched alkane of carbon number 20. The phytyl chain is attached to the central part of the chlorophyll molecule by an ester linkage (see Fig.45).

The central part of the chlorophyll molecule will react rather easily and will be degraded relatively quickly after the death of the organism of which it formed part. The long tail, on the other hand, will, after oxidation and decarboxylation, ultimately yield a hydrocarbon molecule of little chemical activity, e.g. either one of the branched isoprenoids *pristane*, C_{19}, or *phytane*, the C_{20} compound (Fig.72,

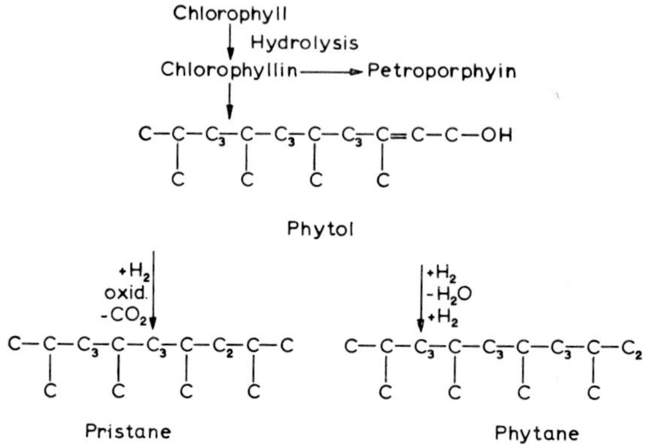

Fig.72. A model for the formation of pristane and phytane from chlorophyll, via phytol. (From Bendoraitis et al., in CALVIN, 1965.)

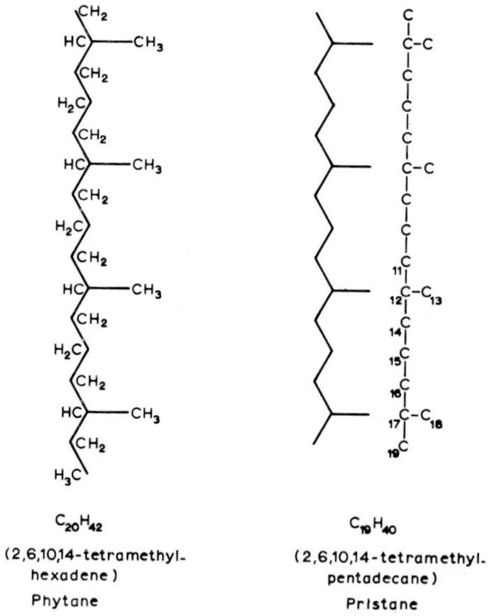

Fig.73. Formulas of the branched isoprenoids phytane and pristane. In chemical shorthand these formulas are often noted in either of the two ways given for pristane. That is, indicating either the carbon skeleton, or a row of carbon atoms. In the latter notation the number of a given carbon atom on the chain may easily be added. In the example given, it is seen how the isoprenoids smaller than pristane will break off at carbon numbers 19, 18 or 16. Carbon number 17 will not be represented in this way, because C_{18} remains attached to C_{17}.

73). These hydrocarbons, being members of the paraffin series, can survive for any length of time in sediments, provided the conditions of diagenesis and metamorphosis do not become too drastic.

It is, of course, possible that further experiments along the lines described in Chapter 6 will indicate that alkanes of this composition can be readily synthesized inorganically under the conditions of the primeval atmosphere. They are, in fact, produced inorganically during the industrial production of ethylene (McCARTHY and CALVIN, 1967; CALVIN, 1968). But this occurs at high temperature and pressure in a carefully controlled cracking technology, and does not bear any relation to a simulated primeval atmosphere. In that case the finding of such molecules in the bituminous extracts of old and very old sediments would not any longer offer proof of a biological provenance of this material. Or, in other words, the existence of molecular fossils would not be proved any longer.

There is, however, one element in the analyses reported of the bituminous extracts of ancient sediments which strongly suggests a biological origin of the alkanes. That is the conspicuous absence of the C_{17} isoprenoid in all of the material analyzed so far. During the diagenesis which takes place after sedimentation phytane and pristane may break at any carbon bound. However, when the break occurs below C_{18}, it will take the C_{17} block with it, leaving a C_{16} molecule as the larger fragment. So, by degradation of the C_{20} and C_{19} molecules of phytane and pristane, the retention of the C_{17} isoprenoid is not favoured. In synthesis it is on the other hand probable that a step-by-step process will be followed and that the formation of the C_{17} will be favoured in about the same way as the formation of the C_{16} and the C_{18} isoprenoids. So, regardless of the fact whether the interpretation of the occurrence of phytane and pristane as break-down products of chlorophyll-like compounds is correct—and BROOKS et al. (1969) have for instance already proposed that these isoprenoids form mainly from the degradation of waxes—the statistical abundance of isoprenoids of various carbon numbers indicates that they represent degradation products from biological material, not products formed by inorganic synthesis.

Returning now to the Fig Tree Series of South Africa, which is over 3.2 billion years old, we saw in the preceding section that there is some hesitation to ascribe a biogenic character to the alkanes found in this sediment. This rests, among other things, on the alkane spectrum as reproduced in Fig.74, in which not only the C_{17} compound is present, but also a relatively high number of higher alkanes. The reason for their presence might be that they represent breakdown products, not only from chlorophyll, but also from a number of more complicated molecules. But it has been argued that it might also be the result of an inorganic synthesis similar to the Fischer-Tropsch process which in modern technology is used to produce paraffins and petrol artificially. ORÓ and NOONER (1967) have concluded that more research is needed before this question can be decided,

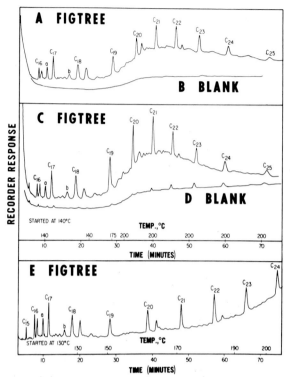

Fig.74. Gas chromatographic separation of cherts from the Fig Tree Series of South Africa. (From ORÓ and NOONER, 1967.)

a conclusion also reached by McLEOD (1968) and with which a mere geologist like myself cannot but agree.

If, on the other hand, the optical activity of Fig Tree amino acids reported by KVENVOLDEN at al. (1969) would be confirmed, this would, in the eyes of most scientists working in the field, constitute indubitable proof in favour of a biological origin of these compounds. But even here, some doubts can be expressed, as will be set forth in section 2 of Chapter 14.

Turning now to the cyclic isoprenoids found in ancient sediments, we have seen in Fig.69 that various compounds of this series in the C_{27}–C_{30} range may exist as molecular fossils. Moreover this picture demonstrates that several isomers of these compounds occur, since compounds of the same molecular weight appear at different times on the gas chromatographic chart. This is due to differences in the interaction between the liquid coating of the tube of the gas chromatograph and these isomers as a result of structural differences between isomers of the same molecular weight.

A further analysis of the cyclic isoprenoids is then carried out with the mass spectrometer. Examples of this type of research are shown in Fig.75 and 77. As stated in section 9 of Chapter 6, the mass spectrometer—or mass-spectrograph, as the case may be—is used in this type of research to determine the fragmentation of a large molecule in its disintegration as a result of the ionization it undergoes upon entering the mass spectrometer (see section 9 of Chapter 3). We see, for instance, in Fig.75 how cyclic isoprenoids of molecular weights 372, 386 and 400, separated in the gas chromatograph, all show a first break caused by the loss of a building block of molecular weight 15 (the peak indicated by M-15). This represents a CH_3 unit (molecular weights of one carbon C = 12 and of three hydrogens H = 1) at the end of the chain. Together with the known absence of atoms of oxygen and nitrogen, and of several other considerations which we cannot discuss here, this indicates that in all these cases we are dealing with fully saturated hydrocarbons.

Other relatively stable fragmentation products are found at $m = 217$ and

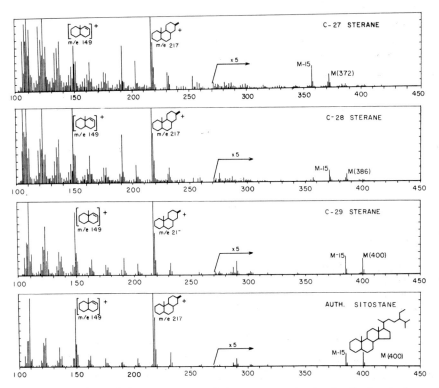

Fig.75. Comparison of the mass-spectra of the sterane molecular fossils C_{27} (mol. wt. 372), C_{28} (mol. wt. 386) and C_{29} (mol. wt. 400) with the laboratory product C_{29} sitostane.

All four compounds first loose a CH_3 group (M-15) and have relatively stable building blocks of molecular weight 217 and 149. The close similarity between the spectra of the C_{29} molecular fossil and of sitostane shows their close relationship. (From BURLINGAME et al., 1965.)

$m = 149$ molecular weight. Further analysis shows that these fragmentation blocks consist of a three ring and a two ring cyclic compound each, such as is indicated by the structural formulas in Fig.75. These spectra compare well with the mass spectrum of the C_{29} compound *sitostane*, as synthesized in the laboratory, indicating a close resemblance between these molecules and sitostane (Fig.76).

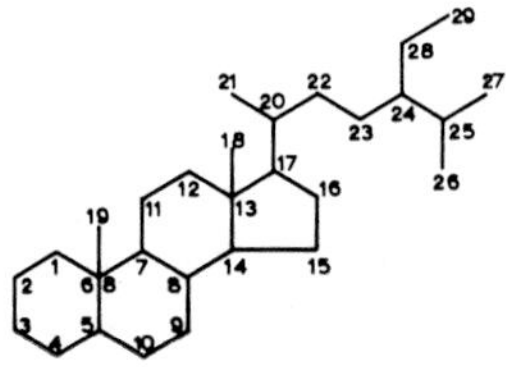

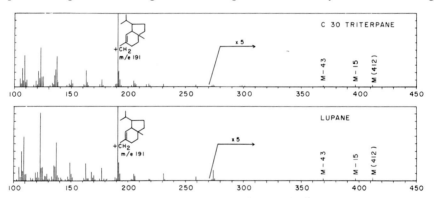

Fig.76. Carbon skeleton of the cyclic isoprenoid C_{29} sitostane.

This strongly suggests that at least this fraction of the bituminous extract of the shales of the Green River Formation is indeed sitostane.

The C_{30} isoprenoid found in the gas chromatographic analysis of the Green River Shales shows a structure which is quite different from the C_{27}–C_{29} isoprenoids represented in Fig.75. Its mass spectrometric analysis is shown in Fig.77.

Fig.77. Comparison of the mass-spectrum of the triterpane molecular fossil C_{30} (mol. wt. 412) with the laboratory-produced C_{30} lupane. The main building block of this compound, with a molecular weight of 191, is quite different from those of the C_{27}–C_{29} steranes, indicating that its basic structure differs from the saturated steranes. (From BURLINGAME et al., 1965, fig.6, 7.)

It has a first strong fragmentation peak at molecular weight 191, which represents a six and a five ring structure, as shown by the structural diagram, and a number of peaks of lower molecular weight. There is a close resemblance to the mass-spectrometric spectrum of artificially synthesized *lupane*, of which the structural skeleton is illustrated in Fig.78.

Fatty acids, finally, have been extracted from the kerogen of the McMinn shale, 1.6 billion years in age. They are represented by compounds ranging in

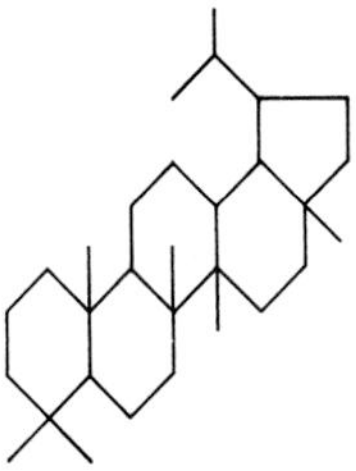

Fig.78. Structural skeleton of the C_{30} cyclic isoprenoid lupane.

carbon numbers from 11 to 21 (ABELSON and HOERING, 1965). Although these compounds also attest the presence of complicated life forms already 1.6 billion years ago, the information supplied by this group of organic materials is not as straightforward as that supplied by the alkanes. The original biological material has been strongly altered during its degradation into kerogen, a fact which is aggravated by more alterations occurring in the laboratory during the extraction of the soluble material from the residue. Nevertheless, it is not doubtful that interesting facts about the history of life will eventually be found by this type of research too. First indications are to be found in the papers by VAN HOEVEN et al. (1969) and HAN and CALVIN (1969), who reported fatty acids from the Gunflint Chert and the Soudan Iron Formation and from the Onverwacht Series.

11. FURTHER EVIDENCE TO BE EXPECTED FROM MOLECULAR FOSSILS

In the two preceding sections we have shown how the presence of certain types of molecules in old and very old sediments may provide evidence that life has already existed at the time of their sedimentation. For those readers who may have found these two sections too technical, we may here summarize their main result: Life has been present on earth already in its very early history, whilst moreover the nature of the molecular fossils themselves may provide a clue as to what sort of metabolism this early life possessed. In the example given, the evidence rests on the presence in these old sediments of several compounds belonging to the isoprenoid branched alkanes, which can be interpreted as fragments of chlorophyll or of chlorophyll-like molecules. This in turn indicates that the capacity of photosynthesis was present at a very early date in the history of life.

Parenthetically, we are as yet not able to determine what type of photosynthesis (photolithotrophic or photoorganotrophic, see section 7 of Chapter 8) was present. For the various chlorophyll molecules found in present-day life in plants and in the various photosynthetic bacteria are very similar in structure. Moreover the differences reside in the central part of molecule, which is, as we saw, more easily destroyed than its paraffin tail.

The existence of molecular fossils opens up a wide vista of what in future can be arrived with the modern methods of organic chemical research now available. Even though the larger original molecules are broken apart during the diagenesis over the long geologic history, it is possible, under favourable circumstances, to tell from the structure of their fragmentation products from what sort of original molecule they have been derived. So, much as a detective builds his case from various scraps of evidence, or an archeologist pictures a vanished city from characteristic bits of rubble, organic geochemists will, with enough evidence at their disposal, in future be able to reconstruct a reasonable model of early life on earth. Amongst the various possible materials, the pigments will, for instance, come under close scrutinity pretty soon. Of this group the *porphyrins*, also derived from chlorophyll or chlorophyll-like molecules (see the "petroporphyrin" in Fig.72), seem to be the most promising (BOYLAN et al., 1969). Another important area will be the further analysis of the soluble parts of kerogen, which, up till now has not proceeded much further than the investigation of the Green River Shale, which is, as we saw, only 0.06 billion years old (BURLINGAME and SIMONET, 1969; BURLINGAME et al., 1969).

Further advances in technology will also give us the possibility for a more refined analysis of Precambrian sediments. As an example, the instrument constructed by MURPHY et al. (1969) may be cited. These authors appended a computer to the gas chromatograph–mass spectrometer illustrated in Fig.40, and automatized the instrument to take a reading every four seconds. In this way minor constituents of the extract from the sediments are detected too. Moreover all data are directly stored in the computer memory, which makes further structural analysis more easy. This technique has, however, also been only applied so far to the Green River Shale.

Fig.79 gives an idea of the variety of the organic compounds which so far have been found in fossil sediments. By far the largest portion are found either in the Phanerozoic—those of $6 \cdot 10^6$ years and younger—and in the Late Precambrian (those between $1.8 \cdot 10^9$ and $6 \cdot 10^6$ years old). Organic remains found in the Early and Middle Precambrian so far only form a fraction of those found in younger beds, but with the active research which is going on in this field at the present time, many more finds will no doubt be reported in the near future. It also follows that the molecular fossils described in some detail in the earlier sections represent only a fraction from what has been described already.

The study of the molecular fossils consequently complements the experiments concerned with the possible origin of life through natural causes, as reviewed in Chapter 6. The experiments indicate the many possible pathways by which "organic" molecules may be synthesized inorganically under the conditions of a simulated primeval atmosphere. The molecular fossils, on the other hand, indicate what pathways have actually been used during the origin of life. Such

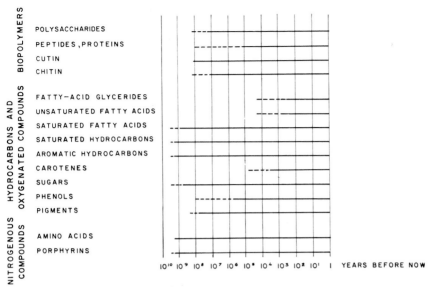

Fig.79. Organic compounds found in ancient sediments. Time is plotted on a logarithmic scale.
The authors state that all of these compounds have been originally synthesized by living organisms, which is, no doubt, a schematization in accordance with the level of the journal in which this illustration appeared. In reality it will become more and more difficult, the older the sediments in question are, and certainly for all sediments belonging to the Early and Middle Precambrian, to differentiate between really organic and "organic" compounds. That is, between compounds synthesized by life and those synthesized inorganically by the shorter ultraviolet radiation of the sun in a primeval anoxygenic atmosphere, and thus belonging to pre-life. (From EGLINTON and CALVIN, 1967.)

a two-fronted attack on the problem of the origin of life has already been proposed in 1937 by the Dutch microbiologist KLUYVER. He deliberately inserted this idea in the title of a public oration, which, but for the fact that it was held in Dutch, would at that early date certainly have received the same fame outside our country, as it did within. The title, *'s Levens Nevels*, of "Life's Nebulae" is a palindrome, spelling forwards as well as backwards, and indicating how our knowledge of life will eventually stem both from experimental research as to what can be synthesized, and from paleontological studies which look backwards as to what actually has been.

12. THE OLDEST REAL FOSSILS

The oldest real fossils, or *fossils sensu stricto*, i.e. morphologically recognizable remnants of former life, are by now well known from the Middle Precambrian. Their estimated age is about 2 billion years. More hypothetical finds have been reported from much older sediments, in fact from the oldest sediments known, of a minimum age of 3.2 billion years.

The fossils from the Middle Precambrian have first been described from the Gunflint Formation of the Animikie Group of the Canadian Shield (see Table XXII). They will be described in the next section. Similar fossils have since been described from other old shields too. The hypothetical fossils which have been described from sediments dating far way back in the Early Precambrian belong to the Onverwacht and Fig Tree Series of the Swaziland System of South Africa and will be described in section 17 of this chapter. As we have seen already in sections 9 and 10 of this chapter, extracts from rocks of these localities contain certain molecules which for the Gunflint Formation have been interpreted as molecular fossils, whereas for the Onverwacht and Fig Tree Series it is still doubtful whether these remains are truly biogenic, or whether they are "organic" (see Table XVI).

All fossils found so far in these ancient sediments occur in the so-called *banded iron formations*. In the next chapter we will go into more detail about these formations, but we have to point out here that they are formed by an alternation of thin laminae of rather pure, extremely finely crystalline, silica with equally thin laminae of iron ore. Iron formations of this type occur only in the Early and Middle Precambrian and are not found in younger formations of the Late Precambrian and the Phanerozoic. The very finely crystalline siliceous material is called *chert* in American and *flint* in English. The material of the Gunflint Formation, which received its name in colonial times, evidently was well suited for the flint-locks of early guns. All of the fossils and hypothetical fossils occur in the siliceous laminae, but are often stained with iron.

Siliceous sediments are at present mainly formed in an oceanic environment. They develop from diatomaceous or radiolarian oozes on the deeper floor of the oceans, or, in exceptional circumstances, in closed basins. As we will see in section 3 of Chapter 14, these more recent siliceous sediments are not comparable to the banded iron formations of the Early and Middle Precambrian, which is one of the reasons why it is postulated that the latter have been formed under the primeval anoxygenic atmosphere. Silicification, however, occurs, albeit in a more restricted way, also on the continents, in swampy environments. In these areas humic acid derived from rotting swamp vegetation attacks and dissolves calcareous and phosphatic organic remains, such as shells and skeletons of animals, but cellulose and other plant substances are relatively resistant to such acid waters. They can be fossilized by replacement, molecule for molecule, of their original organic material by silica. Some of the original material remains at the outer walls of cells, and—much more rarely—at original membranes within cells. Together with iron stain this may colour these replacements in such a way that the original morphology and structure of the organisms is preserved in the minutest detail, as will be evident from the photographs reproduced in the next section. Silicification has provided us with the most beautifully preserved plant remains. *Silicified woods* are, for instance, known from many localities and from various geological formations and are often so impressive to be preserved in specially created national parks.

So we must distinguish between the processes leading to the formation of siliceous sediments, which are clearly different in the Early and Middle Precambrian from those prevailing in the Late Precambrian and the Phanerozoic, and the process of silicification, from which we know examples from more recent times. The process of silicification results from an excess of silica. Regardless of the process which has led to this excess, the important thing in the study of the earliest fossils is that silicification is an optimal method of fossilization.

13. PRIMITIVE FOSSILS FROM THE MIDDLE PRECAMBRIAN, AROUND TWO BILLION YEARS OLD

As stated in the preceding section, the first very old real fossils were described from the Gunflint Formation of the Canadian Shield in the southwestern part of Ontario. A preliminary notice by TYLER and BARGHOORN (1954) has since been followed by a full description by BARGHOORN and TYLER (1965). Similar, but less convincing fossils have been described from other old shields and from both the Early and the Middle Precambrian by LABERGE (1967). But the wealth of forms encountered in the Gunflint Formation has not been met with elsewhere. This wealth of forms has been partly represented here by Plates II–VI.

Barghoorn and Tyler give detailed comparisons between the Gunflint fossils and younger fossils and even living forms. In some cases they equate the Gunflint fossils with these younger organisms, only on the basis of the morphology of the fossils in question. This is, in a way, permissible because the paleobotanical nomenclature is based on morphology only, not on genetic relationship. It is, however, misleading to anyone who is indoctrinated in the common notion that a Latin generic or specific name, and also the names of the higher hierarchy in taxonomics, stand for similar kind, instead of for similar form only. I will return to the interpretation of the Gunflint fossils in section 12 of Chapter 14, but want to state already at this point that such comparisons had better be avoided.

Two important points stand out from the study of BARGHOORN and TYLER (1965). The first is that we may rest convinced that what we see here are real fossils. That is, fossilized remnants of organisms which formerly have been living, either in the water above, or on the sediments into which their remains became entrapped after death. The variations between filamental and spheroidal, between septate and non-septate structure, the queer outlines of, for instance, *Kakebekia*, all point to a varied and intricate morphology, such as is only exhibited by living matter.

PLATE II

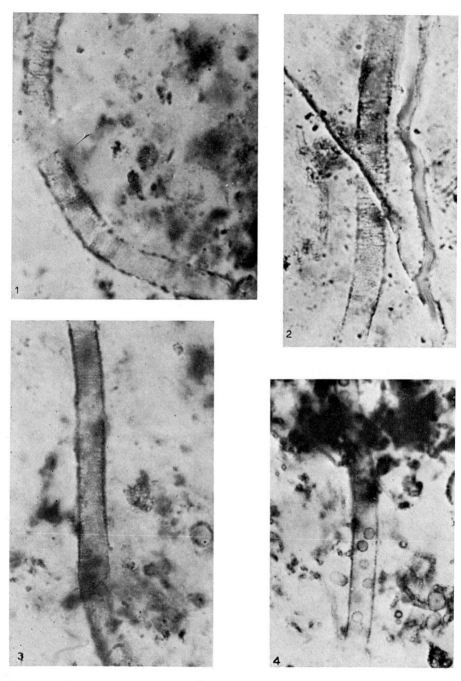

Several specimens of a filamentous fossil called *Animikiea septata* by E. S. Barghoorn. Gunflint Formation of Ontario. The size is indicated by the rod in each microphoto which measures 0.01 mm. Magnification around ×1,000. (From BARGHOORN and TYLER, 1965. Courtesy Professor E. S. Barghoorn.)

PLATE III

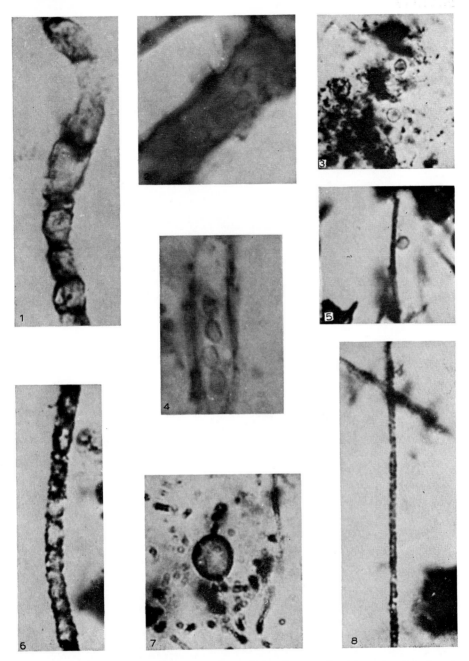

Filamentous and non-filamentous microscopic fossils from the Gunflint Formation. Numbers 1, 4, 6 and 8 represent several species of a group of septate filamentous forms grouped under the new genus *Gunflintia* by E. S. Barghoorn. They are decidedly distinct, both in their outer form, as well as in the position of their septae, from *Animikiea*. Non-septate filaments, called *Entosphaeroides*, contain little spheric bodies, which, after degradation of the filamentous wall, may also occur separately (numbers 2, 3 and 5). Number 7, finally, is thought to be a spore, and called *Huroniospora*. Magnification around ×1,500. (From BARGHOORN and TYLER, 1965. Courtesy Professor E. S. Barghoorn.)

PLATE IV

Typical assemblage of tangled filaments, predominantly *Gunflintia minuta* and spore-like bodies of *Huroniospora*, indicating the local abundance of microfossils in the Gunflint Formation of Ontario. (From BARGHOORN and TYLER, 1965. Courtesy Professor E. S. Barghoorn.)

PLATE V

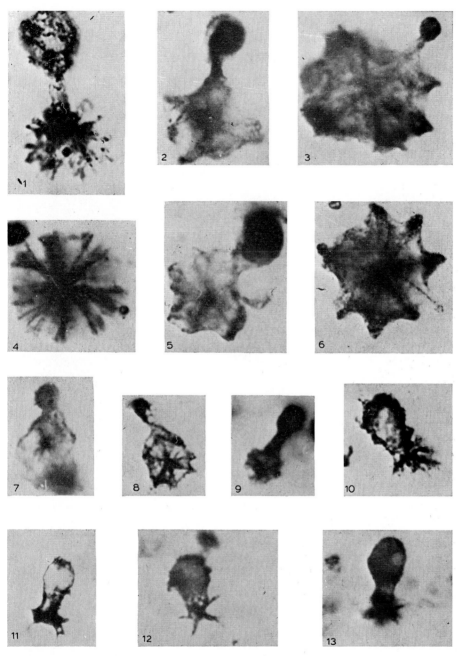

Thirteen different specimens of an organism of uncertain affinities, which has no known fossil or living counterpart. From the locality near Kakabeka Falls it has received the generic name of *Kakabekia* and from its tripartite division into a bulb, a stipe and an umbrella-like crown, its specific name of *umbellata*. This has no further meaning at this point than that the fossil has now been described in formal paleobotanical nomenclature. The specimens 1–13 are thought to represent different ontogenetic stages of a single species. (From BARGHOORN and TYLER, 1965. Courtesy Professor E. S. Barghoorn.)

PLATE VI

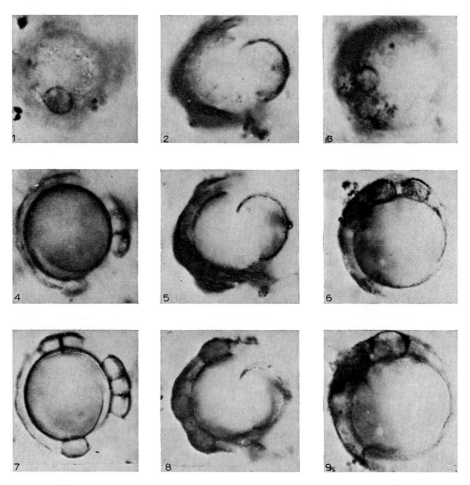

Three specimens of a common bulbose microfossil from the Gunflint Formation of Ontario, called *Eosphaera tyleri* by E. S. Barghoorn.

In each vertical column are three photos of the same specimen, but taken at different focus level, the third row from above representing the equatorial plane of the spheres. This gives some idea of the three-dimensional morphology of these organisms. Their structure is that of a sphere within a sphere, separated by tubercle-like spheroids attached to the wall of the inner sphere. Just as is the case with the segmented filamentous organisms shown in the earlier figures, such complex structure proves conclusively that we have to do with rather advanced organisms. (From BARGHOORN and TYLER, 1965. Courtesy Professor E. S. Barghoorn.)

The other point to be made is that the organisms of the Gunflint Formation, although clearly still primitive in character, must, on the other hand, have had time to evolve into quite a number of different forms of varied morphology and structure. Although the oldest fossil organisms known, they still are a far cry from the earliest life which much have shown much less variation in its make-up.

The interpretation of the structures found in the cherts of the Gunflint Formation as real fossils, that is as morphologically preserved remnants of former life, is strengthened by an interpretation of some of the morphologically different cells as reproductive elements. LICARI and CLOUD (1968) have postulated that differences in diameter and coloration of some cells can be explained by interpreting the larger sized cells as reproductive units of smaller sized cells, and not as different species. This interpretation is based on a comparison with the present-day reproductive cycle of blue-green algae, as illustrated in Plate VII and VIII. If this interpretation is accepted, it proves that at least a part of the Gunflint structures possessed the faculty of reproduction and as such must have belonged without any doubt to contemporary life.

The wealth of forms encountered in the Gunflint Formation is far less apparent in the study of LABERGE (1967) on organic remains found in chert layers of banded iron formations of other old shields and of other ages. With the exception of *Eosphaera tyleri* (Plate VI), which is only reported with certainty from other localities of the Middle Precambrian of the Canadian Shield, and as such does not extend our knowledge much further, the "spheroidal structures" illustrated by Laberge show no inner structure. Nor do they possess such queer outer forms such as does, for instance, *Kakabekia* (Plate V). Remembering the microspheres synthesized by Fox (see section 6 of Chapter 6), which are usually only one order of magnitude smaller than the "spheroidal structures" of Laberge, I can see no irrefutable evidence for the biological origin of the latter.

This is not just a pointless criticism of a valuable paper, but is instead meant as a warning to be very cautious in what to accept as real fossils and what to reject as doubtful. The basis for this caution lies in the possibility of a co-existence of pre-life and early life all through the Early and Middle Precambrian (see section 2 of Chapter 16). If this is so, structures possessing a very simple morphology which can be formed both by pre-life and by early life could occur in beds of the same age, and even in layers of the same formation. Since the environment of pre-life and early life will have been different, even though they were contemporaneous, it is doubtful whether one would find the remnants of both preserved in one and the same bed, although even such a *thanatocoenose* remains possible. But since the environmental conditions must have changed rhythmically to deposit series such as the banded iron formations, a repetitive succession of both forms in successive layers of the same formation is well possible. This will be further detailed in section 18 of this chapter, but at this point we must remember that one has to be very careful in distinguishing between structures which might have been formed by pre-life and also by early life, from those structures which exhibit such a complexity of structure, such a weirdness of outer form, that they can only be visualized as having been formed by life. Pending further evidence, and in accordance with the usage in the study of carbonaceous meteorites (see Chapter 17), the former group of structures had better be cautiously indicated as "*organized elements*".

Both from the Gunflint and from other localities of Middle Precambrian age macroscopic biogenic deposits, in the form of "algal reefs" of Stromatolites have been described (MOORHOUSE and BEALES, 1962). This is to be expected, because organisms forming limestone deposits have been at work from the Lower Precambrian (the MacGregor algal limestones of Southern Rhodesia, described in section 8 of this chapter), all through the Precambrian and the Phanerozoic, up to the present time. In the siliceous environment of the Gunflint chert the biogenic deposits which originally have consisted of limestone and possibly of dolomite have often become silicified. MOORHOUSE and BEALES (1962) describe, however, a carbonate reef 100 ft. long and 20 ft. thick found in Port Arthur, which, probably due to its large dimensions, has escaped silification.

A quite different proposition is the fact that Moorehouse and Beales report the probable presence of animal fossils in the Animikie. The case lies with the finding of what is interpreted as possible sponge spicules in thin sections and of possible sections of complete sponges in acetate replicas of polished rock surfaces. On the one hand I am of the opinion that neither the thin sections, nor the acetate peels on which the interpretation of these authors is based, are at all convincing. They might be organic remains, as claimed by the authors, but they might just as well be anything else. They do not show a structure which proves their biogenic origin. On the other hand, we will see in the next chapter that the formation of

---

PLATE VII

Microphotos of the fossil Gunflint flora and drawings of the present-day blue-green algae to illustrate the assumption of the faculty of reproduction for the Gunflint flora. Bar scale of 20 μ applies to no.1–4 and 7–9, scale of 10 μ to the drawings no.6 and 7.

1.   Microphoto of Gunflint chert showing profusion of *Huroniospora* ellipsoids and *Gunflintia* filaments. Cf. Plate IV.
2, 3. Filaments exhibiting a change in diameter from that characteristic of *Gunflintia grandis* BARGHOORN to that of *Gunflintia minuta* BARGHOORN. Against the opinion of Barghoorn the authors interpret this variation in cell diameter not as a taxonomic one, as a difference between two distinct species, but as a difference in size of the cells produced by one and the same species at different moments of its life cycle.
4.   Differentiated cells of *Gunflintia* filaments interpreted as heterocysts (*h*) and akinetes (*a*). Cf. no.5.
5.   Drawing of the living alga *Anabaena oscillarioides* BORY with differentiated cells. Akinetes (*a*) and heterocysts (*h*).
6.   Heterocyst (*h*) of a trichome of living *Aulosira implexa* BORNET and FLAHAULT.
7–9. *Gunflintia minuta* BARGHOORN, showing probable heterocysts.
(After LICARI and CLOUD, 1968. Courtesy Professor P. E. Cloud Jr.)

PLATE VII

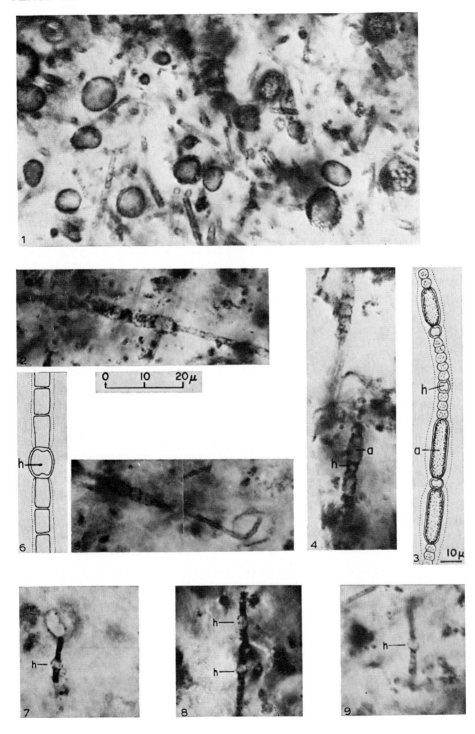

the banded iron formations, to which the Animikie belongs, can be correlated with the presence of the primeval anoxygenic atmosphere. In this atmosphere the level of free oxygen could not rise above about 1% of that of the present-day atmosphere, and this is too low a level for animals to survive (Table XIII). Accordingly, as long as no more convincing evidence for the existence of animal life in the Animikie is produced, we must conclude from the circumstantial evidence offered in the next chapters, that the occurrence of animal life in the Animikie, and in all of the Middle Precambrian at that, is highly improbable.

15. AGE OF THE MIDDLE PRECAMBRIAN FOSSILS

The age of the fossils of the Middle Precambrian is not easily established, because they do of course all occur in sediments. And, as we saw in Chapter 3, absolute dating is normally performed on igneous rocks, where the time of crystallization of a mineral from the molten magma gives us a certain fixed point relating to the "birthdate" of that rock. In sediment similar methods can, as we have seen, only be applied to minerals which form at the time of sedimentation; the so-called *authigenic* minerals, but the results are normally less reliable than for igneous rocks. We will survey the vicissitudes of the age determination of the Gunflint Formation as an example of the difficulties encountered in dating Precambrian sediments.

In the beginning, in 1954, the year when the Gunflint flora was discovered, the age of the Gunflint could only be bracketed in an extremely wide range. It was at that time correlated, on purely geological grounds, with the Negaunee Iron Formation of northern Michigan, across Lake Superior. From magnetite crystals of the Negaunee Iron Formation uranium–helium datings were available, which ranged from 800 to 1650 m.y. and averaged at 1300 m.y. Because helium datings are notoriously untrustworthy, and always give values which are on the low side, due to the loss of the daughter element helium, which easily leaks away, these

PLATE VIII

Microphotos of the fossil Gunflint flora and drawings of present-day organisms to illustrate the assumption of the faculty of reproduction of the Gunflint flora. (Bar scale of 20 μ applies to all figures.)

1.   Microphoto of *Gunflinitia-lake* filaments interpreted as a radiating colony.
2.   Drawing of the living, nonmotile blue-green alga *Synechocystis sallensis* SKUJA showing single cell (*b*) and vegetable cell division (*a* and *c*).
3.   The simple living Dinoflagellate *Desmocapsa geltatinosa* PASCHER, showing a swarmer (*a*) and cell division (*b*).
4.   Differentiated cells terminating a filament of *Gunflintia grandis* BARGHOORN.
5–9.   Double cells of *Huroniospora* interpreted as stages of vegetative cell division.
10, 11. Pore-like openings of *Huroniospora*.
(From LICARI and CLOUD, 1968. Courtesy Professor P. E. Cloud Jr.)

PLATE VIII

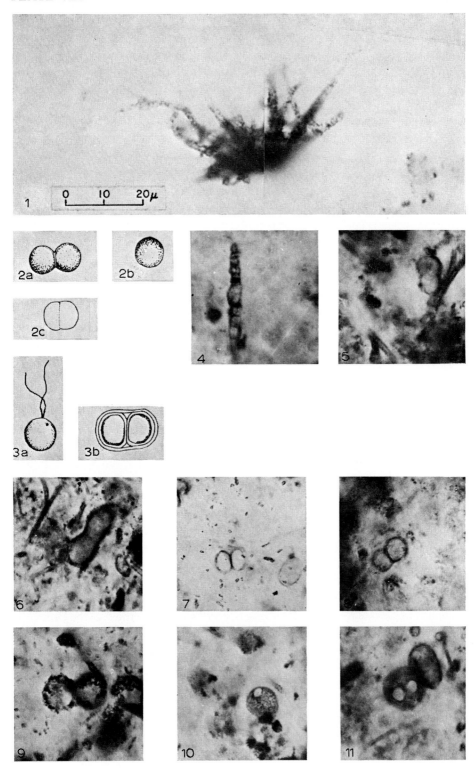

datings were cursorily raised to 2000 m.y. And although this meant a correction of no less than 50%, which was moreover quite arbitrary, this age was happily accepted thereafter. This is the figure, which, of course, caught the headlines, being the "oldest fossils in the world, 2 billion years old". Although this optimistic extrapolation has been corroborated, more or less fortuitously, by later research, it is well to realize how flimsy the basis for this statement has been, all through the period of 1954–1961.

At that date a detailed study by GOLDICH et al. (1961; see also GOLDICH, 1968) of the age of the Precambrian of Minnesota has provided a firmer base for the establishment of the age of the Gunflint. Both the Gunflint and the Animikie Group to which it belongs cross the border from Ontario to Minnesota and here it was established that the whole of the Animikie Group is older than the Penokean orogeny (Table XX). The latter is well dated on the base of igneous rocks belonging to this orogenetic period, and its age is 1.7 billion years. This consequently establishes a minimum age for the Animikie and the Gunflint. A maximum age of 2.5 billion years was found for the next older orogenetic period, the Algoman, whose dated igneous rocks are nonconformably overlain by sediments of the Gunflint. The factual basis for the dating of the Gunflint flora consequently is no more precise than: "Older than 1.7 billion years and younger than 2.5 billion years" (see PETERMAN, 1966).

A further corroboration has been supplied by HURLEY et al. (1962), who analyzed the age of clay minerals of a layer of fine volcanic ash intercalated in the chert. It is supposed that these clay minerals were formed at the time of the volcanic eruption which supplied the ash, so they can be considered as authigenic. They were dated by the K–Ar method at 1600 m.y., which seems to be on the young side. The authors have, however, previously made an extensive study of comparable clay minerals in rocks that were dated by other methods as well. They found that in these particular clay minerals a quite consistent loss of 20% of the radiogenic argon normally occurs. After applying this empirical correction to the Gunflint Formation, they arrived at a probable age of 1900 ± 200 m.y.

FAURE and KOVACH (1969) have, however, recently come to the conclusion that the age of the Gunflint Formation is only 1635 ± 24 m.y. These authors used the rubidium–strontium method of absolute dating (see section 9 of Chapter 3), as applied to "whole rocks" and utilizing isochrons similar to those in the Holmes-Houtermans analysis of uranium-lead ages (see section 16 of Chapter 3). Their evidence is admittedly circumstantial. More important, it gives an absolute age for the Gunflint which is less than that of the Penokean orogeny, which, as we saw, is younger than the Gunflint. Since the ages of the main orogenies can be much more reliably established than those of sediments, it follows that the date supplied by Faure and Kovach must, until further corroboration should become available, be dismissed as being too young.

Accordingly, when I follow the custom in assigning an age of two billion years to the Gunflint flora, this is not a true absolute date, but no more than a considered guess.

Finally it might be stressed that the absolute age of the Gunflint flora is not so important for our problem. This may seem a silly statement after the lengthy discussion of the efforts made to establish the absolute age. But although it would, of course, be nice to know its age exactly, the relative age of the Gunflint flora is the more relevant. From its relative position in the stratigraphical column, as presented in Table XX, we learn that it belongs to the Middle Precambrian. It is contemporaneous with, and in fact occurs within, one of the banded iron formations of the Lake Superior type. These are in turn indicative of a contemporary anoxygenic atmosphere, as we will learn in the next chapter. So, regardless of its absolute age, the relative age of the Gunflint Formation proves that the Gunflint flora represents a group of plants belonging to early life, which thrived under the primeval anoxygenic atmosphere. It is this aspect of the Gunflint flora, which is, to my mind, the most important.

16. METABOLISM OF MIDDLE PRECAMBRIAN ORGANISMS

As will be set forth in the next chapters, we may deduce from the fact that the Middle Precambrian fossils are found in chert layers of banded iron formations that they lived under the anoxygenic primeval atmosphere. It is probable that the level of free oxygen in the primeval atmosphere did not rise above 1% of that in the present-day atmosphere. On the other hand, the organisms have lived aerobically, in swamps or lakes, probably shielded by a water layer of a certain thickness, but still at the earth's surface. But, although aerobic, their metabolism must, by the low level of oxygen in the primeval atmosphere, have been comparable to that of the present-day anaerobic life.

This does not imply that the same organisms were already in existence during the Middle Precambrian that now form the world of the anaerobic microbes, but only that their life functions will be comparable. It seems probable that they had already developed the same major subdivisions as to their feeding habits, as are found in present-day microbes (see section 7 of Chapter 8) and that both chemotrophic and phototrophic organisms have been present. As stated already in section 7 of Chapter 8, there is, however, one possible major exception, because, as we will see in Chapter 16, pre-life has possibly co-existed with early life all through the Early and Middle Precambrian. In this case "organic" molecules must have been readily available for early life to feed on. In this picture of early life browsing on the products formed by pre-life—Hawkin's "aquatic garden of Eden", see section 6 of Chapter 4—it would be probable that photo"organo"-

trophic organisms have occupied an important place, as is indicated in Fig.137.

But this is a detail. In general the Gunflint organisms must have been both aerobic and anoxygenic. They differed from anaerobic organisms in being able to live in free contact with the atmosphere, although their life processes must have been similar to those of the present-day anaerobic world.

17. EARLY PRECAMBRIAN FOSSILS?

Since the succesful find of a varied fossil assemblage in the Middle Precambrian Gunflint an intensive search has started for older fossils. The difficulties in this hunt become larger, the older the fossils one wants to find. As set forth in the preceding chapter, we must search for those areas within the old shields in which a sedimentary series has since its formation escaped later strong mobilization, such as burial in geosynclines, or strong tectonic disturbance in fold belts. Only in those areas in which subsidence and sedimentation were not too active in the first place, will the pile of sediments remain relatively thin and escape serious metamorphism. We must therefore be on the look out for areas that were already relatively stable during sedimentation. That is in areas which are technically called *foreland* or *epicontinental* sedimentary basins. Moreover they must have remained largely untouched by later orogenies, for otherwise the older sediments would have become metamorphosed and the fossils destroyed. Such areas are technically called *cratonic*.

As we saw in the preceding chapter, the big African Shield contains a number of such relatively stable areas. Of these the oldest known to date is that of the Transvaal or Kaapvaal craton, which contains the Swaziland System. Following NICOLAYSEN (1962) and other workers, the Swaziland System can be divided into two series, the older Onverwacht and the younger Fig Tree. The Onverwacht Series is mainly volcanic and consists of a thick pile of ultrabasic to intermediate lavas, followed by rhyolitic tuffs. Fine-grained sediments, mainly cherts, are, however, intercalated in the volcanic series (VILJOEN and VILJOEN, 1967). The Fig Tree Series, on the other hand, consists mainly of sediments, with some volcanics near its base and top. Banded cherts, ironstones and laminated shales characterize the Lower Fig Tree, whereas coarser sediments are found in its upper part. It is in the fine-grained sediments, both in the cherts and in the shales, that fossil-like forms have been found and in fact been interpreted as fossils (BARGHOORN and SCHOPF, 1966; SCHOPF and BARGHOORN, 1966; PFLUG, 1966b, 1967, for the Fig Tree; ENGEL et al., 1968; NAGY and UREY, 1969; NAGY and NAGY, 1969; for the Onverwacht). As this is the most extensively described region of probably fossiliferous sites of the Early Precambrian, I will confine myself to a description of this area (Fig.80).

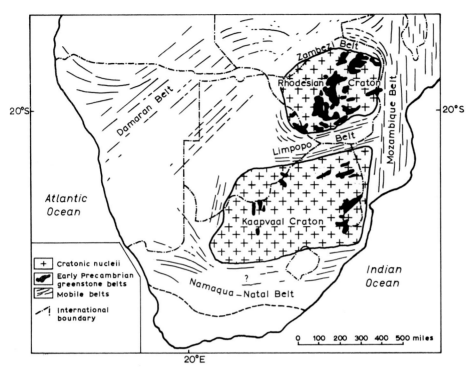

Fig. 80. The southern African crystalline shield stripped of its younger cover showing the ancient greenstone–granite *cratons*, which were not remobilized during their later history, and the encircling, younger, mobile metamorphic belts.

The Kaapvaal craton, called the Transvaal craton by other authors, contains the belts of greenstone and related sediments of the Transvaal System. The Rhodesian craton contains similar volcano-clastic series of the Bulawayan, which contains MACGREGOR's (1940) algal limestones (see section 8 of this chapter). The scattered occurrence of outcrops of both sedimentary series, disconnected not only by the surrounding granites, but also by rocks of the younger fold belts, is the reason why there is as yet no general agreement as to the age and the correlation of the Bulawayan with the Swaziland.

The Kaapvaal and Rhodesia cratons have been schematically united in Fig.56. (From ANHAEUSSER et al., 1968.)

The Swaziland System forms the oldest series of rocks known from South Africa and it has moreover the distinction of being the oldest sedimentary series known to date. At present only a minimum age of "older than 3.2 billion years" can be given (Fig.81, Table XVII). But it is hoped that dating of the Onverwacht volcanics will in future supply us with a more precise absolute age for the fossil-like forms found so abundantly in the Swaziland System.

The volcano-clastic series of the Swaziland System is extremely thick, estimates of over 10,000 m being given (VILJOEN and VILJOEN, 1967; ANHAEUSSER et al., 1967, 1968). However, such estimates are arrived at by adding the thickness

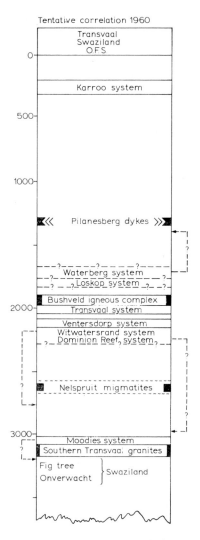

Fig.81. Stratigraphic column of the Precambrian of the Transvaal. The broken lines outside the column indicate major uncertainties in correlation and in dating. Notably the gold–uranium ores of the Witwatersrand and Dominion Reef systems, which we will meet with in the next chapter, are often thought to be nearer 3 billion than 2 billion years old. (Ages given in million-year intervals. From NICOLAYSEN, 1962.)

of local sections and do not necessarily imply that at any one point a pile of rocks that thick has ever been deposited. During sedimentation the subsidence of such sedimentary basins normally shifts, often to a so-called foreland, and younger sediments are not deposited right on top of, but next to, older sediments. Such a development is, for instance, well known from the Tertiary bordering the Gulf of

TABLE XVII

ABSOLUTE AGES, DETERMINED BY THE RUBIDIUM–STRONTIUM METHOD[1] OF THE "WHOLE ROCK" AND MINERAL CONCENTRATIONS FROM SEVERAL SUITES OF THE SWAZILAND GRANITES[2]
(From CLIFFORD, 1968)

Granite	Dated sample	Age (m.y.)
G1	whole rock	3240 ± 300
	microcline	2595 ± 90
	plagioclase	5660 ± 1000
G4	whole rock	2900 ± 60
	biotite	2330 ± 50
	muscovite	3800 ± 100
	microcline	2670 ± 30
G5	whole rock	2070 ± 50
	whole rock	2400 ± 50
	whole rock	2720 ± 340
	feldspar	2375 ± 70

[1] See Chapter 3, section 9.
[2] It is seen that the mineral fractions separated from the rocks give ages which are discordant from those obtained from the whole rock. It is postulated that this is due to diffusion of ^{87}Sr from mineral to mineral during the later history of the rocks. If this assumption holds true, the "whole rock" determinations would indicate the real age of these granites and this forms the basis to assign to the Fig Tree and Onverwacht Series an age of "more than 3.2 billion years".

Mexico and from the Upper Carboniferous in the Ruhr Basin (PAPROTH and TEICHMÜLLER, 1961).

Nevertheless, the thickness of the Swaziland System is imposing and this has led some authors to imply that the structures described from the basal beds of the Onverwacht Series might be much older than those found in the Fig Tree Series. This is, however, not necessarily the case. Series of similar thickness, such as found in the continental deposits of the Upper Carboniferous mentioned above, were formed during periods of the order of 50 m.y. That is of 0.05 billion years. This is considerably less than the uncertainty still prevailing at present in the dating of the Swaziland System, which is variously given as "older than 3.1 or 3.2 billion years". Consequently, even although the authors cited above are of the opinion that these Early Precambrian sedimentary series are in several ways distinct from the later geosynclines, we may still consider the Fig Tree and the Onverwacht Series together in our study of the oldest vestiges of early life.

Even so, the age of the Swaziland System is still not well known, as follows from the survey paper on radiometric dating of African rocks by CLIFFORD (1968). Absolute dates are now known from many of the granites surrounding the schist belts of the Swaziland System, but these dates vary widely inter se, whilst it is moreover in many cases only assumed that these granites developed during the post-Swaziland orogeny, which consequently only suggests that the Swaziland volcano-clastic series of sediments is older than the oldest known of these granites. That is, according to Table XVII, older than 3240 ± 300 m.y. The only direct relation established so far, is that the G4 granite of Table XVII with a whole-rock age of 2.9 billion years, is intrusive into the Moodies Series, the uppermost division of the Swaziland System, and as such younger than both the Fig Tree and the Onverwacht Series.

The structures described from the Swaziland System comprise small, spherical and filamentous forms, which range in size around 20μ. They contain carbonaceous material, at least in their outer sections, and have been located both in thin sections and in maceration concentrates. Moreover, much smaller structures, less than 1μ in diameter, have been found by electron microscopy in replicas of polished rocks. There is no doubt that these structures are real. Nor can it be doubted that they belong to and are included within the rocks studied. The techniques used preclude all possibility of contamination by outside matter. The main forms of the spherical structures from the Fig Tree Series, in which, according to PFLUG (1967) several types, indicated as A_{1-3} and B can be distinguished, are illustrated in Plate IX. Filamentous forms from the Fig Tree Series are illustrated in Plate X, whilst the smaller, rod-like structures can be found in Plate XI. An organized structure from the Onverwacht Series is illustrated in Fig.82.

There is consequently no doubt that organized structures of various types occur both in the Fig Tree and in the Onverwacht Series of the Swaziland System. It has, however, not yet been proved that these structures are biogenic. Or, in other words, that they are real fossils, remnants of life of the Early Precambrian. The material of the structures, at least of their outer walls, has been analyzed by various refined techniques, and has uniformly been proclaimed to be *organic*. However, the reason for this lies only in the fact that it is built up by carbon compounds.There is as yet no indication whether this material is really organic, that is of biological origin, and not "organic" and thus of abiological origin; a point already discussed in section 10 of this chapter. Nor is there any hint to be taken from the morphology of these structures which could decisively point to a biological origin. There is none of the delicate inner structures, none of the weirdness of outer form, so characteristic for at least a part of the Gunflint fossils.

Until further research has supplied us with more convincing proof for the biogenic origin of these structures, it behoves us to indicate them with a strictly neutral term, such as "organized elements".

Positive results from either the chemical composition or the morphology of these structures will, of course, sustain the claim that they are real fossils. If the morphology of newly found structures undoubtedly points toward a biological origin, then it is the more probable that the carbon compounds are really organic.

PLATE IX

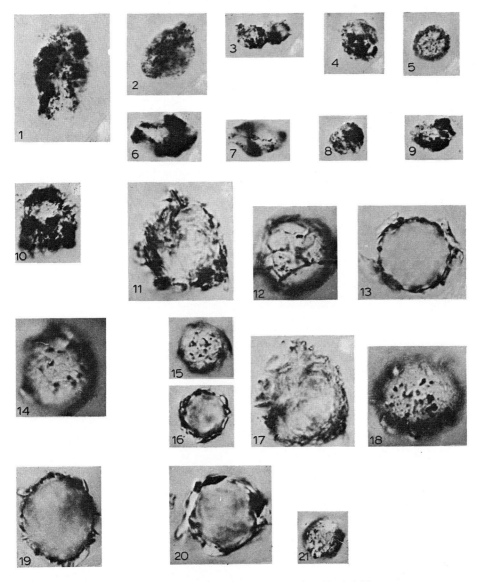

Globular structures from the Early Precambrian Fig Tree Series, South Africa.
1–10. Globular type A₃ structures. (×425.)
11. Globular type A₃ structure. (×1,100.)
12–16. Globular type B structures. (×1,100.)
17. Globular type A₃ structure. (×1,100.)
18–20. Globular type B structures. (×1,100.)
21. Globular type B structure. (×425.)
Similar globular structures have been described by SCHOPF and BARGHOORN (1967). Without applying the finer distinctions cited by Pflug, they have classified them together somewhat rashly as *Archaeosphaeroides barbertonensis* new genus, new species. It remains doubtful, whether science is served with such rather meaningless names.
(From PFLUG, 1967. Courtesy Professor H. D. Pflug.)

PLATE X

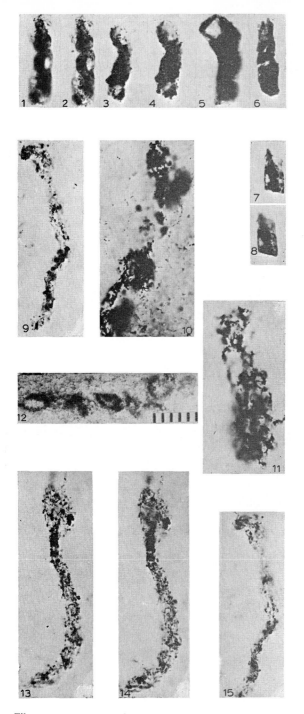

Filamentous structures from the Early Precambrian Fig Tree Series, South Africa.
1–11.   Filamentous structures of various types. (×375.)
12.   Colony of globular type A_2 structures. (×150.)
13–15. Filamentous structures of various types. (×375.)
(From PFLUG, 1967. Courtesy Professor H. D. Pflug.)

PLATE XI

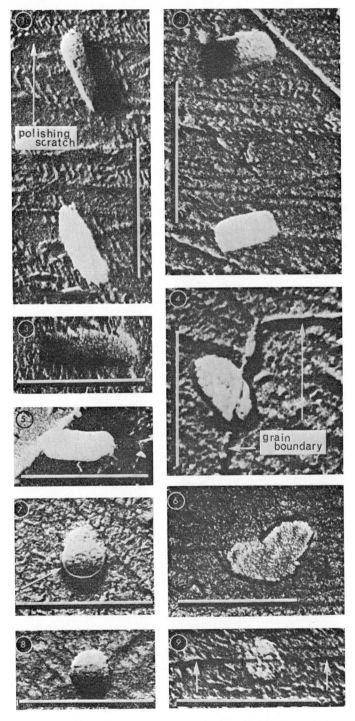

Negative prints of electron micrographs of platinum-carbon surface replicas of chert from the Fig Tree Series. Showing small rod-like structures, which, according to the authors, represent fossil bacteria and have been called *Eobacterium isolatum*, new genus, new species. Line in each figure is 1 μ. (From BARGHOORN and SCHOPF, 1966. Courtesy Professor E. S. Barghoorn.)

Or if, vice versa, undoubted molecular fossils prove the existence of life during the sedimentation of the Onverwacht and Fig Tree Series, then it becomes the more likely that these "organized elements" really represent the fossilized remnants of early life. But up to now the question must remain undecided.

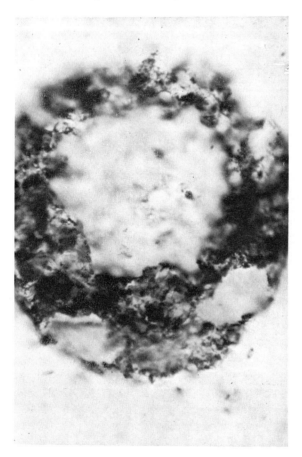

Fig.82. Organized structure of carbonaceous material from the Onverwacht Series of South Africa. The structure is 106 μ across. Although we can, as yet, not be sure that this structure represents a real fossil, it has the distinction of being the oldest "fossil-like" structure known. From ENGEL et al., 1968. Courtesy Professor B. Nagy.)

18. CONTEMPORANEITY OF "ORGANIZED ELEMENTS" AND OF FOSSILS: OF PRE-LIFE AND OF EARLY LIFE?

Unfortunately the question of distinguishing between "organized elements" synthesized inorganically, and "organic elements" which are fossilized remnants of early life, becomes more complicated by the probability mentioned already that pre-life and early life have co-existed over a time span of the order of 2 billion years. We will go into the arguments in favour of this co-existence of pre-life and early life in Chapters 16 and 18, whereas here only one possible example of such a co-existence will be discussed.

As indicated earlier, even if pre-life and early life have been co-existent for longer periods, this does not mean that the "organic" molecules of pre-life were formed at the same localities where early life has established itself. Inorganic synthesis of the "organic" molecules of pre-life and the existence of life at the same locality, that is under the same environmental conditions, is, as we have shown over and over again, impossible. But we may well visualize the possibility of early life striving already in lakes, sheltered by a layer of water, and leaving fossilized remnants on the lake bottom, whilst elsewhere on the continent the conditions for pre-life were fulfilled. Taking two different deposits of similar age, it would thus be possible to find real fossils in one, "organized elements", synthesized as pre-life, in the other.

An example which might provisionally be used to illustrate this possibility lies in the comparison between the fossils of the Gunflint Formation of Ontario described earlier in this chapter, and "organized elements" described from the Witwatersrand System of South Africa by SCHIDLOWSKI (1965, 1966). The structures described by the latter author were first called "probable life-forms", but more prudently indicated as "cellular structural elements" in his second publication. Nothing in these small, round circles (Fig.83) proves that they are real fossils. Their age, however, is given as 2.15 billion years, that is only somewhat older than the Gunflint. So we might expect the existence of life forms complicated enough to be capable of producing fossils at the time of formation of the Witwatersrand System.

Although it is no more than an assumption, we might explain this difference by the mode of formation of the two formations, which will be discussed in more detail in the next chapter. It is probable that the cherts of the Gunflint formed on the bottom of lakes, where life was sheltered from the shorter ultraviolet rays of the sunlight by the overlying water. The sands and gravels of the Witwatersrand, on the other hand, formed in rivers and deltas, on the shores of lakes, and perhaps the oceans. In this unsheltered environment the ultraviolet sunlight could strike the surface of the earth, and "organized elements" could be synthesized inorganically. Moreover, even if the environment in which pre-life was formed differed from that in which early life developed, a contamination will have occurred because the materials formed by pre-life would be swept into lakes and oceans by the rivers flowing off the continents, in this way still contributing to the "thin soup" (see Fig.124). So, although pre-life and early life will have existed in different environments, "organized elements" formed by pre-life, which

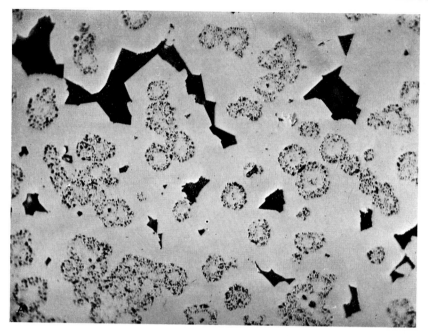

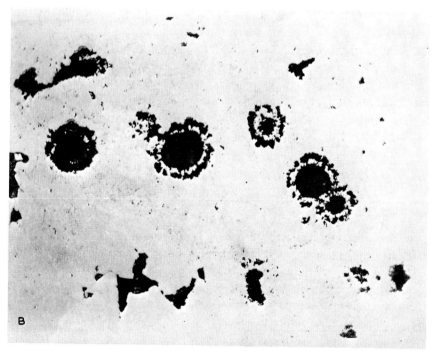

Fig.83. "Organized elements" from the Witwatersrand System, South Africa; ×1150. Polished sections seen in reflected light. (From SCHIDLOWSKI, 1965, 1966. Courtesy Dr. M. Schidlowski.)

could stand the transport by the rivers, would get mixed with contemporaneous organized forms of early life, and would in this case even stand a change of becoming fossilized in the same bed. For the Phanerozoic we are quite familiar with such a mixture of fossils derived from different habitats or *biocoenoses* into the same burial ground or *thanatocoenose*, so there is no reason at all why a similar mixing could not have occurred during the Precambrian.

19. PSEUDOFOSSILS?

In finishing this chapter we have to take note of the fact that there are quite a number of "fossil" finds of the Precambrian of which we are not sure whether these really represent former living organisms, or whether they only resemble the remains of such organisms, and as such must be called *pseudofossils*. I will not try to give a full review of the voluminous literature, which would only result in a tedious enumeration of uncertainties and would not bring us any further. Instead I will cite two examples, which might serve as case histories for this whole group of speculative structures. It so happens that these examples comprise in *Corycium enigmaticum* SEDERHOLM (1911) from Finland the earliest described structure, and in the form genus *Rhysonetron* HOFMANN (1967) from Canada one of the most recently described members of this group.

Notwithstanding its specific name, *Corycium enigmaticum* has been generally considered to be of organic origin. The literature has been carefully reviewed by MATISTO (1963). However, much of the evidence for a biogenic origin of *Corycium* is based on the occurrence of trace elements and on the ratio of the stable isotopes of carbon. As we will see in section 9 of Chapter 14 such arguments can no longer be accepted as proof, either for, or against a biogenic origin.

Corycium in itself is not very convincing as a fossil, as may be seen from Fig.84. It consists of sack-like structures, of the order of one or several centimeters thick and one or more decimeters long, which are enveloped by a thin film, either of graphite, or of graphite-rich material. It is not surprising that a sedimentologist like VAN STRAATEN (1950) has once proposed an inorganic genesis of these structures, by a slumping process of clay sediments at the time of their sedimentation.

The sedimentary series containing *Corycium* is older than the Svecofennian orogeny, now dated at 1.8–1.9 billion years. The age of *Corycium* consequently is now estimated to be over 2 billion years (Dr. A. Matisto, in lit., 1968). It consequently falls in the same age bracket as Barghoorn's Gunflint flora reviewed earlier in this chapter. Geologically the difference is that the Gunflint Chert of the Canadian Shield is much less metamorphosed than the phyllites containing *Corycium* of the Baltic Shield, whilst the conservation of the organic material is much better in the Gunflint. Since we are primarily interested in a world wide study of the

Fig.84. Section of Precambrian phyllite, a slightly metamorphosed clayey sediment, showing two specimens of *Corycium enigmaticum* SEDERHOLM. Mylyniemi, Aitolakti, 7 km northeast of Tampere, Finland.

origin of life and not in a regional study of the Baltic Shield, we may do well to concentrate on the North American finds and leave the subject of *Corycium* as undecided, until better material becomes available.

The other doubtful Precambrian fossils which will be cited here, belong to the newly coined form genus *Rhysonetron* HOFMANN. In reality they do not belong to the true fossils at all, but are sandstone structures, which are interpreted as fossilized casts or molds of contemporaneous organisms. Their form, which is depicted in Fig.85, is described as "discrete, long, curved, cylindrical and tapering rods or spindles". Their most fascinating character is the longitudinal median marking or sculpturing and the oblique, lateral, crescentic corrugations. Their maximum size reaches 1 dm and over.

HOFMANN (1967) is convinced that such an intricate pattern must not only be biogenic in origin but even represent casts or molds of animals (or, to put it slightly more prudently, of Metazoans). But he thinks it is at present still highly speculative to assign them to a more definite taxonomic position. Annelids, sponges and coelenterates might be held responsible for these structures, or perhaps something entirely new to science; a group which has since died out and of which these casts are the only remnants known so far. The latter possibility cannot be ruled out, because of the great age of these structures. *Rhysonetron* was found near Elliot Lake in Ontario in sandstones belonging to the Upper Huronian of the regional stratigraphy, that is to the Middle Precambrian of our subdivision. They are thought to be older than 2155 ± 80 m.y. but are certainly younger than the underlying pre-Huronian granites dated at 2500 m.y.

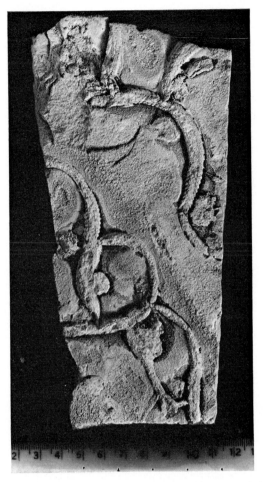

Fig.85. Huronian sandstone of the Middle Precambrian near Elliot Lake, Ont., Canada, containing spindles of the pseudofossil *Rhysonetron*. (From HOFMANN, 1967. Courtesy Geological Survey of Canada and Dr. H. S. Hofmann.)

This age implies that the casts called *Rhysonetron* are contemporaneous with the primeval anoxygenic atmosphere. Which in turn signifies that if and when these structures are fossilized casts or molds of animals (or Metazoans), we had at that time a primitive group of animals capable of respiration at a level of atmospheric oxygen of 1% of that of our present atmosphere. In this case Hofmann's speculation, cited above, that these structures represent an up till now entirely unknown group of animals becomes plausible. It would consist of primitive animals, with a respiratory system functioning already at very low levels of oxygen. It would then in the course of time have become superseded by younger groups with a better respiratory system, which developed when a higher level of atmospheric oxygen became available.

On the other hand, I have strong objections to accept these spindle-like structures as fossilized casts of animals. Or in fact, to accept them as biogenic at all. The reason for this is that they occur at one base of sandstone beds. Now, all during the Phanerozoic, the base of a sandstone bed is not exactly the best site for fossilization. Fossils are found scattered in sandstone beds, but not "glued" to the base of a sandstone layer, as are the *Rhysonetron* spindles. Instead, we know of thick series of alternating clay and sand, the so-called flysch facies, in which the base of the sandstone beds show all sorts of casts, which, at least in part, are of inorganic origin. The weight of the heavy sand, presumably transported in rapid flow over the underlying mud, has scooped out the clay and in this way any number of rather regular casts of sand form when the sand is deposited. True, the spreading of the sand layer may also fossilize tracks of animals that lived on or in the clay, but, with the exception of the casts aligned perpendicularly to the bedding, these are far more irregular than the *Rhysonetron* spindles, which show more resemblance to casts of inorganic nature. Consequently, until more material becomes available, I prefer to interpret these spindles as inorganically formed pseudofossils.

20. OPTIMISTIC OUTLOOK

Less than a decade ago I closed a review of the scanty remains of life known at that time from the Precambrian with the optimistic outlook that because the interest in the study of the Precambrian had quickened considerably over the last few years, new results were bound to follow soon. This has been amply justified so far. Not only have the fossils of the Middle Precambrian become much better known, but the whole new group—although still controversial—of the Early Precambrian Swaziland System has been discovered since. Moreover, apart from these classic fossil or pseudofossil finds, the entirely new notion of the existence of molecular fossils has sprung up in the meantime.

There is now no longer any doubt possible that life has already existed under the anoxygenic conditions of the primeval atmosphere.

Of course, the remnants of early life on earth will forever remain scanty. This is to be expected for two reasons. For one thing, early life will have been microscopic and we have seen that microbes and other small organisms without hard parts do not easily fossilize. On the other hand, most of the rocks dating back from the early history of the earth have become metamorphosed, due to the vicissitudes of later orogenetic periods, whereby every possible evidence of early life has become destroyed.

Nevertheless, I think I may well close this chapter with the same optimism. I believe that the discoveries awaiting us in the next decade will be at least as many and as important as those in the last decade. I have a feeling, however, that there will be a change in the quality of these discoveries. Up till quite recently there has been a record-seeking hunt to find older and still older vestiges of life. Every peculiar structure and every carbon compound found, has been hailed as a new proof of the antiquity of life. With the advance of the experimental checks into the possibility of inorganic synthesis of "organic" molecules, it has, on the other hand, become of prime importance to decide whether the remnants found are really organic, of perhaps only "organic"; whether they were formed by life or by pre-life. It is reasonable to expect that the next decade will see an intensification of research into a proper distinction between products formed by life and by pre-life.

REFERENCES

ABELSON, P. H. and HOERING, T. C., 1965. Fatty acids from the oxydation of kerogen. *Carnegie Inst. Wash., Yearbook*, 64: 218–223.

ABELSON, P. H. and PARKER, P. L., 1962. Fatty acids in sedimentary rocks. *Carnegie Inst. Wash., Yearbook*, 61:181–184.

ANHAEUSSER, C. R., ROERING, C., VILJOEN, M. J. and VILJOEN, R. P., 1967. The Barberton Mountain Land: A model of the elements and the evolution of an Archean fold belt. *Econ. Geol. Res. Unit, Univ. Witwatersrand, Johannesburg, Inform. Circ.*, 38:31 pp.

ANHAEUSSER, C. R., MASON, R., VILJOEN, M. J. and VILJOEN, R. P., 1968. A reappraisal of some aspects of Precambrian shield geology. *Econ. Geol. Res. Unit, Univ. Witwatersrand, Johannesburg. Inform. Circ.*, 49:30 pp.

BARGHOORN, E. S. and SCHOPF, J. W., 1966. Microorganisms three billion years old from the Precambrian of South Africa. *Science*, 152:758–763.

BARGHOORN, E. S. and TYLER, S. A., 1965. Microorganisms from the Gunflint Chert. *Science*, 147: 563–577.

BARGHOORN, E. S., MEINSCHEIN, W. G. and SCHOPF, J. W., 1965. Paleobiology of a Precambrian shale. *Science*, 148:461–472.

BELSKY, T., JOHNS, R. B., McCARTHY, A. L., BURLINGAME, A. L., RICHTER, W. and CALVIN, M., 1965. Evidence of life processes in a sediment two and a half billion years old. *Nature*, 206:446–447.

BERTRAND, J., 1969. Les édifices stromatolitiques précambriens de la "série à stromatolithes" du Nord-Ouest de l'Ahaggar (Sahara). *Bull. Soc. Géol. France*, 7 (10):168–178.

BOYLAN, D. B., ALTURKI, Y. I. and EGLINTON, G., 1969. Application of gas chromatography and mass spectrometry to porphyrin microanalysis. In: P. A. SCHENK and I. HAVENAAR (Editors), *Advances in Organic Geochemistry 1968*. Pergamon, London, pp.227–240.

BROOKS, J. D., GOULD, K. and SMITH, J. W., 1969. Isoprenoid hydrocarbons in coal and petroleum. *Nature*, 222:257–259.

BURLINGAME, A. L. and SIMONET, B. R., 1969. High resolution mass spectrometry of Green River kerogen. *Nature*, 222:741–747.

BURLINGAME, A. L., HAUG, P., BELSKY, T. and CALVIN, M., 1965. Occurrence of biogenic steranes and pentacyclic tripertanes in an Eocene shale (52 million years) and in an Early Precambrian shale (2.7 billion years): A preliminary report. *Proc. Natl. Acad. Sci. U.S.*, 54: 1406–1412.

BURLINGAME, A. L., HAUG, P. A., SCHNOES, H. K. and SIMONET, B. R., 1969. Fatty acids derived from the Green River Formation oil shale by extractions and oxidations. A review. In: P. A. SCHENK and I. HAVENAAR (Editors), *Advances in Organic Geochemistry 1968*. Pergamon, London, pp.85–130.

CALVIN, M., 1965. Chemical evolution. *Proc. Roy. Soc. (London), Ser. A*, 288:441–466.

CALVIN, M., 1968. Molecular paleontology. Bennet Lecture—*Trans. Leicester Lit. Phil. Soc.*, 62:45–69.

CALVIN, M., 1969. *Chemical Evolution*. Clarendon, Oxford, 278 pp.

CALVIN, M. and BASSHAM, J. A., 1962. *The Photosynthesis of Carbon Compounds*. Benjamin, New York, N. Y., 127 pp.

CLIFFORD, T. N., 1968. Radiometric dating and pre-Silurian geology of Africa. In: E. I. HAMILTON and R. M. FARQUAR (Editors), *Radiometric Dating for Geologists*. Interscience, New York, N. Y., pp.299–416.

CLOUD, P. E., 1968. Pre-Metazoans and the origins of Metazoa. In: E. T. DRAKE (Editor), *Evolution and Environment*. Yale Univ. Press, New Haven, Conn., pp.1–72.

CLOUD, P. E., GRUNER, J. W. and HAGEN, H., 1965. Carbonaceous rock of the Soudan Iron Formation (Early Precambrian). *Science*, 148:1713–1716.

EGLINTON, G., 1969. Hydrocarbons and fatty acids in living organisms and recent and ancient sediments. In: P. A. SCHENK and I. HAVENAAR (Editors), *Advances in Organic Geochemistry 1968*. Pergamon, London, pp.1–24.

EGLINTON, G. and CALVIN, M., 1967. Chemical fossils. *Sci. Am.*, 216 (1):32–43.

EGLINTON, G., SCOTT, P. M., BELSKY, T., BURLINGAME, A. L. and CALVIN, M., 1964a. Hydrocarbons of biological origin from a one-billion-year-old sediment. *Science*, 145:263–264.

EGLINTON, G., SCOTT, P. M., BELSKY, T., BURLINGAME, A. L., RICHTER, W. and CALVIN, M., 1964b. Occurrence of isoprenoid alkanes in a Precambrian sediment. In: G. D. HOBSON and M. C. LOUIS (Editors), *Advances in Organic Geochemistry*. Pergamon, London, pp.41–74.

ENGEL, A. E. J., NAGY, B., NAGY, L. A., ENGEL, C. G., KREMP, G. O and DREW, C. M., 1968. Alga-like forms in Onverwacht Series, South Africa: Oldest recognized lifelike forms on earth. *Science*, 161:1005–1008.

FAURE, G. and KOVACH, J., 1969. The age of the Gunflint Formation of the Animikie Series of Ontario, Canada. *Geol. Soc. Am. Bull.*, 80:1725–1736.

GLAESSNER, M. F., 1961. Precambrian animals. *Sci. Am.*, 204:72–78.

GLAESSNER, M. F., 1962. Precambrian fossils. *Biol. Rev. Cambridge Phil. Soc.*, 37:467–494.

GLAESSNER, M. F., 1966. Precambrian paleontology. *Earth-Sci. Rev.*, 1:29–50.

GLAESSNER, M. F., 1968. Biological events and the Precambrian time scale. *Can. J. Eearth Sci.*, 5:585–590.

GLAESSNER, M. F. and WADE, M., 1966. The Late Precambrian fossils from Ediacara, South Australia. *Paleontology*, 9:599–628.

GLAESSNER, M. F., PREISS, W. V. and WALTER, M. R., 1969. Precambrian columnar *Stromatolites* in Australia. Morphological and stratigraphical analysis. *Science*, 164: 1056–1058.

GOLDICH, S. S., 1968. Geochronology in the Lake Superior region. *Can. J. Earth Sci.*, 5:715–724.

GOLDICH, S. S., NIER, A. O., BAADSGAARD, H., HOFMAN, J. H. and KRUEGER, H., 1961. The Precambrian geology and geochronology of Minnesota. *Minnesota Geol. Surv., Bull.*, 41: 193 pp.

GOLDRING, R. and CURNOW, C. N., 1967. The stratigraphy and facies of the Late Precambrian at Ediacara, South Australia. *J. Geol. Soc. Australia*, 14:195–214.

GRAVELLE, M. and LELUBRE, M., 1957. Découverte de Stromatolithes du groupe des Conophytons dans le Pharusien de l'Ahaggar occidental (Sahara Central). *Bull. Soc. Géol. France*, 6 (7):435–442.

HAN, J. and CALVIN, M., 1969. Occurrence of fatty acids and aliphatic hydrocarbons in a 3–4 billion-year-old sediment. *Nature*, 224:576−577.

HARE, P. E., 1965. Amino acid artifacts in organic geochemistry. *Carnegie Inst. Wash., Yearbook*, 64:232–235.

HÄRME M. and PERTTUNEN, V., 1963. Stromatolite structures in Precambrian dolomite in Tervola, North Finland. *Compt. Rend., Soc. Géol. Finlande*, 35:79–81.

HOERING, T. C., 1962. The isolation of organic compounds from Precambrian rocks. *Carnegie Inst. Wash., Yearbook*, 61:184–187.

HOERING, T. C., 1965. The extractable organic matter in Precambrian rocks and the problem of contamination. *Carnegie Inst., Wash., Yearbook*, 64:215–218.

HOERING, T. C., 1967. The organic chemistry of Precambrian rocks. In: P. H. ABELSON (Editor), *Researches in Geochemistry*, II. Wiley, New York, N. Y., pp.87–111.

HOFMANN, H. J., 1967. Precambrian fossils (?) near Elliot Lake, Ontario. *Science*, 156:500–504.

HURLEY, P. M., FAIRBAIRN, H. W., PINSON, W. H. and HOWER, J., 1962. Unmetamorphosed minerals in the Gunflint Formation used to test the age of the Animikie. *J. Geol.*, 70: 489–492.

JOHNS, R. B., BELSKY, T., MCCARTHY, E. D., BURLINGAME, A. L., HAUG, P., SCHNOES, H. K., RICHTER, W. and CALVIN, M., 1966. The organic geochemistry of ancient sediments. II. *Geochim. Cosmochim. Acta*, 30:1191–1222.

KLUYVER, A. J., 1937. 's Levens nevels. *Handel. 26e Nederl. Natuurk.-Geneesk. Congr.*, 82:82–106. (Translated as "Life's fringes" in: A. F. KAMP, J. W. M. LA RIVIÈRE and W. VERHOEVEN, *Albert Jan Kluyver*. North-Holland, Amsterdam, pp.329–348.

KVENVOLDEN, K. A., PETERSON, E. and POLLOCK, G. E., 1969. Optical configuration of amino-acids in Precambrian Fig Tree chert. *Science*, 221:141–143.

LABERGE, G. L., 1967. Microfossils and Precambrian iron-formations. *Geol. Soc. Am. Bull.*, 78:331–342.

LICARI, G. R. and CLOUD, P. E., 1968. Reproductive structures and taxonomic affinities of some nannofossils from the Gunflint Iron Formation. *Proc. Natl. Acad. U. S.*, 59:1053–1060.

MACGREGOR, A. M., 1940. A pre-Cambrian algal limestone in Southern Rhodesia. *Trans. Geol. Soc. S. Africa*, 43:9–16.

MACLEOD, W. D., 1968. Combined gas chromatography–mass spectrometry of complex hydrocarbon trace residues in sediments. *J. Gas Chromatog.*, 6: 591–594.

MATISTO, A., 1963. Über den Ursprung des Kohlenstoffes in *Corycium*. *Neues Jahrb. Geol. Paläontol., Monatsh.*, 1963:443–441.

MCCARTHY, E. D., 1967. A treatise in organic geochemistry. *Lawrence Rad. Lab., Berkeley*, *Rept.* UCRL-17758:290 pp.

MCCARTHY, E. D. and CALVIN, M., 1967. Organic geochemical studies, I. Molecular criteria for hydrocarbon genesis. *Nature*, 216:642–647.

MEINSCHEIN, W. G. and BARGHOORN, E. S., 1964. Biological remnants in a Precambrian sediment. *Science*, 145:262–263.

MOORHOUSE, W. W. and BEALES, F. W., 1962. Fossils from the Animikie, Port Arthur, Ontario. *Trans. Roy. Soc. Can.*, 56:97–110.

MURPHY, R. C., DJURICIC, M. V., MARKEY, S. P. and BIEMANN, K., 1969. Acidic components of Green River Shale identified by a gas chromatography–mass spectrometry-computer system. *Science*, 165:695–697.

NAGY, B. and NAGY, L. A., 1969. Early Pre-Cambrian Onverwacht microstructures: Possibly the oldest fossils on earth? *Nature*, 223:1226–1229.

NAGY, B. and UREY, H. C., 1969. Organic geochemical investigations in relation to the analysis of returned lunar rock samples. *Life Sci. Space Rev.*, 7:31–46.

NICOLAYSEN, L. O., 1962. Stratigraphic interpretation of age measurements in southern Africa. In: A. E. J. ENGEL (Editor), *Petrographic Studies—Geol. Soc. Am., Buddington Vol.*, pp.569–598.

ORÓ, J. and NOONER, D. W., 1967. Aliphatic hydrocarbons in Pre-Cambrian rocks. *Nature*, 213: 1082–1085.

ORÓ, J., NOONER, D. W. and ZLATKIS, A., 1965. Hydrocarbons of biological origin in sediments about two billion years old. *Science*, 148:77–79.

PAPROTH, E. and TEICHMÜLLER, R., 1961. Die paläogeographische Entwicklung der subva-rischischen Saumsenke in Nordwestdeutschland im Laufe des Karbons. *Compt. Rend. Congr. Intern. Stratigraphie Géol. Carbonifère, 4e, Heerlen, 1961*, 2:471–491.

PETERMAN, Z. E., 1966. Rb–Sr dating of metasedimentary rocks of Minnesota. *Geol. Soc. Am. Bull.*, 77:1031–1044.

PFLUG, H. D., 1966a. Einige Reste niederer Pflanzen aus dem Algonkium. *Paleontographica*, 117B: 59–74.

PFLUG, H. D., 1966b. Structured organic remains from the Fig Tree Series of the Barberton Mountain Land. *Econ. Geol. Res. Unit, Univ. Witwatersrand, Johannesburg, Inform. Circ.*, 28:14 pp.

PFLUG, H. D., 1967. Strukturierte organische Reste aus über 3 Milliarden Jahre alten Gesteine Südafrikas. *Naturwissenschaften*, 54:236–241.

PFLUG, H. D., MEINEL, W., NEUMANN, K. H. and MEINEL, M., 1969. Entwicklungstendenzen des frühen Lebens auf der Erde. *Naturwissenschaften*, 56:10–14.

PRASHNOWSKY, A. A. and SCHIDLOWSKI, M., 1967. Investigations of Pre-Cambrian tucholite. *Nature*, 216:560–563.

RUTTEN, M. G., 1966. Geologic data on atmospheric history. *Paleogeography Paleoclimatol. Paleoecol.*, 2:47–57.

RUTTEN, M. G., 1969. Sedimentary ores of the Early and Middle Precambrian and the history of atmospheric oxygen. *Proc. Inter-Univ. Congr., 15th, 1967, Leicester*, pp.187–195.

SALOP, L. I., 1968. Pre-Cambrian of the U.S.S.R. *Proc. Intern. Geol. Congr., 23rd, Prague*, 4:61–73.

SCHIDLOWSKI, M., 1965. Probable life-forms from the Precambrian of the Witwatersand System (South Africa). *Nature*, 205:895–896.

SCHIDLOWSKI, M., 1966. Zellular strukturierte Elemente aus dem Präkambrium des Witwatersrand Systems (Südafrika). *Z. Deut. Geol. Ges.*, 115:183–786.

SCHIDLOWSKI, M., 1970. Untersuchungen zur Metallogenese im südwestlichen Witwatersrand-Becken (Oranje-Freistaat-Goldfeld, Südafrika). *Beih. Geol. Jb.*, 85:80pp.

SCHOPF, J. W. and BARGHOORN, E. S., 1967. Alga-like fossils from the Early Precambrian of South Africa. *Science*, 156:508–512.

SCHOPF, J. W., KVENVOLDEN, K. A. and BARGHOORN, E. S., 1968. Amino acids in Precambrian sediments. An assay. *Proc. Natl. Acad. U. S.*, 59:639–646.

SEDERHOLM, J. J., 1897. Über eine archäische Sedimentformation im südwestlichen Finnland und ihre Bedeutung für die Erklärung der Entstehungsweise des Grundgebirges. *Bull. Comm. Géol. Finlande*, I, 6:254 pp.

TYLER, S. A. and BARGHOORN, E. S., 1954. Occurrence of structurally preserved plants in the pre-Cambrian rocks of the Canadian Shield. *Science*, 119:606–608.

VAN HOEVEN, W., 1969. Organic geochemistry. *Lawrence Radiation Lab., Berkeley, Rept.*, UCRL-18690: 242 pp.

VAN HOEVEN, W., MAXWELL, J. R. and CALVIN, M., 1969. Fatty acids and hydrocarbons as evidence of life processes in sediments and crude oils. *Geochim. Cosmochim. Acta*, 33: 877–882.

VAN STRAATEN, L. M. J. U., 1950. Occurrence in Finland of structures due to subaquatic sliding of sediments. *Bull. Comm. Géol. Finlande*, 144:9–18.

VILJOEN, M. J. and VILJOEN, R. P., 1967. A reassessment of the Onverwacht Series in the Komati River valley. *Econ. Geol. Res. Unit, Univ. Witwatersrand, Johannesburg, Inform. Circ.*, 36:35 pp.

WADE, M., 1969. Medusae from the uppermost Precambrian or Cambrian sandstones, Central Australia. *Paleontology*, 12:351–356.

YOUNG, R. B., 1940a. Further notes on algal structures in the Dolomite Series. *Trans. Geol. Soc. S. Africa*, 43:17–22.

YOUNG, R. B., 1940b. Note on an unusual type of concretionary structure in limestones of the Dolomite Series. *Trans. Geol. Soc. S. Africa*, 43: 23–26.

Chapter 13 | The Contemporary Environment

1. RECONSTRUCTION OF THE CONTEMPORARY ENVIRONMENT FROM FOSSILS AND FROM OTHER GEOLOGICAL EVIDENCE

It seems important at this point, now that we have passed in review the fossil remains of the early stages of life, to make clear what kind of evidence on early life one may derive from fossils and from other geological data.

Precambrian fossils, both the fossils sensu stricto and the biogenic deposits, normally do not tell us much more than that there has been life on earth. In the later history of the earth fossils will be able to tell us considerably more, for instance whether they are the remains of organisms that lived in the sea, in lakes or on land. Or, whether it was a warm or a cold, a wet or a dry environment in which they became deposited. Such information is, however, not at the moment forthcoming from the remains of early life. The organisms of that period were so primitive that they might have well been at home in any environment. Even the finding of molecular fossils indicating capabilities of organic photosynthesis does not tell us whether the environment was oxygenic or anoxygenic. For this gives no indication about the oxygen balance, the ratio of oxygen production and consumption.

The only clear biological indicator about the oxygen level of the contemporary atmosphere is the existence of a continental flora or fauna. As we will see in Chapter 15, it is probable that such organisms could not exist in an atmosphere with less than one tenth of the present level of oxygen, because of the lethal action of the shorter ultraviolet sunlight reaching the earth's surface at any lower level of free oxygen. But, as we will further see in Chapter 16, this level has in all probability only been reached with the Silurian and therefore it is not of great importance in the study of early life.

An answer concerning the composition of the early atmosphere can, on the other hand, be given by rocks which formed whilst in contact with that atmosphere. At present this is only possible somewhat categorically, telling us if at a given date we still had an anoxygenic, or already an oxygenic atmosphere. The most important of such rocks are the sediments formed as the result of a cycle of weathering–transportation–sedimentation. Such processes, which take place at the surface of the earth, are called *exogenic* processes, in contrast to the *endogenic* processes taking place within the crust.

This chapter will go into the evidence that can be derived from the contemporary rocks, starting with short notes on the difference between endogenic and exogenic processes and on the weathering of rocks.

2. EXOGENIC AND ENDOGENIC PROCESSES

In geology processes which take place on the surface of the land (that is on top of the so-called *lithosphere*) or in water (in the *hydrosphere*), or in the air (in the *atmosphere*), are grouped together as *exogenic processes*. They only affect the surface and the outer skin of the solid earth. They are, at least in part, directly affected by the atmosphere. Processes occurring within the crust of the earth are called *endogenic processes*. Typical endogenic processes are, for instance, mountain-building, metamorphism of rocks at depth and the formation of granites. Typical exogenic processes are weathering, soil formation, erosion, transportation by wind or by rivers, and sedimentation.

For our topic we are concerned only with rocks formed during exogenic processes in which the material comes into regular contact with the contemporary atmosphere. It is hoped that the composition of rocks formed during exogenic processes will in some way be a witness to the composition of the contemporary atmosphere.

3. WEATHERING OF ROCKS

In the most important of the exogenic processes in regard to our problem, in the sequence weathering–erosion–transportation–sedimentation, weathering is not necessarily confined to the beginning of this sequence. Depending upon the individual steps, such as the speed of erosion and transportation, and the environment of sedimentation, such as either lowlands or oceans, weathering may take place during the later parts of this sequence too. Weathering is closely bound up with the composition of the contemporary atmosphere, because during weathering the rock and its constituent minerals is permanently in contact with the atmosphere, whereas it might be cut off from contact with the atmosphere during a smaller or larger part of the rest of the sequence.

To be able to evaluate the effect of weathering it is necessary to say a few words on the composition of the rocks of the crust of the earth. The most common elements of the crust of the earth are silicium, Si, aluminium, Al, and oxygen, O. The two first elements have given their first letters to the technical term of *sial* for the continental crust of the earth.

Most rock-forming minerals of the earth's crust are compounds of these three elements, with other elements added. Quartz is silicium oxide, SiO_2. Most other minerals are silicates, that is compounds of Si, O and Al, incorporating in various ways the alkaline earth metals, Ca and Mg; the alkaline metals, K and Na; base metals, such as Fe; hydrogen H; the halogens, Cl and Br, and other elements. Silicates containing only alkalis and calcium combined with aluminium and silica form light-coloured minerals, such as the felspars. Dark minerals, like biotite, augite, hornblende and olivine, contain magnesium, iron and other metals in addition to aluminium and silica. Quartz, together with the groups of the felspars and the dark minerals, make up by far the bulk of all crustal rocks. In addition ore minerals occur, mainly in plutonic and metamorphic rocks and in related veins. These contain sulphides, such as pyrite, FeS_2, and metal oxides, such as magnetite, Fe_3O_4. In general the sulphides are low-temperature minerals, whereas the oxides form at a higher temperature.

Returning now to the process of weathering, we find that the rocks exposed at the surface of the earth are attacked in two different ways, by physical and by chemical processes.

Purely physical weathering is rare at present. It is, for instance, found both in extremely cold and in extremely hot and dry climates, that is in the high-arctic tundras and in deserts. In cold climates rocks disintegrate mainly through frost splitting, in deserts by sun blasting. Everywhere else on earth we find a combination of physical and of chemical weathering. In most cases the chemical weathering is much more active than the physical weathering. The influence of the latter is often but slight, and commonly one speaks of either physical or chemical weathering, neglecting the part played by the other process (REICHE, 1950).

4. MINERALS UNSTABLE IN PRESENT WEATHERING

Chemical weathering, under the circumstances of our present oxidizing atmosphere, will attack all minerals with the exception of the oxides. Both the felspars and the dark minerals of normal rocks, as well of the sulphides of ore veins, will be oxidized and form soluble ionic compounds. So the only minerals left are the oxides: the common quartz and the much rarer fully oxidized metal oxides such as hematite. The solutions carrying the ions derived from the decomposition of the felspars, the dark minerals and the ore minerals, are transported by the flow of groundwater and by the rivers into lowlands, lakes and oceans. There, mostly below the watertable, and often to the exclusion of free oxygen, new compounds will form. These newly formed minerals predominantly belong to the group of the clay minerals.

In a simplified version this is the normal sequence of weathering–transportation–recombination in our present oxidative atmosphere. It illustrates the main points important for our study: (*1*) that all minerals except oxides are at present unstable in chemical weathering, and, (*2*), that the ions derived from chemical weathering, upon being transported into areas of sedimentation, will recombine, mostly forming clay minerals.

This is the reason why we now have only three common types of sediment: sand, clay and limestone. The sands are almost exclusively quartz sands, left-over from chemical weathering, although often eroded, transported and re-sedimented several times. The clays are newly formed by recombination of ions derived through chemical weathering, whilst the carbonate material of the limestones is mainly biogenic in origin, derived from the shells of animals and from limestone-secreting plants.

Only under exceptional environmental circumstances will other minerals stand the chance of survival through a cycle of weathering–transportation–sedimentation. Sand may, for instance, locally still contain an appreciable amount of felspar. But this occurs only when detritus from igneous rocks is deposited quite near its source area, and buried so speedily by continuing sedimentation, that further chemical weathering is stopped. Sulphides. which do not only weather more quickly than felspars, but are also readily attacked through biochemical action by sulphur bacteria, nowadays are preserved even more rarely. There are some examples from high up in the tundras, where the fiercely cold arctic climate is prohibitive of chemical erosion, or from the rapidly subsiding Indus valley, where younger sediments so quickly cover older deposits, that the air supply is effectively sealed off, even before the pyrite grains have had time to weather.

These exceptions, by their scarcity, and by the extreme conditions of their environment, only confirm the rule that normally, in our present oxidative atmosphere, all sands are quartz sands; and that the only stable minerals are the oxides.

5. MINERALS STABLE UNDER AN ANOXYGENIC ATMOSPHERE

This will not have been so under a primeval atmosphere of reducing character. Under it, felspars, dark minerals and sulphides could have lain on the surface of the earth for a much longer time before they finally disintegrated. They have been taken up in new cycles of erosion–transportation–sedimentation, for instance, when through slight crustal movements, or a changing of river courses, or a lowering of the sea level, their original sedimentation area became attacked by erosion. They would weather mainly by mechanical attack, through physical weathering, and much less than at present by chemical weathering. In repeated sequences of erosion–transportation–sedimentation, they would become well-rounded and get wellsorted as to hardness, size and specific weight, in a way entirely comparable to the quartz sands of today.

Under such a reducing atmosphere sands of all sorts could be formed. For instance, coarser and finer sands of the light-weight minerals, such as quartz and felspar (not quartz alone). Or sands of medium-weight minerals, such as horn-blende and augite. But also sands of the heavy ore minerals such as sulphides, the not fully oxidized metal oxides, such as magnetite and uraninite. The size, form and specific weight of the individual mineral grains, that is their physical proper-ties, not their chemical properties, would determine their ultimate place of sedimentation.

6. STUDIES BY RANKAMA: DETRITUS OF GRANITES

A first attempt to use this difference in exogenic processes under oxygenic and anoxygenic atmosphere was undertaken already in 1955 by RANKAMA in Finland. He tried to evaluate the character of the atmosphere by studying ancient detrital rocks surrounding a granite from which the material of the sediments was thought to have been derived.

The granite, actually a quartz diorite, contains about four percent ferrous iron and two percent of ferric iron, measured by weight of the oxides. Now, in oxidative weathering the relative amount of ferric iron will always be greater in sediments derived from a granite, because of oxidation of ferrous iron during weathering, transportation, and sedimentation. In the Finland example this is not the case, the ratio Fe_2O_3/FeO in the sediments being comparable to that of the neighbouring quartz diorites, and even lower than that of a diorite pebble inclu-ded in the sediments (Table XVIII). Rankama concluded that this particular quartz diorite weathered under a reducing atmosphere.

This has given us the first indication that a primeval anoxygenic atmosphere has really existed on earth in the Precambrian. As such it was an important break-through in the search for a possible origin of life through natural causes.

Rankama's studies would have supplied us even with a fixed date for the existence of an anoxygenic primeval atmosphere, if a reliable age for the sediments surrounding the quartz diorite could have been established. Unfortunately this is not the case. An age of 1800 m.y., originally given, has turned out to be probably the age of a younger period of metamorphism, so that really the sediments may be much older. Moreover, since the studies of Rankama were published, other authors have doubted whether the metamorphosed sediments surrounding the granite were indeed derived from the erosion of that particular granite. His whole study is

TABLE XVIII

FERROUS AND FERRIC IRON CONTENT (wt.%) IN GRANITES (QUARTZ DIORITES) AND IN DERIVED
SEDIMENTS OF PRECAMBRIAN AGE NEAR TAMPERE (FINLAND)
(From RANKAMA, 1955)

	Quartz diorite pebble	Quartz diorites					
Fe_2O_3	1.98	0.79	0.61	0.75	1.46	0.50	1.09
FeO	3.67	6.70	4.06	6.23	7.99	4.35	5.65
Fe_2O_3/FeO	0.54	0.12	0.15	0.12	0.18	0.11	0.19
		Derived sediments					
Fe_2O_3		0.16	1.73	0.49	0.65	1.43	
FeO		2.94	5.19	5.06	6.31	6.71	
Fe_2O_3/FeO		0.05	0.33	0.10	0.10	0.21	

therefore full of doubtful points. Nevertheless, the underlying idea was sound, and the application of this idea has since led to much more trustworthy results to be dealt with in our next sections.

7. STUDIES BY RAMDOHR: PYRITE SANDS WITH GOLD–URANIUM REEFS

In 1958 RAMDOHR drew attention to the fact that ancient Precambrian deposits from the old shields of South Africa, Brazil and Canada, which are of economic importance for their content of gold and/or uranium, and are consequently well known, could be considered as ancient sediments laid down under an anoxygenic atmosphere.

Historically, the South African deposits have already been mined over a long time for their gold content, hence they have been called gold reefs. In more recent decades their uranium content has been exploited too, which has, of course, added considerably to their economic importance. The quite similar Brazilian and Canadian deposits, however, are so poor in gold, that they have never been called gold reefs, and only recently acquired economic importance as uranium ores. The deposits described by Ramdohr belong to the following districts: Dominion and Witwatersrand Reefs, South Africa; Serra de Jacobina, Bahia, Brazil; and Blind River, Ontario, Canada. There is consequently a wide variation in their geographical and geological setting, whilst their age is also quite different. They have, however, in common that they all belong to the Early and Middle Precambrian, and are not found either in the Late Precambrian or in the Paleozoic.

Quite apart from their economic importance, they have a general significance for geology, and more specially for our own subject in that they are formed by ancient sands and gravels, and represent sediments laid down on the surface of the earth in an exogenic process. Consequently their composition was influenced by that of the ancient contemporary atmosphere. These deposits are formed not only by grains of quartz, but also of sulphides, notably pyrite, and by pitchblende. Pitchblende is a complex mineral with a composition varying between UO_2 and U_3O_8. Its varied composition is, however, due to later oxidation, and originally it consisted entirely of the least oxidized form of uranium, the mineral uraninite, UO_2.

Notwithstanding the separation in geographical location, in geologic setting and in age, the composition of these quartz-pyrite sands, with a variable, but small, amount of gold and uranium, is so strikingly similar, that even experts often cannot tell from hand samples from which of these widely separated districts they have been derived. The mineral composition shows, of course, its variation from layer to layer, whilst also lateral variations in the relative abundance of the constituent minerals may occur within a single horizon. But the variations all fall within sharp limits and the overall picture is remarkably consistent. And—most important—this overall picture, so strikingly similar for all these ancient deposits, so far removed in place and time, contrasts strongly with that of all younger deposits of gravel and sand of the later history of the earth, from the Late Precambrian to the present time.

8. SANDS WITH PYRITE, URANINITE AND ILMENITE

The deposits mentioned in the preceding section are ancient conglomerates and sands, cemented to a very hard rock, which forms the "reefs". Apart from quartz they carry pyrite, FeS_2, ilmenite, $FeTiO_3$ and pitchblende, derived from the primary mineral uraninite, UO_2, in considerable quantities. They show all the features characteristic for deposits originally laid down as superficial gravels and sands in river beds and in lakes. When rich in valuable minerals such sediments are called *placer deposits*. The roundness of the individual grains, the fact that they are so well sorted as to grain size, the differences in grain size and in mineral composition between successive layers cannot be accounted for by any other process of deposition. Moreover RAMDOHR (1958) describes evidence of erosion of earlier beds, and re-deposition of fragments of these older sands and gravels, together with individual grains, in younger beds. There is, accordingly, evidence for repeated cycles of weathering–erosion–transportation–sedimentation, which proves that these beds must have been in repeated contact with the contemporaneous atmosphere.

In these ancient deposits, with grains of such a different mineralogical composition, and hence of different specific weight, as quartz, pyrite and pitch-blende, one finds the physical processes of classifying and sorting individual mineral grains far better expressed than in the younger sands and gravels, which almost only contain grains of quartz. In the ancient deposits in one single horizon the lighter quartz grains are always much larger than the heavier pyrite grains, whereas the much heavier grains of pitchblende are by far the smallest.

Apart from the features mentioned already, these ancient sands and gravels are also very similar to those of younger sedimentary basins in their macro-distribution. As an example we may take the Witwatersrand Basin of South Africa (SCHIDLOWSKI, 1965). As we see from Fig.86, this forms at present a saucer-shaped basin, some 300 km by 100 km, lying on a much older basement. In the middle of the basin later uplift and erosion has again uncovered the old basement in the Vredefort Dome.

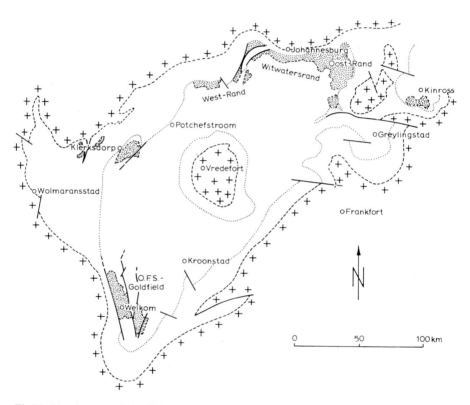

Fig.86. Sketch map of the Witwatersrand Basin in South Africa. Crosses: Older basement, about 3.1 billion years old, surrounding the basin and also cropping out in the Vredefort Dome. White: Witwatersrand System. The dotted line gives the boundary between the lower and the upper parts of the system. All younger rocks are thought to have been removed. Stippled: "Gold fields"-areas where gold content of the gravels is high enough for mining.

The maximum thickness of the Witwatersrand Series of sediments is about 8 km, but we must not think of an original basin 8 km deep, which was gradually filled in. Instead, the characters of the deposits indicate continental and/or shore conditions, deltas and such, which have persisted all through the deposition of the Witwatersrand System. What we must visualize is a slowly sinking crustal basin, in which sinking was offset by sedimentation during all the time of the formation of these sediments. The earth surface was always near sea level, no matter how deep the basement, the original crust had sunk. Only after sedimentation had stopped, and presumably at a much later date, discontinuous with the deposition of the Witwatersrand System, did further crustal sinking produce the present saucer-shaped "basin", in which the Witwatersrand Series is covered by much younger rocks over a greater part of its area.

During the sedimentation of this pile of clastic sediments variations in the rate of crustal sinking and in that of sedimentation have led to so-called *sedimentary cycles* such as shown in Fig.87 and 88, which are entirely comparable to those found in the younger sedimentary basins of the world.

It follows that, as to their mode of deposition, of re-working and of re-deposition, there is a striking similarity between these ancient gravels and sands and those formed in later times. There is but one difference, but that a paramount one, i.e. all newer sands and gravels are formed predominantly or even exclusively by quartz, whereas in these ancient deposits we find large amounts of other minerals, such as sulphides and uraninite, which are not stable under the present oxygenic atmosphere.

To compare the structure of ancient and modern sands, Fig.89 and 90 are given. The first picture is from a section of an ancient sand of the Witwatersrand System, over two billion years old (cf. Fig.80), the second from a recent black sand from the coast north of Buenos Aires. The ancient sand consists mainly of pyrite, the recent sand of magnetite, which have comparable specific weights. The similarity in structure is evident.

Fig.91 is again from an ancient sand of the Witwatersrand System, which is formed mainly by quartz and pyrite. In this picture too, we see that the amount of sulphide grains is large compared to that of the quartz. The pyrite grains do not form a negligible minority. Instead they form one of the two main constituents. Moreover, in the upper half of the picture, there is a re-deposited lump of an older sandstone, consisting almost entirely of pyrite. The rounded borders of the individual pyrite grains cemented together to form the older pyrite sandstone, can still be seen faintly.

Such re-cycling of earlier sediments is not just an exception, as follows, for instance from the detailed sedimentological investigation of the Kimberley Reefs in the East Rand Goldfield by ARMSTRONG (1968). This author even concluded that the bulk of the material of the Upper Kimberley 9 A reefs was derived through

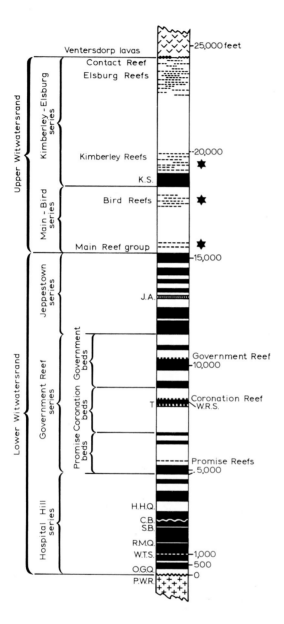

Fig.87. Stratigraphic column of the Witwatersrand System near Johannesburg, South Africa, showing the main sedimentary cycles. Black: shale. White: sand. Stippled: conglomerates. The Dominion Reef, forming the base of the system, is not developed in this area. Thicknesses in feet. (From Schidlowski, 1965.)

erosion from the older Upper Kimberley 9 B reefs, which in turn are thought to be derived from various horizons of the Middle Kimberley Series.

Fig.92 shows a sedimentary rhythm, in all aspects comparable to rhythms in sedimentation found in younger sands and gravels. On top of a bed consisting

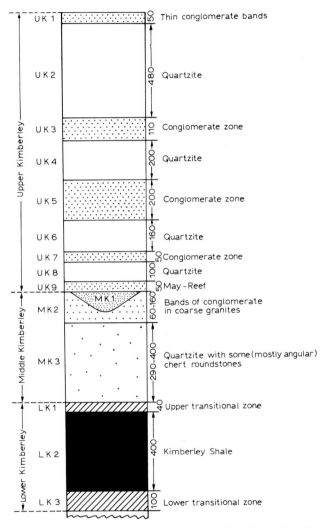

Fig.88. More detailed stratigraphic column of the Kimberley Cycle, a part of the Witwatersrand System, South Africa. Black and hatched: shales, former clays. White: quartzite, former sand. Stippled: conglomerate, former gravel. At MK 1 an erosional channel is indicated.

Variations in grain size in the sediments may have been influenced mainly by the rate of crustal movements. When movements were slow, erosion was at its minimum and fine-grained shales were deposited. The strongest crustal movements, resulting in a rapidly rising hinterland attacked by strong erosion, would then have led to the deposition of the conglomerates. (From SCHIDLOWSKI, 1965, based on a sketch by SHARPE, 1949.)

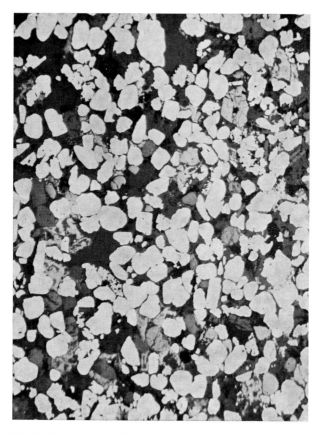

Fig.89. Microphotograph of ancient sand, over 2 billion years old, from the Witwatersrand System, South Africa; ×70. White: grains of pyrite. Grey: grains of various oxides. Dark: grains of quartz. (From RAMDOHR, 1958.)

almost entirely of quartz grains lies a thin horizon composed of grains of pyrite and pitchblende, which is again covered by an almost pure quartz sand. The grain boundaries of the quartz are not seen in the reproduction, but in the original microphotograph they stand out well and indicate that the quartz grains are at least twice as large as those of the pyrite. The pitchblende grains, on the other hand, are not only the smallest, but they moreover show the strictest classification in size.

The uraninite grains themselves, when seen under a larger magnification, also show the typical roundness of clastic grains transported by water, which have, in the process lost all the edges the original crystals possessed (Fig.93).

The conclusion drawn by RAMDOHR (1958), and accepted by most geologists, is that these deposits are sediments, deposited under an anoxygenic primeval atmosphere, which did not allow the chemical weathering of sulphides, uraninite and similar minerals, such as would take place under an oxygenic atmosphere.

Not every geologist agrees with this conclusion. The most outspoken critic has been the late Professor DAVIDSON of St. Andrews (1963, 1965). He did, however, himself concisely summarize the arguments in favour of Ramdohr's hypothesis as follows: "I can only say that there is no known geological environment from which detrital gold, uraninite and pyrite devoid of other heavy minerals can be derived as alluvial deposits. If (one) can suggest how such deposits could form syngenetically in the Dominion Reef, in the south-easth of Finland, Blind River and so forth—if he can suggest any geological environments from which such associations can be derived syngenetically,—then he will have made a great step forward." (DAVIDSON, 1965, p.1205.) This is exactly what Ramdohr has done, in proposing a sedimentary origin under anoxygenic atmospheric conditions for these gold–uranium reefs.

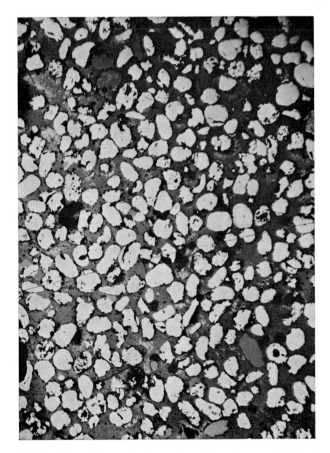

Fig.90. Microphotograph of recent sand from the coast of the Argentine; ×45. White: grains of magnetite. Grey: grains of quartz. Compare with structure of ancient pyrite sand of Fig.89. (From RAMDOHR, 1958.)

It moreover again stresses the point I have made already and will go on making, that the difference between the primeval and the present-day atmosphere is fundamental. It is a difference of black and white. It is so all-embracing that almost everybody, reared as he is in the present oxygenic atmosphere, has difficulties in visualizing all aspects of the primeval anoxygenic atmosphere.

Since the question whether the deposits of the Witwatersrand System are ancient sediments, or whether their gold and uranium content has been brought by hypogene, i.e. endogenic solutions, is so important for further exploration of the ore, ever more refined methods have been applied, which have all resulted in favour of the primary sedimentary origin of these deposits. As an example, two approaches will be cited. JENSEN and DECHOW (1964) and DECHOW and JENSEN

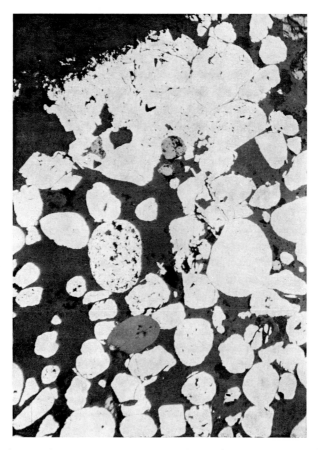

Fig.91. Microphotograph of Precambrian pyrite–quartz sand, Witwatersrand System, South Africa; ×45. White: grains of pyrite. Dark: grains of quartz. In the upper half of the picture a lump of re-deposited older pyrite sand. Rounded borders of original pyrite grains within the older sandstone are still visible. (From RAMDOHR, 1958.)

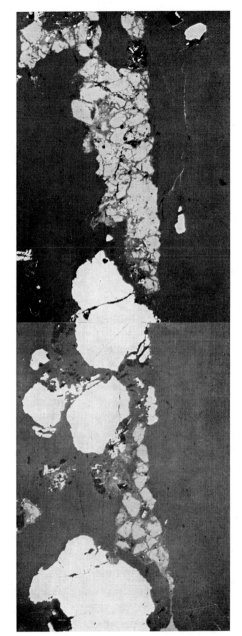

Fig.92. Microphotograph of a quartz–pyrite/pitchblende–quartz sedimentary rhythm in Precambrian sand. Blind River, Canada; ×22.5. Black: quartz, individual grain boundaries not visible in reproduction. White: pyrite. Grey: small grains of pitchblende. (From RAMDOHR, 1958.)

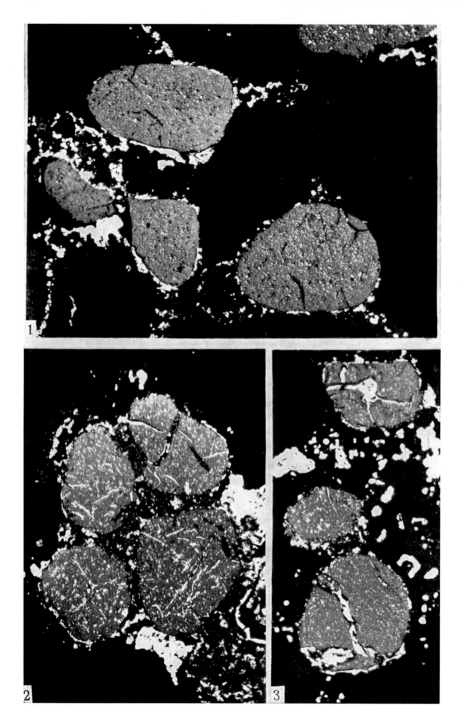

(1965) studied the relation of the stable isotopes of sulphur. In igneous rocks, and in sulphur deposits derived from hypogene solutions, the ratio of the stable sulphur isotope is rather constant within a single body of igneous rocks, a fact probably due to a sort of fractionating process in the magmatic stage. In sediments the ratio of the sulphur isotopes, on the other hand, varies much more widely, because sediments derive their material from many and varied sources. The sulphur isotope ratios of the pyrites of the Witwatersrand System were found to fall in the group of sediments. The scatter is far too wide for igneous rocks.

The other study was by RAMDOHR et al. (1965), who studied the approximate age of individual grains of uraninite by measuring the relative amount of uraninite and galena (i.e. lead, see Fig.93), both optically and by electron microprobe scanning. The "ages" of the individual grains of uraninite were found to scatter widely, such as could be expected in sediments with a provenance from various different older igneous rocks. Moreover, the ages of the individual grains of uraninite are normally higher than that of the formation of the Witwatersrand System, which precludes the possibility of their formation through younger hypogene solutions.

Although there are, of course, many more specialistic points in this whole question, I think that for our topic we may rest convinced that the pyrite–uraninite deposits of the Early and Middle Precambrian are sediments formed under an anoxygenic atmosphere.

9. LATER HISTORY OF PRECAMBRIAN GOLD–URANIUM REEFS

The conclusion reached in the last section expressly relates to the origin of these deposits. During the couple of billion years of their later history changes have, however, often occurred which make the deciphering of their original genesis difficult.

One of them is the mobility of gold. Gold has almost certainly moved in younger times through the sedimentary pile. Moreover, gold might, at least in part,

---

Fig.93. Grains of uraninite with rounded outline, typical of clastic material transported by rivers or along shores of lakes and oceans. Basal Reef, Witwatersrand System, Orange Free State; × 315. Black: groundmass of quartz. Gray: uraninite. White: galena, PbS.

This lead is radiogenic, formed through the decay of the uranium atoms in the uraninite, which has combined with sulphur and was deposited in small holes and cracks in the parent mineral. Volumetric estimates of the relative amounts of uraninite and of galena give an approximate value, either for the absolute age of the uraninite grain, or for the latest period of metamorphism, it has gone through coupled with a loss of Pb (RAMDOHR et al., 1965). Normally this age lies considerably higher than the 2.15 billion years estimated for the time of deposition of the Witwatersrand System. Evidently the grains of uraninite carried at least part of the earlier radiogenic lead with them, all through the cycle of weathering–erosion–transportation–sedimentation which led to the formation of the sediments of the Witwatersrand System. (From SCHIDLOWSKI, 1966a.)

be derived from younger hypogene solutions. And although a later enrichment in gold of a particular tract of a particular gold reef has no consequences at all for our topic, as long as the original structure of the sediments remains visible, it may be a question of fame or failure for the mining geologist concerned. The history of gold, moreover, is difficult to assess, because the old adage "gold leaves no traces" holds true in ore mineralogy too. Nevertheless, ARMSTRONG (1968) in his sediment-ological study of the Upper Kimberley reefs in the East Rand gold field, cited already in the preceding section, stresses the sedimentological control of the distribution of the gold.

A better picture of the various changes brought about during the later history of these deposits can be drawn from the two minerals pyrite and uraninite.

In the case of pyrite the original rounded grains will serve as crystallization

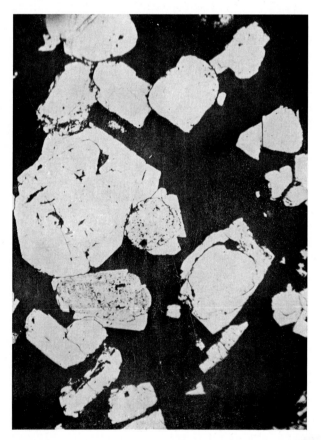

Fig.94. Microphotographs of grains of pyrite and quartz in Precambrian sands of the Witwaters-rand System, South Africa; ×70. The rounded original grain shows later growths of pyrite, formed after the deposition of the sands, which follow crystal planes, and tend to obliterate the original roundness of the clastic pyrite grains.

centers for younger solutions of iron and of sulphur, which may percolate the rocks. The new crystallization will, of course, follow the rules of crystallography, and the new growths will be bounded by crystal planes, which will appear in the sections as straight lines and sharp corners. These may ultimately obliterate the rounded form of the original clastic grains, and thus conceal the original sedimentary character of these deposits (Fig.94, 95; see also SAAGER, 1970).

Fig.95. Microphotograph of a rounded pyrite grain surrounded by a younger deposit which shows the crystal form of pyrite. Witwatersrand, South Africa; × 180. (From HOEFS et al., 1968.)

Uranium, on the other hand, is much less easily transported in solutions, the result being that the uraninite will not show crystal growth such as shown by the pyrite, but will combine with titanium present in the rock to form the mineral brannerite UTi_2O_6. In this reaction the original homogeneous structure of the uraninite disintegrates into a composite mass of very small fibres, but the original rounded outline of the primary grains is normally preserved (Fig.96), and the evidence for the original sedimentary character of the individual grains is not destroyed.

These examples show that these deposits have quite a complex history since the time of their formation a couple of billion years or more ago, and make it understandable why some ore mineralogists of different lineages are still contesting the details of the formation and later alteration of these deposits.

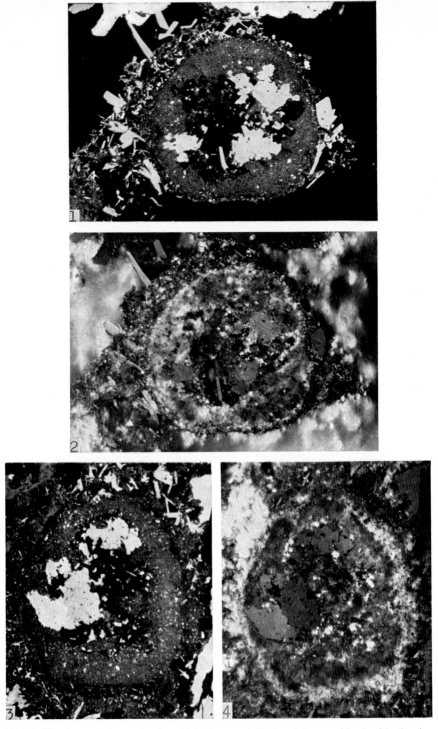

Fig.96. Clastic uraninite grains in which the uraninite has later combined with titanium to brannerite, UTi_2O_6. The rounded outline of the original uraninite grains has been preserved, although their inner structure has been completely altered. These "uraninite ghosts" are called *Uranpecherzgeister* in German; $\times 315$. (From SCHIDLOWSKI, 1966b.)

10. AGE OF GOLD–URANIUM REEFS FORMED UNDER THE ANOXYGENIC ATMOSPHERE

We have seen in Chapter 3, why in general, it is not possible to arrive directly at the age of sediments through absolute dating. Consequently, the ages given in the literature, and also in Table XIX, are still to be regarded as tentative.

TABLE XIX

TENTATIVE AGES OF GOLD–URANIUM REEFS FORMED UNDER AN ANOXYGENIC ATMOSPHERE

Gold–uranium reefs	Age (m.y.)	Reference
Blind River, Ontario, Canada	1,800	DERRY (1960)
Witwatersrand, South Africa	2,000–3,000	CLIFFORD (1968)
Dominion Reef, South Africa	3,000	HALES (1961)

But we also saw how a sediment can be used, in exceptional cases, in absolute dating. This is possible when a mineral containing a radioactive element has been formed during sedimentation. Such minerals are called *authigenic*. In the Blind River uranium ores of Ontario, for instance, DERRY (1960) cites the difference in ages of the various minerals in a single rock sample. Instead of analysing the uranium-lead content for the whole sample—the "whole-rock" analysis—, the various mineral grains from a sample are separated first, and then their ages are determined individually.

The sands and gravels contain grains of the minerals monazite and zircon, which are certainly detrital. Their "absolute ages" vary around 2.5 billion years, but this age has nothing to do with the age of the sediments. It is the age of crystallization of older granites, that is the time when these granites intruded the crust and solidified.

On the other hand, in the Blind River deposits several secondary uranium minerals occur, comparable to the brannerite described from Witwatersrand. It is postulated that at least part of these uranium minerals have been formed during the time of sedimentation, so these authigenic minerals may indicate the age of the sediments. However, there has also been certain later recrystallization, which might have led to the loss of lead or of uranium, or to both. The calculated age of various uranium mineral grains is found to vary from 1680 m.y. to 2000 m.y. DERRY (1960) supposes that uranium loss has occurred in some minerals, leading to a too high apparent age, and lead loss in others, leading to a too low apparent age. His tentative age for the Blind River deposits is 1.8 billion years.

From their geological position in the Lower Huronian, these deposits might, however, well be much older, with a maximum age of about 2.4 billion years.

(See Table XX where the Blind River is tentatively correlated with the Pokegama of the base of the Middle Precambrian.) The granites from which the grains of mona-zite and zircon have been derived, 2.5 billion years old, would in this case have intruded during the Algoman orogeny. This separates the Lower and the Middle Precambrian, and a generally accepted tentative age for this orogenetic period is 2.5 billion years. The sands and gravels of the Blind River deposits, which form the basal post-orogenetic series of the Middle Precambrian, are derived from the subsequent erosion of this fold belt, and the age of these sediments does not have to be much younger than the orogeny itself.

The example of the Blind River shows why the ages assigned to these ancient deposits are still tentative. Similar uncertainties are found in relation to the age of other comparable deposits, elsewhere in the old shields.

Many more and much more refined age determinations will have to be made, before we can assign somewhat more precise ages to these deposits. But at present it seems already well established that sediments formed under anoxygenic conditions up to about 2 billion years ago. And if one bears in mind the tentative character of these dates, one is even justified to use the 1.8 billion year figure.

11. PRECAMBRIAN IRON FORMATIONS

Another type of ancient sediments has in later years also become of impor-tance as a possible indicator for an anoxygenic atmosphere, i.e. the Precambrian iron formations. Just as the Precambrian gold–uranium reefs, the iron formations are of great economic importance, in this case as iron ore.

Iron formations are superficially formed deposits of iron, always associated with sediments. There are other concentrations of iron, rich enough to have served or to serve as iron ore, which occur in veins. In this case the iron is deposited by solutions originating from deep-seated sources, the so-called hypogene solutions. In fact, most of the economically important iron ores of early human history and of mediaeval times are found in veins. Their iron content may reach quite high percentages, which, in the early technologies, was the prime consideration. In contrast, the iron percentage of the sedimentary iron formations is normally lower, averaging below 30%

At present a still active district of iron mining based on veins is found in the Siegerland in Western Germany. Elsewhere, most iron mining has turned to the sedimentary iron formations. Their disadvantage of a low grade has become offset by the advantage of great volume. But the difference in grade of the hypogene iron ores from veins and the supergene iron ore found in the sedimentary iron formations still persists in the names of the latter. Those of Lorraine, in Europe, have of old been called "minette", that is rocks not good enough for a mine. And the North American Precambrian iron formations have long been considered as a proto-ore only. In the case of the minette the introduction of the blast furnace has made possible their exploitation, whilst the North-American deposits had to await the development of modern concentration techniques.[1]

Precambrian iron formations are typically very thin-bedded, iron richer beds alternating with beds either poor in iron, or containing no iron at all. In contrast to the younger minettes, this alternation is commonly so rapid in the Precambrian iron formations, the individual layers are so thin, that they are called laminations (Fig.97).

Fig.97. Polished slab of a Precambrian banded iron formation from Ishpeming, northern Michigan.
The extremely fine lamination barely visible in the thin shaly beds, is possibly the result of cyclic yearly deposition, as described in the next section. The thicker bed in the middle is a sandy intercalation. The irregular whitish streaks are formed by replacements dating from the later history of this rock. (Courtesy Mr. Tsu-Ming Han, Cleveland-Cliffs Iron Company.)

This is, however, not the only difference between all younger iron formations and those of the Precambrian. A much more fundamental difference lies in the fact that the former are always associated with carbonate, mainly with $CaCO_3$, whereas the latter are always associated with chert, that is with very fine-grained

---

[1] We are, of course, not concerned here with the iron ores of higher concentration, of up to 60% Fe, which up to a short time ago formed the bulk of the iron ore actually mined. These have been derived from the original sedimentary banded iron formations through oxidation and concentration processes during their later history.

siliceous beds. Siliceous beds occur in the younger iron formations too, but always as a secondary phenomenon, in which silica has at a later date replaced the original carbonate.

All younger, predominantly calcareous, iron formations are nowadays called by their old European name and designated as the "Minette type". Whereas the Precambrian, siliceous representatives are called the "Lake Superior type," regardless of their geographical situation. The difference in composition between these two types is evident from Fig.98 and 99.

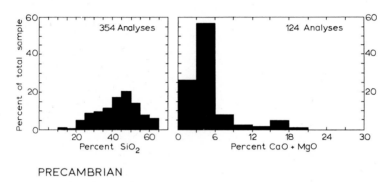

PRECAMBRIAN

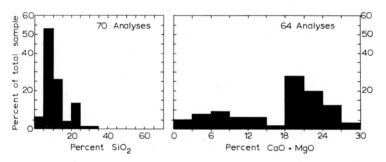

Fig.98. Distribution of SiO_2 and $CaO + MgO$ in Precambrian (upper) and in younger iron formations (lower). (From LEPP and GOLDICH, 1964.)

The iron of the Precambrian iron formations occurs largely in the form of the primary minerals siderite, $FeCO_3$; pyrite, FeS; magnetite, Fe_3O_4; and as iron silicates. There is no direct evidence how much of the hematite, Fe_2O_3, which represents the most oxidized form of iron, and which nowadays occurs widely in these iron formations, is primary. There is a general feeling, however, that most of this hematite is secondary, and has been derived from primary magnetite through later oxidation. On the other hand, it is also conceded that a minor part of the hematite might be of primary origin.

Although we thus find a similar situation as encountered in our study of

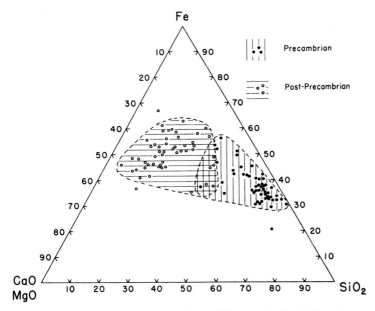

Fig.99. Variations in the content of FeO, SiO₂ and CaO + MgO in Precambrian and in later iron formations.

It is evident that, apart from a slight overlap the composition of the Precambrian iron formations is quite different from that of the younger ones.

There are minor, but quite as consistent, differences in the amount of other elements too. Alumina, phosphorus and titanium are much lower; manganese, on the other hand, is higher in the Precambrian iron formations. Alumina averages only 1.6% in the older, as against 6.1% in the younger formations; phosphorus 0.26% versus 0.86%; titanium 0.15% versus 0.45%, whereas manganese shows an average of 1.0% for the older, as against 0.34% for the younger ormations. (From LEPP and GOLDICH, 1964.)

the Precambrian gold–uranium reefs, the distinction between the iron formations of the Precambrian and those of younger times is not as clear-cut as in the case of the gold-uranium reefs. There we found, in the ancient sands, reduced iron in the form of pyrite as against oxidized iron, such as hematite, in the more recent sands.

In the iron formations, on the other hand, the iron is oxidized both in the ancient and the more recent beds. True, it occurs in distinctly less oxidized forms, such as magnetite, in the Precambrian deposits, as against its most oxidized form, hematite, in the younger beds. The clear antithesis between Precambrian and younger iron formations rests, however, not in the iron minerals, but in the association of iron with silica in the older, versus iron and carbonate in the younger beds.

So there is a distinction in the difference between Precambrian and younger sediments of various sorts. The reason for this cannot be found in a difference of age. As Table XX shows, both gold–uranium reefs and siliceous iron formations have been formed during both the Lower and the Middle Precambrian. We will return to this distinction in the next section, when discussing the genesis of the iron formations.

Turning now to the Precambrian iron formations for a closer look, we may note that most of our up to date information comes from North America. Here the iron formations of the Canadian Shield have been intensively studied, both in

TABLE XX

STRATIGRAPHIC SUCCESSION AND CHRONOLOGY OF THE PRECAMBRIAN OF MINNESOTA[1]
(From GOLDICH, 1968)

Era	Group	Formation	Event
Paleozoic			
———————— — — — — — — — — *Unconformity* — — — — — — — —			
600 m.y.		Hinckley Sandstone	
		— — — — — — — — — — — — — — —	
		Fond du Lac Sandstone	
		— — — — — — — — *Unconformity* — — — — — — — —	
Late	North Shore	undivided flows, tuffs, and sediments	Keweenawan igneous
	Volcanic Group		activity
Precambrian —			(1000–1200 m.y.)
		Puckwunge Conglomerate (?)	
		Unconformity — — — — — — — —	
		Sioux Quartzite (?)	
———————— — — — — — — — — *Unconformity* — — — — — — — — — —			
1800 m.y.		Rabbit Lake = Virginia = Rove =	Penokean orogeny
		Thomson	(1600–1900 m.y.)
Middle		— — — — — — — — — — — — — —	
	Animikie Group	Trommald = Biwabik = Gunflint	
Precambrian		— — — — — — — — — — — — — —	
		Mahnomen = Pokegama = Kakabeka	
———————— — — — — — — — — *Unconformity* — — — — — — — — — —			Algoman orogeny
2600 m.y.		undivided, slate, graywacke, con-	(2400–2750 m.y.)
	Knife Lake Group	glomerate, tuffs, lavas	
Early	— — — — — — — — *Unconformity* — — — — — — — — — —		Laurentian orogeny
		Soudan Iron-Formation	(age?)
	Keewatin Group	— — — — — — — — — — — — — —	
Precambrian		Ely Greenstone	
	— —		
	Coutchiching (?)	Metasedimentary rocks (?)	
	— —		
		Older rocks	(?)
			(3300–3550 m.y.)

[1] The three major successions of iron formations in Minnesota are from older to younger, the Soudan-, the Agawa- (in the middle of the Knife Lake Group) and the Biwabik (or Gunflint) iron formations. The Blind River gold–uranium reefs of Ontario, found in what is locally called the Lower Huronian, are the approximate equivalent in age of the Pokegama Quartzite of Minnesota. The Keweenawan orogeny, a local name for Minnesota, can be correlated with the Grenville orogeny of the Canadians (see Fig.127).

Canada and in the northern states of the U.S.A. Similar iron formations occur, however, in the Precambrian of the other old shields. And local names such as *taconite* and *itabirite* have gained world-wide fame in the geological literature. Just as we have found it to be the case with the Precambrian gold–uranium reefs, there is amazing overall similarity, one could even say congruence, between the Precambrian iron formations of the various old shields. Moreover, there is a world of difference between any Precambrian and any younger iron formation. There is, consequently, no objection in taking the best studied example of the Precambrian iron formations, those of the Canadian Shield, as typical for the whole group. An idea of the extent of the iron formations of the Canadian Shield can be gained from Fig.100. Their stratigraphic position follows from Table XX.

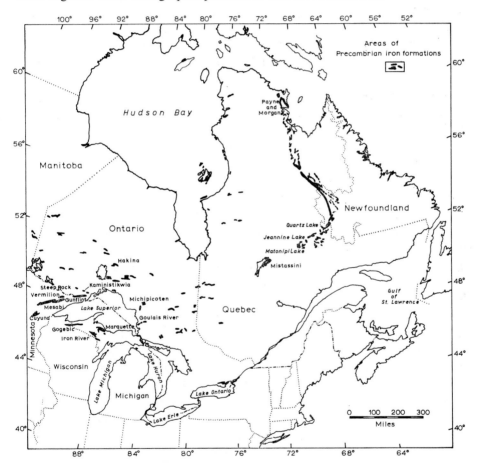

Fig.100. Outline map of the Canadian Shield indicating, in black, the occurrences of Precambrian iron formations of the Lake Superior type. The iron formations shown mainly belong to three distinct stratigraphic horizons of the Lower- and Middle Precambrian (see Table XX). (From LEPP and GOLDICH, 1964.)

12. PYRITE SANDS AND PRECAMBRIAN IRON FORMATIONS

Returning now to the apparent contradiction in the degree of oxidation of sediments laid down at about the same time and thus under the same atmosphere, we must look more closely both into the difference between the anoxygenix and the oxygenic atmospheres and into the difference between the sedimentary histories of pyrite sands and iron formations.

Regarding the difference between the two atmospheres, we must now further qualify the term *anoxygenic*. This does not necessarily mean "absolutely without any free oxygen", but only "with insufficient oxygen, to sustain modern life and weathering". Similar definitions must be applied to all merely quantitative antitheses. It would, for instance, seem to be fairly easy to distinguish between light and dark, But, as everyone knowns, one sees more and more, even in a really dark night, the longer one's eyes get accustomed to it. A sensible definition is, when one calls dark an environment in which the energy of the light striking a square centimeter falls below a certain value. Although there is still "light" below that value this is neglected and the environment is called "dark". Such a definition is, as we will see in Chapter 15, commonly applied in the study of the extinction of sunlight.

It is difficult to estimate what the level of free oxygen in the anoxygenic atmosphere has been. We will return to this topic in Chapters 15 and 16, where the anoxygenic atmosphere will be provisionally defined as an atmosphere containing not more than one percent of the free oxygen contained in the present atmosphere. Purists might prefer to use the term "oligo-oxygenic" for such an atmosphere (JONG, 1966), but I think this to be unnecessarily complicated. Usage of anoxygenic versus oxygenic is in accordance, not only with the example cited of black versus light, but also with fresh versus salt water, and many other similar antitheses.

Consequently, having acquired a sharper understanding of the difference between the anoxygenic and the oxygenic atmospheres, we realize that even in the earliest atmosphere some oxygen has always been present. This knowledge can now be applied to the difference in sedimentary histories between the pyrite sands and the iron formations. We must at this point recall the properties of the orogenetic cycle, as discussed in Chapter 9, and realize that the pyrite sands were deposited during the post-orogenetic periods; the iron formations during the geosynclinal periods, of successive orogenetic cycles during the Early and Middle Precambrian. During a post-orogenetic period vertical crustal movements, adjusting the disequilibrium resulting from the preceding orogeny, are still relatively strong; erosion, transportation and sedimentation relatively quick. Coarse-grained sediments, such as sands and gravels, will be formed predominantly. During a geosynclinal period, on the other hand, crustal movements are weak, especially outside the geosynclinal belts. Weathering, erosion, transportation and sedimentation will all be slowed down, and fine-grained sediments, such as shales and sediments deriving from chemical precipitation, are formed predominantly.

On an earlier page in this chapter we have seen that the weathering of minerals and rocks must have been quite different under the oxygenic and the anoxygenic atmospheres. Whereas under the oxygenic atmosphere chemical weathering predominates, physical weathering was prevalent under the anoxygenic atmosphere. We have also met with a clear-cut example of the latter process in the formation of the pyrite sands of the Early and Middle Precambrian. This distinction holds good, however, only under more or less normal environmental circumstances. Extreme conditions may override it. For the present oxygenic atmosphere we have, for instance, already seen how in the extreme climatic conditions of arctic regions and of deserts physical weathering may even now predominate, because it either is too cold, or because there is not enough moisture for chemical reactions to proceed. A reversed situation will occur under an anoxygenic atmosphere if the crustal movements are so small that the continents are base-levelled completely, and erosion comes to a standstill. In such cases chemical weathering may predominate over mechanical weathering, even under an anoxygenic atmosphere. The chemical weathering will mainly lead to leaching, leaving behind the insoluble residue which cannot be transported owing to the lack of topographical relief and the consequent lack of superficial running water. The transported materials will be mainly in a dissolved, and not in a granular, state and sedimentation will take place by precipitation, and not by mechanical deposition.

It follows that during the geosynclinal period of each successive orogenetic cycle, when the continents have become base-levelled, the sequence weathering–transportation–sedimentation will not only take much longer than during the post-orogenetic periods, but that the material also comes into longer and closer contact with the contemporary atmosphere. In this way partly oxidized oxides, such as magnetite, may form even under the "anoxygenic" atmosphere, which are never completely devoid of free oxygen. Whereas, because of some as yet not properly understood chemical mechanism, iron silicates precipitated, instead of the iron carbonates formed under and oxygenic atmosphere.

The effects of the processes active during the sequence weathering–erosion–transportation–sedimentation thus are due, not only to the amount of free oxygen in the atmosphere, but also to the speed and the intensity of these processes. If their rate is normal, fully oxidized materials will form under the oxygenic, as against non-oxidized materials under the anoxygenic atmosphere. If the rate is high, non-oxidized materials may be deposited even under the oxygenic atmosphere. Whereas, when this sequence proceeds slowly, some oxidation may take place even under the anoxygenic atmosphere. The nature of sediments laid down under the anoxygenic and the oxygenic atmospheres consequently depends upon the balance between the rate of oxidation at any level of free atmospheric oxygen and the rate of the exogenic processes which led to the formation of these sediments.

We have at present no ways of estimating definite values for this balance. For one thing, we do not know the absolute rate of the exogenic processes in a post-orogenetic period or in a geosynclinal period. We have only some idea that the difference has been large, but no data on which to base estimates of exact values. Neither do we have any idea about the absolute values of the rates of oxidation at different levels of atmospheric oxygen.

As regards the latter point, it must be noted that geochemistry is almost exclusively interested in the final equilibria reached by mineral reactions, not in their kinetics. So the normal geochemical interpretations cannot be applied to our problem, which depends on the balance between the rate of several non-connected processes.

We will return to this question in the next chapter, but may already at this stage of our investigation draw some conclusions about the behaviour of free oxygen. The potential for oxidation by free oxygen—not the rate of this process—is so strong that when the equations of thermodynamics are extrapolated towards an atmosphere really devoid of oxygen, the non-sensical answer is found that already a single atom of oxygen in the entire atmosphere of the earth would be instable. This is also indicated by the equilibrium diagram of the system iron–water–oxygen of Fig.101. When we only look at the equilibria, we find that the shift from the present atmosphere to one containing only one percent of its amount of oxygen would hardly alter anything in the diagram. Still, we have seen that the properties of sediments laid down under the primeval atmosphere are quite distinct from those formed at present. It follows that arguments in favour of an anoxygenic atmosphere cannot be based on the equilibria of mineral reactions, which in the case of free oxygen evidently cannot be reached at all, but on their kinetics.

13. GENESIS OF PRECAMBRIAN IRON FORMATIONS

Among the authors who have studied the iron formations of the Lake Superior type there are, as always, differences in detail as to how they think these deposits were formed. There is, however, general consensus about their contemporary environment. They must have originated from continents which were in a morphologically mature state, characterized by extensive lowlands and without any marked mountain ranges. The materials from which the iron formations were formed were derived by leaching from these lowlands and precipitated either in large lakes or in shallow seas.

This type of weathering by leaching from flat, extensive low lands is compa-

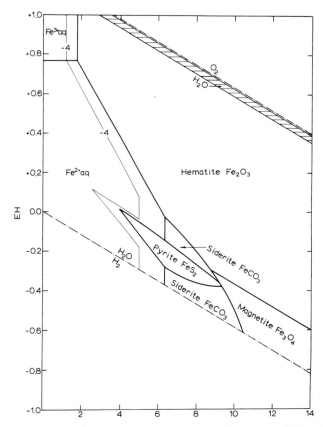

Fig.101. Stability fields of iron compounds in water at 25° and 1 atm total pressure, when $\Sigma S = 10^{-6}$. The barred area in the upper part indicates the shift of the boundary between the fields of O_2 and H_2O between 1 PAL O_2 (upper line), and 0.01 PAL (lower line). It does hardly affect the stability field of hematite. (From GARRELS, 1960. Courtesy Dr. W. C. Kelly.)

rable to the weathering which at present forms our lateritic soils. In recent lateritic soils a so-called cementation zone is found below the zone of leaching, in which, depending upon the topography and other environmental factors, the elements leached from above may become concentrated. In this way iron-lateritic and aluminium-lateritic (bauxite) deposits may be formed which are of economical value, and are often called laterites—or laterites sensu stricto. In our case, we are, however, concerned with the active lateritic weathering, the leaching, (see, for instance, GERASSIMOV, 1968) of the upper layers. Upon the further erosion of the continents this will proceed downwards and in its turn attack the underlying cementation zones, which accordingly are of a transitory nature only.

Lateritic weathering develops in tropical and subtropical climates, preferably with monsoonal variations. Apart from rainfall, humic acid derived from the

cover of vegetation is at present an important factor in the leaching which leads
to lateritic soils. In this context it is important to note that LEPP and GOLDICH
(1964) report frequent intercalations of graphite in the iron formations. As we will
see in Chapters 15 and 16, it is improbable that any vegetation, even of the most
primitive kind, covered the continents under the anoxygenic atmosphere during
the Early and Middle Precambrian, because the lethal rays of the shorter ultraviolet
sunlight penetrated that atmosphere. On the other hand, we have seen that these
same rays are able to synthesize "organic" molecules inorganically. For the
weathering process it is not important whether the acids are really organic, or
only "organic", so the presence of graphite in the iron formations is indeed an
indication of a certain similarity between the Precambrian weathering on these flat
extensive low lands and the present lateritic type of weathering.

LEPP and GOLDICH (1964) developed a model which explains how solutions
containing iron, silica, calcium and sodium were transported to the sea, whereas
other elements, notably alumina and phosphorus, remained as insoluble residue in
the source areas. Under an anoxygenic atmosphere silica and iron will then
precipitate in about the same regions of sedimentation whereas calcium, magnesium
and sodium are carried to the open sea (Fig.102). Under an oxygenic atmosphere,
on the other hand, iron and calcium will fall out in the same general area, whereas
silica and sodium will be transported to the open sea.

The simple model of LEPP and GOLDICH (1964) does not, however, explain
the lamination, the ultra-thin alternation of thin bands of iron minerals and of
chert, so typical for many of the Precambrian iron formations. HOUGH (1958) has

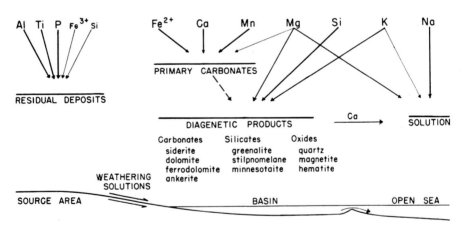

Fig.102. Schematic model for the lateritic weathering in a low lying continental area and deposition of iron formations in a shallow sea under an anoxygenic atmosphere.
Thick arrows indicate stable residues and deposits. Thin arrows indicate transportation.
Al, Ti and P will be left behind as residual deposits in the source area. A number of intermediate
products will form in the sedimentary basin, but ultimately Ca, Mg, K and Na will be transported
to the open sea, leaving Fe and Si to be deposited in the basin. (From LEPP and GOLDICH, 1964.)

pointed out that this might be the result of deposition, not in shallow seas or in lagoons, but in large, relatively deep lakes. Such closed waterbodies show a yearly cycle in the mixing of their waters (Fig103). Consequently the lamination shown by the iron formations could well be the result of a yearly rhythm in deposition connected with this mixing cycle of lake waters.

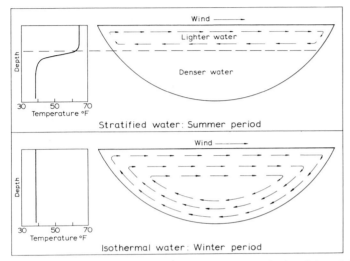

Fig.103. Schematic representation of water circulation in deep lakes of the so-called *monomictic* type, occurring in sub-tropic and warm zones. (From HOUGH, 1958.)

Lakes show either a yearly cycle, and such lakes are called *monomictic*, or a mixing every half-year, the so-called *dimictic* lakes. Dimictic lakes at present occur only in the temperate and more northern zones, where the influence of the seasons and of the freezing over in winter is of great importance. Monomictic lakes occur in the sub-tropic and warm-temperate zones. For our problem it does not matter so much if the cycle is yearly or twice yearly, but since there is no evidence for any real cold climates during the formation of the Precambrian iron formations, the model of the monomictic lake seems the more appropriate.

In monomictic lakes a stratification develops during summer, when a body of lighter and warmer water, the *epilimnion* rests on the deeper, colder water, the *hypolimnion*. The temperature of the latter is at 4°C at about the highest density of fresh water. In winter, on the other hand, the superficial water, cooling to below 4°C, will attain about the same density as the deeper water, and currents originating from wind and waves at the lake's surface will tend to mix all of the lake water, down to the lake bottom.

In summer the upper strata are well aerated and well oxygenated, but in the lower stagnant water body the oxygen is quickly used up in the decomposition of down sinking organic (or "organic") material, and reducing conditions will prevail. In winter, when there is no stratification, the whole water body will be well aerated throughout. Now we have to remember that iron under reducing conditions is much more soluble than under oxidizing conditions. Although actual solubility will be governed by a number of factors, a relation of 100,000 times more soluble in reducing conditions than in oxidizing conditions is probable (HOUGH, 1958).

Iron will precipitate as soon as circulation begins to bring oxygen in the lower water body. Moreover, this reaction will start already at a low level of oxygen. Finally, the precipitates of iron will be relatively heavy. They are not greatly influenced by the circulation of the lake water, but will be deposited on the lake bottom.

The deep water of a large fresh water lake—the hypolimnion—will therefore act as a trap for all the iron brought in solution into the lake. Moreover, it will not precipitate the iron continuously, all through the year, but only seasonally. Although less is known about the behaviour of silica, it is well possible that this element would be deposited in alternation with iron.

The ideas of HOUGH (1958) have since been taken up by GOVETT (1966), who introduced a somewhat more refined model, the details of which are not relevant to our topic. The important point in the deep lake model proposed by both Hough and Govett lies in the fact that only very little oxygen is needed to precipitate oxidized iron during the seasonal times of circulation of the lake water. In this way we may understand how it could be possible that partly oxidized iron formations formed under a similar atmosphere as the pyrite sands that are characterized by reduced iron. The latter result from a predominantly physical cycle of weathering, transportation and sedimentation, in which fragmentation, rapid transport towards the sedimentary basin and mechanical sedimentation prevailed. The former depend upon a chemical cycle of weathering, transportation, and sedimentation, in which leaching of the source rocks and chemical precipitation in the sedimentary basins dominate.

Returning now for a final look at the difference between the Precambrian banded iron formations and the younger iron formations, it seems well to realize that there is a difference in mode of formation too. The younger iron formations are never banded, whereas on the other hand they are always associated with marine fossils: They were definitely not laid down in lakes. Stratification of water, and a seasonal circulation rhythm, thought to be the controlling agents in the formation of the banded iron formations, is at present only found in lakes, never in the sea.

One aspect in the genesis of the banded iron formations is the possible influence of biochemical reactions in the formation of the cherts. As we will see in the next chapter, most of the chert of the Phanerozoic is thought to be deposited at the time of sedimentation of a given layer, and not to be due to much later influences during the diagenesis of a rock series, as was formerly thought to be the case. Instead, they are now thought to form, either directly, or by silicification of already existing matter on the sea floor. In the preceding chapter we saw how organic remains are preserved in the banded iron formations of the Middle, and possibly of the Early Precambrian. Moreover, the presence of graphite, sometimes even in macroscopic quantities, has variously been cited, which makes a biochemical influence on the deposition of the chert of the banded iron formations a possibility. On the other hand, it has already been stated that pre-life and early life may have coexisted during the deposition of the banded iron formations (see section 12 of Chapter 12 and section 2 of Chapter 18). In that case it becomes difficult to say, whether these cherts have formed under the cooperation of organic and/or "organic" compounds. Or, to say it in other words, through biochemical, or through purely inorganic reactions. So a clear-cut answer to this problem cannot, as yet, be given.

14. MARINE OR CONTINENTAL (OR LIMNIC)?

One of the basic aspects of today's environment is, whether it is below or above the sea level. As the distinction between lakes and seas plays a part in the unravelling of the history of the Precambrian iron formations, we may do well to take a closer look at this distinction. Very broadly, we speak of a marine environment in the first case, of a continental one in the second. Both major groups are then further subdivided.

In the marine environment, we may distinguish open seas, as against lagoonal seas. In the latter category we may further distinguish between brackish and saline, depending upon the balance of inflow of rain or river water against evaporation. In the open sea, depth is an important factor. We may note four zones, the first being the oceans, with depths ranging from some 10 km to 200 m. Then follow the shallow seas, commonly called the shelf seas, less than 200 m deep. The upper part of the shelf sea, ranging from 50 m to low tide, comprises the phototrophic zone, in which enough sunlight penetrates to allow organic photosynthesis. The last category then is the tidal, or littoral zone, between low and high tide.

For the continental environment altitude and climate are the most important factors. In geology, we are, however, not so much concerned with altitude. For all higher levels will eventually be attacked by erosion, and thereby vanish from the geological record.

So we normally find preserved only the environment of former lowlands. In wet climates marshes, swamps and larger and smaller lakes will develop. This is the environment called limnic. It stands opposed to the environment of the arid low lands.

In geology the distinction between marine and limnic environment rests on the fossils. It is in most cases easy to assign certain groups of plants and animals to a marine or to a limnic environment. In non-fossiliferous series it is in most cases impossible to tell whether a given stratum was deposited in a marine or in a limnic environment. All of our recent insight acquired through the intensive study of sediments is of no avail here. There is no infallible sedimentological criterion to distinguish between sedimentation in the sea or in a fresh-water lake.

We have, however, a good criterion for continental environment—provided we can find it—in the former soils, or paleosoils. The widest known paleosoils are the so-called *Stigmaria* horizons, or underclays, or seat rocks, found beneath the coal seams in continental coal series. They not only contain roots—called *Stigmaria*—of the former trees which lived in the coal swamps, but they also have a particular blocky structure, in which the original stratification has been lost. But this does not help us very much in our analysis of the environment of the Precambrian. For in the Precambrian plants with roots did not yet exist. Whilst the other criterion, the blocky structure of these paleosoils, which is due to the homogenization resulting from the activity of a burrowing fauna, which destroyed the original stratification (RUTTEN, 1963), is also of no avail, because such a fauna had not yet developed during Early and Middle Precambrian times.

As an illustration of the difficulty to distinguish between a marine and a limnic environment in the Precambrian, a study of the paleobiology of the Nonesuch Shale (BARGHOORN et al., 1965) may be cited. The authors conclude to deposition in shallow water, to a deltaic environment, near the shoreline. They do not, however, state whether the Nonesuch Shale was deposited above the shore line, in a limnic, or below the shore line, in a marine environment. Failure to make this distinction—one would say of prime importance in a study on paleobiology— is the result of the lack of available criteria.

Apart from the macroscopic criteria of a *Stigmaria* horizon, as cited above, there are other criteria for recognizing soils. These are the leaching or accumulation of certain elements during soil formation. They are well known in pedology, in relation to modern soils, but this knowledge has as yet hardly been applied to the study of paleosoils. It seems to me that here lies a promising field for detailed geochemical studies of Precambrian sediments. We find ourselves at present in the annoying position of complete ignorance. To take an example, we have no way of knowing whether the voluminous deposits of algal limestones of the Precambrian were developed in a limnic or in a marine environment, or if, perhaps, both types have occurred.

15. AGE OF THE PRECAMBRIAN IRON FORMATIONS

Knowledge about the age of the Precambrian iron formations is necessarily vague, because we have to do with sediments, in which the authigenic minerals, iron oxides and silica, have not incorporated radioactive elements at the time of their formation. So, the age of Precambrian iron formations, which also do not contain index fossils, can only be given as: "younger than the igneous rock they overlap", and "older than the igneous rock by which they are intruded".

To obtain absolute age determinations from the sediments directly, LEPP and GOLDICH (1964) cite a number of absolute dates determined by "whole rock" analyses of slates alternating with the iron formations. The idea being that in such fine-grained rocks clastic grains of parent and daughter elements of earlier radio-active decay processes will have become washed far apart, and not incorporated into the same sediment. A "whole rock" analysis consequently will approximate the analysis of the new state of affairs which originated at the time of deposition of such a fine-grained sediment.

Such dating methods are required when one wants to correlate one single period of deposition of an iron formation in a part of, or across a single old shield, with a view to find more outcrops of a given good iron ore. As such it is rather conspicuous in the economically slanted literature on the iron formations, though it can probably not be used for world-wide correlations from one old shield to the other.

For our topic we are not so directly concerned with the exact age of a given iron formation, but with the period of time when iron formations of the Lake Superior type were formed all over the world. As a first approximation we may therefore use the well-known ages of the intrusive rocks which bracket the sedimentary series. In general, this boils down to the age of the preceding and the succeeding orogenetic periods.

For the Precambrian of Minnesota, that is for part of the Canadian Shield, we learn from Table XX that iron formations occur during the Early and the Middle Precambrian, ranging in age from more than 2.5 billion years to 1.8 billion years. No typical banded iron formations are known from rock series younger than the Penokean (= Grenville) orogeny.

The ages of Precambrian iron formations from all over the world, as collected by GOVETT (1966, see Table XXI), confirm the data arrived at by the study of the Minnesota iron formations, extending the earliest occurrence to more than 3 billion years. I am, however, under the impression that not all of the dates listed in Table XXI are as reliable as those from Minnesota.

In relation to the age of Precambrian iron formations there remains one important aspect to be mentioned. As we have seen in Chapter 9, the history of the earth shows a rather strict sequence in each orogenetic cycle. Each begins with

TABLE XXI

AGE AND DESCRIPTION OF PRECAMBRIAN IRON FORMATIONS
(From GOVETT, 1966)

Age (billion years)	Formation, location, description[1]
Late Proterozoic	*Late Proterozoic, Northern Territory, Australia.* Oolitic and pisolitic; hematite–siderite, greenalite (Edwards, 1959)
1.7–2.5	*Huronian, North America.* Banded silicate–carbonate, silicate–chert, granule-bearing. Banded hematite, oolitic hematite (James, 1954; Goldich and others, 1961)
1.8–1.9	*Karelides-Svecofennides, Eastern Baltic Shield* (Polkanov and Gerling, 1960)
≃2.0	*Transvaal, South Africa.* Banded hematite–chert; oolitic and pisolitic chamosite–hematite–magnetite–siderite (Wagner, 1928; Nicolaysen, 1962)
2.0–2.5	*Proterozoic, Western Australia.* Banded chert–carbonate, ferruginous chert (McKinstry, 1945; Lord, personal communication)
>2.1	*Krivoi Rog, Russia.* Banded iron formation (Vinogradov, 1960)
≃2.5	*Dharwar, iron ore series,* India. Banded hematite–chert (Krishnan, 1960)
2.5	*Saamides, Baltic Shield* (Polkanov and Gerling, 1960)
Archaean (>2.5?)	*Middleback, South Australia.* Banded hematite–chert (Edwards, 1953)
>2.5	*Archaean, Western Australia.* Banded hematite–quartzites, jaspilites (McKinstry, 1945; Lord, personal communication)
>2.5	*Soudan Iron Formation, North America.* Banded hematite–chert (Goldich, 1961)
>2.5	*Shamvian, Southern Rhodesia.* Banded hematite–chert (MacGregor, 1951; Nicolaysen, 1962)
>2.6	*Bulawayan, Southern Rhodesia.* Banded hematite–chert (MacGregor, 1951; Nicolaysen, 1962)
Underlies Bulawayan	*Sebakwian, Southern Rhodesia.* Banded hematite–chert (MacGregor, 1951; Nicolaysen, 1962)
>2.8	*Nyanzian, Kenya.* Banded hematite-chert (Saggerson, 1956)
>3.0	*Swaziland.* Banded hematite–chert (Way, 1952; Nicolaysen, 1962)
>3.0	*Swaziland, South Africa.* Banded hematite–chert; magnetite–siderite-slate (Wagner, 1928, Nicolaysen, 1962)

[1] For a bibliography, consult the original publication.

a post-orogenetic phase, characterized by strong erosion and thick clastic sedimentation, whereas during the middle and later parts of the succeeding geosynclinal phase crustal stability and chemical erosion and sedimentation prevail outside the actual geosynclinal basins. In regard to the iron formations, there are as yet hardly enough data to check whether this simple scheme holds good all over the world, in the various old shields in which iron formations of the Lake Superior type occur. But, at least for Minnesota, it is well developed. All three main periods of iron formation occur either in the middle, or at the end of a geosynclinal period.

In contrast, the Blind River gold–uranium reefs, found in ancient sands and gravels of the Lower Huronian, typically belong to the post-orogenetic sedimentary facies of the Algoman orogeny.

16. SEDIMENTS FORMED UNDER THE OXYGENIC ATMOSPHERE: THE RED BEDS

From the studies reviewed above it follows that under favourable circumstances, in localities which have not undergone too severe changes since their formation, old sediments may supply indications that they were laid down under reducing atmospheric conditions. The opposite is true, too. Sediments may, by the state of oxidation of their constituent minerals, indicate that they were formed under an oxygenic atmosphere.

This distinction has, however, to be handled with care. As we saw, not every sediment with a few grains of sulphides is indicative of a contemporaneous anoxygenic atmosphere, nor does a sediment formed by grains of oxides automatically proclaim that it was formed under an oxygenic atmosphere. The most common mineral of present-day sands, quartz, for instance, does not give an indication for either an anoxygenic or an oxygenic atmosphere. As we observed in the preceding sections, sands can be so well classified, so well sorted out acccording to the specific weight of their grains, that under an anoxygenic atmosphere true quartz sands may also develop. Another instance is the chert layers of the banded iron formations, which also dominantly consist of SiO_2.

For the purpose of recognizing the influence of an anoxygenic or an oxygenic atmosphere on the type of the contemporaneous sediments, elements which may combine with oxygen in different states of oxidation are perhaps the best indicators. We have already shown how the banded iron formations of the Early and Middle Precambrian predominantly contain iron in its least oxidized form, in the mineral magnetite, whereas iron formations of the Late Precambrian and the Phanerozoic contain iron in its most strongly oxidized form, in the mineral hematite.

Another group of sediments with iron, which although containing much less of the metal is as widely distributed as the iron formations, are the *red beds*. Red beds are fine-grained quartz sediments of the silt type, which means that their grain size lies somewhere in between that of normal sands and normal clays. Moreover, most series of red beds carry intercalations of thin beds which consist of clay, and also coarser beds of sands and/or of conglomerates. They are of continental origin. Their colour, varying between bright red and reddish brown, is due to iron, which is, however, present in a small percentage by weight only. It occurs either in small flakes, or as a coating on the quartz grains. It is in the ferric, highly oxidized state, and normally present as hematite, Fe_2O_3.

The formation of red beds is rather complicated, and is not yet well understood. Hematite is the stable form of iron oxide in a dry climate with relative humidity below 60%. In wetter climates the greyish mineral goethite, FeO(OH) is stable (SCHMALTZ, 1959). It has accordingly been thought that red beds formed in deserts and that ancient red beds were indicative of a desert climate. It has turned out, however, that in recent deserts very few red beds are developed. On the basis of the clay minerals occurring in the ancient red beds a two-step origin is now generally thought more likely. The material of the red beds would originate in a savanna climate in upland regions, but the ultimate deposition was in adjacent basins with a desert climate. The transport from the source area to the area of deposition could have been effected by intermittent rivers for the coarser members, and by the same rivers, or by wind action, for the fine-grained members, so typical for an unequivocal red bed series. The possibility that, on the other hand, most of the hematite formed only after deposition, during early diagenesis of the sediments, has also been proposed.

We will not go further into the problem of the origin of red beds, and refer the reader to VAN HOUTEN (1963). But we will retain the fact that this is a group of sediments possessed of a number of special characters, which run true-to-type all over the globe and over all of the Phanerozoic, with the inclusion of the later part of the Precambrian. They are detrital sediments, formed one way or another under dry climates, in which the iron is in its most oxidized form.

17. AGE OF RED BEDS

We know typical red beds from all of the later, "normal" geologic history, that is from the Phanerozoic. There are for instance, the "Old Red" and the "New Red" series of Britain, the former of Devonian, the latter of Triassic age, some 400 m.y. and 200 m.y. old respectively. There are comparable red beds in the Silurian of North America, which are over 400 m.y. old. And during the Permian epoch, around 250 m.y. ago, red beds developed quite extensively in many places scattered all over the earth.

There are also Precambrian red beds. But, just as was the case with most Precambrian fossils, these belong to Late Precambrian times. The Torridonian Sandstones of Scotland and the Jotnian Sandstones of Skandinavia form a case in point. Both are Precambrian in age, but both belong to the orogenetic cycle which started with its geosynclinal period during the later Precambrian, and lasted into the Lower Paleozoic. They form part of a single sedimentary sequence.

Neither the Torridonian Sandstones of Scotland, or the Jotnian Dala Sandstones of Skandinavia can be dated directly. However, within the latter series intercalations of dolerite are found, which by a preliminary estimate have an age of 1300–1400 m.y. (WELIN et al., 1966). Moreover the underlying Dala

Porphyries are dated as 1405 ± 30 m.y. (PRIEM et al., 1968). The latter age determination has made use of the Rb/Sr ratio and seems the more trustworthy, and a maximum age of 1400 m.y. can thus provisorily be used for the oldest red beds.

It is tentative in two ways. On the one hand we might find still older red beds, but on the other hand we might learn to distinguish between red beds and red beds and be able in future to separate the older, coarser and more greyish red beds from the finer-grained and much more reddish newer ones, using this difference for a further distinction between succeeding levels of oxygen in the former atmosphere.

For as we saw in the preceding section, not every coarse-grained sandstone series with a coating of hematite on the quartz grains is the exact replica of the typical red beds of the Phanerozoic. CLOUD (1968) has expressed this in calling the latter "unequivocal" red beds. A highly subjective term, to be sure, but still something conveying a meaning to the field geologist.

The qualification about "unequivocal" red bed series, introduced above, has to be made, because there exist earlier reddish sediments, different from the red beds. But under the anoxygenic atmosphere, some oxidation is also always possible, as we have seen in the case of the banded iron formations. However, just as the typical banded iron formations are confined to the Early and Middle Precambrian, the typical red beds are only found from the later Precambrian onwards through the Phanerozoic.

A case in point of earlier reddish sediments is the Roraima Formation in the Guyanas. This is a series of coarse-grained reddish—but not very red—sandstones. They differ from the typical red beds in that no finer-grained intercalations seem to occur. The age of the Roraima Formation has been estimated as between 1.8 billion years and 2.5 billion years, which would make it the contemporary of the pyrite sands of the gold–uranium reefs. Superficial coating of coarse sand grains, either during sedimentation and early diagenesis, or during the later history, is of course easier to accomplish in coarse-grained sandstones in which ground water may circulate freely, than in fine-grained silt layers. In this context it must be noted that the history of the Roraima Formation is more complex than it would seem at first sight. It is found to contain pollen and spores of presumably Tertiary age washed into cracks, a fact indicating the amount of post-depositional groundwater circulation possible in such coarse sandstones. So, even if the Roraima Formation is correctly dated, a fact doubted by some authors (STAINFORTH, 1966), it is not necessarily a typical red bed series.

A general impression I gained from the Dala Sandstones, for instance, would also indicate that they are on the average much coarser than the red beds of the Phanerozoic and that they lack the fine-grained silt members, so typical for the latter. Curiously, no serious sedimentological research has been made, as to what is characteristic for an "unequivocal" red bed of the Phanerozoic. Nor have the Late Precambrian red sandstone series, such as the Torridonian or the Dala Sandstones, ever been studied in this respect. Red beds have no particular economic importance, and have escaped detailed attention so far.

So it might well be possible that we will learn to distinguish between red beds and red beds in future. We might then acquire a means to distinguish between, for instance, an "oxygenic" and a "more fully oxygenic" atmosphere. But at present we can do no more than lump all red beds of Late Precambrian and Phanerozoic age together and take them as indicative of the existence of a contemporary oxygenic atmosphere.

Consequently the tentative date for the oldest atmosphere leading to the formation of red beds is 1.4 billion years.

18. CONCLUSIONS

To cut a long story short, we have learned in this chapter about the possibility of ancient sediments supplying us with information about the atmosphere at the time of their formation. From the Precambrian quartz-pyrite sands of the gold-uranium ores and from the banded iron formations we have drawn the conclusion that the earth indeed has had a primeval anoxygenic atmosphere. Moreover, we have arrived at a preliminary set of dates, indicating till how long ago the earth still had this primeval anoxygenic atmosphere, and from what age onwards the oxygenic atmosphere of the present has become established.

We have also noted that it is, as yet, not possible to decide from these data at what level of free atmospheric oxygen the boundary between the "anoxygenic" and the "oxygenic" atmospheres may be drawn. This is a consequence of the fact that our deductions are predominantly based on the state of oxidation of the iron included in the sediments. That is, whether these sediments contain wholly reduced iron—FeS—, or partly oxidized Fe_3O_4, or fully oxidized Fe_2O_3. We observed that the geochemistry of iron, and for that of most the rock-forming minerals, has as yet only been studied in regard to its equilibrium reactions, whereas the more important kinetics are virtually unknown. The amount of oxidation of iron in sediments depends not only on the amount of oxygen in the contemporary atmosphere, but also upon the relation between the rates of oxidation and of transportation and sedimentation. In Chapter 15 we will use the faculty many present-day microbes possess to change their metabolism from fermentation to respiration (the Pasteur Point, see section 5 of Chapter 7) to assign an upper level of free oxygen to the primeval anoxygenic atmosphere.

As to the dating of the transition between the two atmospheres, we have seen that both the quartz-pyrite sands and the banded iron formations occur in the Early and Middle Precambrian, but are not found in the Late Precambrian.

The boundary between Middle and Late Precambrian at 1.8 billion years may therefore be taken provisionally (see section 12 of Chapter 14) as the youngest date for the occurrence of the anoxygenic atmosphere. The age of 1.4 billion years of the Dala Sandstones of central Sweden has, on the other hand, been provisionally accepted for the earliest date at which we may speak of an oxygenic atmosphere. It follows that between 1.8 billion years and 1.4 billion years ago the transition from the anoxygenic primeval atmosphere to the present oxygenic atmosphere took place.

We are now in a position to draw up a very schematic history of atmospheric oxygen on earth, and with it an equally schematic history of the origin and the early development of life. This will be done in Chapter 16 in which the main narrative will be continued. But before doing so we must insert in the next chapter a discussion of a number of diverse geological problems, the knowledge of which, though not directly related to our main story, is still important for a better understanding of this story.

REFERENCES

ARMSTRONG, G. C., 1968. Sedimentological control of gold mineralisation in the Kimberley Reefs of the East Rand Goldfield. *Econ. Geol. Unit. Univ. Witwatersrand, Johannesburg., Inform. Circ.*, 47:24 pp.

BARGHOORN, E. S., MEINSCHEIN, W. G. and SCHOPF, J. W., 1965. Paleobiology of a Precambrian shale. *Science*, 148:461–472.

CLIFFORD, T. N., 1968. Radiometric dating and pre-Silurian geology of Africa. In: HAMILTON, E. I. and FARQUAR, R. M. (Editors), *Radiometric Dating for Geologists*. Interscience, New York, N.Y., pp.299–416.

CLOUD, P. E., 1965. Significance of the Gunflint (Precambrian) microflora. *Science*, 148:27–35.

CLOUD, P. E., 1968. Pre-metazoans and the origins of the Metazoa. In: E. T. DRAKE (Editor), *Evolution and Environment*. Yale Univ. Press, New Haven, Conn., pp.1–72.

DAVIDSON, C. F., 1963. The Precambrian atmosphere. *Nature*, 197:893–894.

DAVIDSON, C. F., 1965. Geochemical aspects of atmospheric evolution. *Proc. Natl. Acad. Sci. U.S.*, 53:1194–1205.

DECHOW, E. and JENSEN, M. L., 1965. Sulfur isotopes of some Central African sulfide deposits. *Econ. Geol.*, 60:894–941.

DERRY, D. R., 1960. Evidence of the origin of the Blind River uranium deposits. *Econ. Geol.*, 55:906–927.

GARRELS, R. M., 1960. *Mineral Equilibria*. Harper, New York, N.Y., 254 pp.

GERASSIMOW, I. P., 1968. Recent laterites and their formation. *Intern. Geol. Congr. 22nd, New Delhi, 1964, Rept.*, 14:116–125.

GOLDICH, S. S. 1968. Geochronology in the Lake Superior Region. *Can. J. Earth Sci.*, 5:715–724.

GOVETT, G. J. S., 1966. Origin of banded iron formations. *Geol. Soc. Am. Bull.*, 77:1191–1212.

HALES, A. L., 1961. An upper limit to the age of the Witwatersrand System. *Ann. N. Y. Acad. Sci*, 91:524–529.

HOEFS, J., NIELSEN, H. and SCHIDLOWSKI, M., 1968. Sulfur isotope abundances in pyrite from the Witwatersrand conglomerates. *Econ. Geol.*, 63:975–977.

HOUGH, J. L., 1958. Fresh-water environment of deposition of Precambrian banded iron formations. *J. Sediment. Petrol.*, 28:414–430.

JENSEN, M. L. and DECHOW, E., 1964. Bearing of sulfur isotopes on the origin of South African ore deposits. *Geol. Soc. Am., Progr., Ann. Meeting*, 1964:101.

JONG, W. J., 1966. Actualism in geology and geography. *Tijdschr. Koninkl. Ned. Aardrijkskundig Genoot.*, 58:238–248.

LEPP, H. and GOLDICH, S. S., 1964. Origin of Precambrian iron formations. *Econ. Geol.*, 59: 1025–1060.

PRIEM, H. N. A., MULDER, F. G., BOELRIJK, N. A. I. M., HEBEDA, E. H., VERSCHURE, R. H. and VERDURMEN, E. A. T., 1968. Geochronological and paleomagnetic reconnaissance survey in parts of central and southern Sweden. *Phys. Earth Planetary Interiors*, 1:373–380.

RAMDOHR, P., 1958. Die Uran- und Goldlagerstätten Witwatersrand Blind River District, Dominion Reef, Serra de Jacobina: Erzmikroskopische Untersuchungen und ein geologischer Vergleich. *Abhandl. Deut. Akad. Wiss. Berlin, Kl. Chem., Geol. Biol.*, 1958 (3):35 pp.

RAMDOHR, P., OTTEMANN, J. and SCHIDLOWSKI, M., 1965. *Mikroskop, Mikrosonde und Schätzung der Uran-Blei Alter*. Max. Planck Inst. Kernphysik, Heidelberg, 13 pp.

RANKAMA, K., 1955. Geologic evidence of chemical composition of the Precambrian atmosphere. *Geol. Soc. Am., Spec. Papers*, 62:651–664.

REICHE, P., 1950. *A Survey of Weathering Processes and Products*. (Revised edition). Univ. New Mexico Press, Albuquerque, N.M., 95 pp.

RUTTEN, M. G., 1963. Biological homogeneization in fossil underclays and seat rocks. *Geol. Soc. Am. Bull.*, 74:91–96.

SAAGER, R., 1970. Structures in pyrite from the Basal Reef in the Orange Free State goldfield. *Trans. Geol. Soc. S. Africa*, 73:29–46.

SCHIDLOWSKI, M., 1965. Das Witwatersrand Becken. In: *Max Richter Festschrift*. Pieper, Clausthal-Zellerfeld, pp.127–147.

SCHIDLOWSKI, M., 1966a. Beiträge zur Kenntnis der radioaktiven Bestandteile der Witwatersrand-Konglomerate. I. Uranpecherz in den Konglomeraten des Oranje-Freistaat-Goldfeldes. *Neues Jahrb. Mineral., Abhandl.*, 105:183–202.

SCHIDLOWSKI, M., 1966b. Beiträge zur Kenntnis der radioaktiven Bestandteile der Witwatersrand-Konglomerate. II. Brannerit und "Uranpecherzgeister". *Neues Jahrb. Mineral., Abhandl.*, 105:310–324.

SCHMALTZ, R. F., 1959. A note on the system Fe_2O_3–H_2O. *J. Geophys. Res.*, 64:575–579.

SHARPE, J. W. N., 1949. The economic auriferous bankets of the Upper Witwatersrand Beds and their relationship to sedimentation features. *Trans. Geol. Soc. S. Africa*, 52:265–300.

STAINFORTH, R. M., 1966. Occurrence of pollen and spores in the Roraima Formation of Venezuela and British Guiana. *Nature*, 210:292–294.

VAN HOUTEN, F. B., 1963. Origin of red beds—some unsolved problems In: A. E. M. NAIRN (Editor), *Problems in Palaeoclimatology*. Interscience, New York, N.Y., pp.649–672.

WELIN, E., BLOMQVIST, G. and PARWEL, A., 1966. Rb/Sr whole rock age data on some Swedish Precambrian rocks. *Geol. Fören. Stockholm Förhandl.*, 88:19–28.

Chapter 14 | Miscellaneous Geological Notes

1. INTRODUCTION

This chapter deals briefly with a number of rather loosely connected geologic topics, which all bear on our understanding of the origin of life, but which could not well be incorporated in the two preceding chapters.

We will start with a note on optical activity and the importance of quartz, which, together with the clay minerals, is the most important constituent of the sediments. A note on the various possible ways of formation of chert follows. Cherts, consisting of extremely fine-grained silica, are, as we have discussed in Chapter 12, the main depositories of microfossils of the Early and Middle Precambrian. An understanding of the meagre data pertaining to their formation is therefore essential. Then follow two sections on geochemical inventories, which tell us that geochemical data, as known at present, are in themselves insufficient to give a decisive answer to the question whether the earth ever had an anoxygenic primeval atmosphere.

The clay minerals come next. Insight into their structure and their large chemical variability leads to the modern ideas on the composition of ocean waters. It will be seen how under present atmospheric conditions the composition of ocean waters must have remained remarkably constant throughout geologic history. The reason being that the concentration of the base elements, Na, Ca, Mg and others, is to a large extent buffered by the clay minerals.

This has in 1965 led the Swedish chemist Professor L. G. Sillèn to speak of the "myth of the prebiotic soup" because thermodynamical considerations predict that, when equilibrium concentrations are taken into account only, the concentration of free radicals and of "organic" molecules postulated for the "thin soup" will always remain neglegibly small in ocean waters such as they are known at present. This reason probably holds true even for an ocean in contact with a primeval anoxygenic atmosphere, although its equilibrium reactions are still poorly known. But, just as salt deposits form at present in lagoons and lakes, although the amount of salt in the oceans stays well below crystallization level, similar environments have in the past probably also been more propituous for the development of life than the open ocean.

Apart from the phenomenon of optical activity discussed in section 2 of this chapter, we will touch upon a number of other aspects which deal with the distinction between biogenic and abiogenic formation. This distinction, which is rather facile in our present-day environment, often has all too easily been extrapolated to the early days of the earth's history. For example, the relative abundance of the stable isotopes of a given element may at present be different, depending upon the fact whether this element occurs in biogenic or abiogenic material. But this does not mean that we can, without further ado, use this difference in trying to distinguish between biologically formed organic compounds and "organic" material, synthesized inorganically during the Early and Middle Precambrian.

Another topic, which will be treated in rather detail, is the mechanism of limestone deposition during their metabolism by various organisms. Under our present-day oxygenic atmosphere this is largely due to photosynthetic activities of Algae. But even in our present world, as follows from information kindly supplied by Professor C. B van Niel, there are many other possible pathways of limestone deposition, particularly by anaerobic microbes, so it is not surprising we found in section 8 of Chapter 12 that biogenic deposits of limestone occur already during the Early Precambrian.

A point often not well understood is the process of fossilization. We are truly impressed by the wonderfull morphological preservation of the microfossils of the Precambrian. But we must never forget, that, especially in cherts, this is no more than the preservation of the outer shape of the organisms. The original cell material has not been "biologically preserved", although this has so often been maintained recently. Nor has, in most fossils, anything remained of the inner structure of the former cells. It is, therefore, impossible, in these fossils, to distinguish between procaryotic and eucaryotic cells.

Also a word of warning about the meaning of the nomenclature of the algae in paleobotany will be given. In contrast to the nomenclature of the higher plants and the animals, which is genetically based, the algae, especially in paleobotany, are no more than a so-called *form group*. Their nomenclature does not tell us anything more than that a certain resemblance exists between different species, genera and higher taxonomic units. For our topic, the origin and development of life, the nomenclature based on a form group has no meaning.

We then come to a rather delicate question, that is the supposed relation between the development and the evolution of life on the one hand, and periods of strong disturbance of the earth's crust on the other. I do not believe that such a correlation between so-called "faunal crises" and "major orogenetic periods" exists, but I am aware that many fellow-geologists do not agree with me. The question is still of academic nature for the Precambrian anyhow, since the absolute dating of the Precambrian uses mainly the rocks formed during the major orogenies, whilst we have as yet no means to arrive at a more detailed world wide time scale.

Finally, attention is drawn to the fact that the gradual evolution of life over the last 3 billion years or so indicates that the earth must have had a rather uniform surface temperature during all this time.

2. OPTICAL ACTIVITY AND THE IMPORTANCE OF QUARTZ

Optical activity is the property of certain forms of matter to influence polarized light. It is all-prevalent in living matter, and as such it has been stated that "No other chemical characteristic is as distinctive of living matter, as is optical activity" (WALD, 1957). This is an oversimplification, since some inorganically formed crystals, of which quartz is the most common example, also possess optical activity. It is safer to say that on a molecular scale optical activity is characteristic of many biogenetic compounds, whereas it is never or hardly ever found in inorganically formed substances.

Optical activity depends on a lack of symmetry, either in the molecules themselves, or in the crystal lattice. Matter which possesses a plane of symmetry is never optically active. Matter which does not possess a plane of symmetry cannot be superimposed on its mirror image. Matter may, however, still have some symmetry of a lower order, for instance, symmetry around a single point. In crystallography crystals showing such lower order symmetry are called *triclinic*, whereas crystals possessing a plane of symmetry are either *monoclinic*, or, if they show an even higher order of symmetry, belong to other groups of the crystallographic system. Matter which does not possess a plane of symmetry, but still shows some lower order of symmetry is technically called *disymmetric*, whereas matter without any symmetry is called *asymmetric*. These terms are, however, often used rather loosely.

On the molecular level optical activity depends upon the existence of two stereochemically distinct isomeric forms of the same composition, called *stereoisomers*. On the level of the crystals optical activity depends on the crystals belonging to the triclinic group, and showing right and left hand forms called *enantiomorphs*. It is not necessary for a compound to have stereoisomeric molecules to form optically active crystals. Quartz, SiO_2, for instance, apparently has symmetric molecules, and still solidifies into optically active crystals[1]. Although this

---

[1] The well known "crystal clear" prisms of quartz only superficially resemble hexagonal crystals. In reality they are pseudo-hexagonal, the angles between the prism faces varying slightly from the 60° required for a true hexagonal prism.

is perhaps an oversimplification, one might state that in optically active crystals the light is influenced because the dissymmetry of the crystal lattice results in a sort of corkscrew arrangement of the molecules. In a solution optical activity results from a similar corkscrew influence exerted by the individual dissymmetric molecules.

Although known already since the beginning of last century, the main interest in optical activity goes back to the work of Louis Pasteur in the middle of last century. Pasteur studied the crystals of two stereoisomers of the compound $HO_2C—CHOH—CHOH—CO_2H$, racemic and tartaric acid respectively. Racemic acid produces symmetric crystals, whereas tartaric acid solidifies into dissymmetric crystals with right and left hand enantiomorphs. Racemic acid has since given its name to all optically inactive substances, whereas if in a solution the original optical activity is destroyed, the process is called *racemisation*, regardless of the composition of that solution.

Optical activity was first found as the ability to rotate plane polarized light. This aspect of optical activity is called *optical rotatory dispersion* (ORD). It has since been learned that in circularly polarized light the coefficient of extinction for left- and right hand circularly polarized light is altered by optically active substances. This *circular dichroism* or CD is of course only apparent near absorption bands, where it shifts the position of these bands. This is called the *Cotton effect*. Both optical rotatory dispersion and circular dichroism are valuable analytical tools in organic chemistry. They are specially important because they permit a continuous analysis of the synthesis or the degradation of organic substances, and because they are extremely sensitive. But notwithstanding their apparent difference, they are but two aspects of the same property of optical activity, depending both on the dissymmetry of either stereochemically dissymmetric isomers, or crystallographically dissymmetric enantiomorphs. The two aspects are analytically interconvertible by quantum mechanic equations of the Kronig–Kramer type, and therefore equivalent (SCHELMAN, 1968; GRATZER and COWBURN, 1969).

The importance of optical activity in relation to life has been highlighted ever since the work of Louis Pasteur. He found already, and this has been confirmed ever since, that of all organic compounds containing building blocks capable of stereoisomerism, and thus being able to rotate plane polarized light either to the left (*laevo*-rotatory or L) or to the right (*dextro*-rotatory or D), only a single type is present in living matter. This is not an overall choice, all living matter being, for instance in the L-form; for whilst the common amino acids are all in the L-form, several of the rarer amino acids and most of the sugar phosphates are in the D-form. But for a given compound the distinction is absolute. It is present either in the L-or in the D-form and no mixing of both stereoisomers of a single compound occurs in living matter.

Pasteur has already shown that it is fairly easy to arrive at a comparable strict separation of either the L- or the D-isomer of a given compound, but only if one uses, somewhere in the chain of reactions, a solution which already shows optical activity. Attempts to produce optically active substances directly from racemic solutions have, on the other hand, largely failed. The prevalent opinion therefore is that only active substances can generate other active substances.

This conclusion is similar to that quoted in Chapter 4 in regard to life as a whole, which stated that "only living matter can produce other living matter". We have nevertheless seen how in a reducing environment "organic" molecules can be produced at will inorganically, which proves that if not life, at least the compounds of life can be produced without the interference of life. With S. W. Fox (as cited by CLOUD, 1968) we may submit that "any remaining discontinuity between non-life and life should be regarded as not yet understood, rather than as hopelessly incomprehensible". A similar position has been taken by WALD (1957), who stated: "I am as reluctant as the reader to quarrel with a concept as well founded as the one that holds that spontaneous crystallization of L- and D-forms should be equally probable. The trouble is that the only data I have found indicate the contrary." And further: "I think that a few more experiments in this area would not be amiss, for perhaps we are assuming too much too lightly."

In fact the situation is not as black and white as it is normally presented. Optical activity can be attained either by the preferential synthesis or the preferential destruction of one of the isomers or enantiomorphs. Both pathways have indeed yielded results in specially designed experiments. The experimental difficulties are great and the differences attained are admittedly small. Biochemists, used to the perfect separation between L- and D-isomers in living matter, have tended to regard these results as inconclusive. But here again the tenet holds that in the early history of life even a small distinction may have been sufficient to provide a "survival" value, whereas the present refined state was only arrived at by the interaction of mutations and selection pressure over three billion years. As to preferential synthesis, HAVINGA (1954) showed how either L- and D-enantiomorphs could crystallize from a supersaturated racemic solution. As to preferential destruction, A. Cotton showed already in 1896 that optical isomers have different coefficients of absorption for left and right circularly polarized light (see the Cotton effect, mentioned earlier) and that, theoretically at least, a racemic mixture of dissymmetric molecules might lose one of its components preferentially and thus become optically active, when exposed to such light. This has, however, only been confirmed experimentally in 1929 (WALD, 1957), which may serve as an indication of the experimental difficulties encountered in this field. Another pathway was investigated by GARAY (1969) who in 1961 began to study the possible effect of radiation by β-particles and the accompanying left circularly polarized "Bremsstrahlung" of γ-radiation. He also took seven years before he could publish positive results and found that D-tyrosine was preferentially destroyed over L-tyrosine. It seems indeed that "a few more experiments in this area might not be a miss".

Another possibility to "generate optically pure material without the intervention of a microbe or a man", which has been proposed by CALVIN (1969), is that of autocatalysis of stereospecific compounds. If in a reaction in which a stereospecific compound is formed the new product is autocatalytic—that is catalytic for this reaction—, and if furthermore the reaction rate of this autocatalysis is many powers of 10 greater than the original non-catalytic reaction, then, Calvin reasons, the first molecule of the new compound, be it in the L- or in the D-form, will, by its so much more vigorous autocatalytic reaction rate, suppress the formation of the other form. An example is cited in the experiments of ALLEN and GILLARD (1967), in which peptide complexes on an octohedral cobalt compound generate either L- or D-compounds.

Apart from the fact that not only cobalt, but also the related ambivalent metals are rare on the surface of the earth, we encounter a much larger difficulty in that such reactions may well lead to the results quoted in a testtube but not necessarily in nature. In a testtube the first molecule formed may indeed, if only its rate of autocatalysis is sufficiently large, induce a wholesale formation of compounds of similar structure. But in nature we must assume that the processes of pre-life and of biopoesis have been going on at the same moment in many places scattered all over the earth. Even if the proposed mechanism of autocatalysis should obtain in one or more of these processes, there must have been a great number of foci, where, without outside influence, a "first" molecule of the new compound, either in the L- or the D-form, will have developed. Such molecules will then have induced the formation of similar forms in their more or less immediate neighbourhood. But these areas will eventually have met and a checkboard pattern of local "colonies" of L- and D-forms will have developed. Statistically there will have been no preference for one or the other stereoisomers. To enable a statistical preference of either of the two forms to have developed, a factor must have been present which exerts a global selection pressure in this type of reactions.

In this context I should like to point out the possible importance of quartz. We have met with quartz sands and cherts already several times in our discussion of the fossils and the environment of the Precambrian, and it is clear that quartz, then as now, is one of the most common minerals of the crust of the earth. TERENT'EV and KLABUNOWSKII (1959) and KLABUNOWSKII (1959) have published papers on the subject of optical activity of quartz, to which the reader is referred for details. The results are that quartz crystals not only are optically active for transmitted light, but also have optically active surface properties. Selective adsorption of L- or D-isomers consequently is theoretically possible at the crystal surfaces of L- or D-enantiomorphs. But the overall balance is of course not altered, and no optical activity is induced, when L- and D-enantiomorphs occur on a fifty-fifty basis in nature. We do not know if this is so, the state of the art being summed up by WALD (1957) in a discussion between the German crystallographer

V. M. Goldschmidt, who learned from a manufactor of optical instruments that "large clear crystals were right handed ten times more often than left handed" and the British chemist N. W. Pirie who stated that according to British manufactors right handed crystals are only a little more common.

As there is, consequently, no sound knowledge on the statistical distribution of L- and D-enantiomorphs of quartz in nature, we have to assume that they occur on a fifty-fifty basis, and as such do not contribute to the establishment of optical activity. There is, however, one possibility in that the sunlight reaching the earth is plane polarized by a filtering process in the atmosphere. And although as yet only the influence of circularly polarized light in relation to the genesis of optical activity has been mentioned, we have also seen that plane and circularly polarized light on optical activity are but two aspects of the same property. It seems feasible that the interaction between the ubiquituously present plane polarized light and optically active crystal surfaces might have an influence on the preferential synthesis of either L- or D-isomers of a given compound.

The often heard statement that optical activity has never been found in the inorganically synthesized "organic" molecules is rather meaningless, for no experiments have ever been expressly set up to test the possibility of such a development. The experiments in which "organic" molecules are synthesized run over periods of days or weeks, not over years, such as is needed in experiments cited above on the formation of optical activity. Nor has β- or γ-radiation been applied to these syntheses, or have they been carried out in the presence of optically active quartz crystal surfaces.

It follows that I am rather chary of accepting the statement that because optical activity is at present only found in living matter, the presence of optically active carbonaceous substances in early sediments proves the contemporaneous existence of life. Just as the presence of carbonaceous compounds no longer proves the presence of life, because these may have been synthesized abiogenically, it is quite possible that "a few more experiments" will prove that optical activity may also originate inorganically under the primeval atmosphere. Of course, a distinction might be made in whether such optical activity is found in smaller and more simply built carbonaceous material, or in more complex molecules. For instance, the optical activity reported from amino acids of the Fig Tree Series (KVENVOLDEN et al., 1969, see section 17 of Chapter 12) is less convincing than would be the report of optical activity in isoprenoid alkanes where, on other grounds, also a biogenetic origin is probable[1]. But such activity has to my knowledge only been found in the Eocene Green River Shales (MACLEAN et al., 1968), for which we are

---

[1] A certain doubt about the reality of optical activity in amino acids from the Early Precambrian can, moreover, be expressed. KROEPELIN (1969) found that in amino acids in contact with montmorillonite—one of the common clay minerals, see section 6 of this chapter—racemization is accomplished already by an exposure to 100°C over 100 hours.

quite certain anyhow that life existed already. To my mind the existence of optical activity in fossil carbonaceous material does not in itself prove the existence of contemporaneous life, but it may be taken, in conjunction with other indications, as a strong argument in favour of such life. Just as with the carbonaceous materials and the molecular fossils themselves, the boundary between non-living and living is becoming blurred, in geology too. This has already been indicated in section 2 of Chapter 12, and we will return to it in section 3 of Chapter 18.

3. THE FORMATIONS OF CHERTS

Cherts consist of extremely fine-grained silica. The size of the individual crystals is so small that in most cases they cannot be detected by the light micros- cope. Hence cherts were formerly thought to be only partly crystallized, and amorphous for the other part. Roentgen analysis later revealed that all cherts are crystalline throughout, so nowadays they are described as *cryptocrystalline*.

At present cherts form in deep ocean basins, below a depth of 4 km or more. There the so-called *siliceous oozes* assemble, formed by the tests of micro- organisms carrying a siliceous shell. The commonest forms are the diatoms and the Radiolaria. Diatoms and Radiolaria live as plankton in the upper 50 m or so of the oceans, together with other plankton which may have calcareous (Fora- minifera) or chitinous tests, or has no tests at all. Upon the death of these organisms this material slowly sinks down, being all the time chemically attacked by the ocean water. The calcium carbonate of Foraminifera and the chitin of other members of the plankton dissolves more readily than the silica of the diatoms and the Radiolaria, and below a certain depth the latter material is the only remaining substance. It is not, however, in equilibrium with ocean water, as the many solution phenomena found on these delicate tests clearly proclaim. The deep-sea siliceous oozes consequently consist of foreign material, produced in the upper reaches of the ocean, which can only accumulate thanks to the fact that the rate of sedimenta- tion is higher than the rate of solution by the ocean water.

Chert formation is at present negligible both in shallow oceans and in lakes and swamps. It has, however, been of importance all through the Phanerozoic in shallow seas when crustal movements were dormant. There it normally occurred in the form of strings of irregular *silex nodules* which are now found in the lime- stone series dating from these shallow seas. The most widely known example are those found in the chalk. These have, it may be remarked parenthetically, served as the basic material for the first human heavy industry, i.e. the manufacture of stone tools and weapons.

In contrast to the deep sea siliceous oozes, the silica of the silex nodules, *or flints,* is not a primary deposit. It originates from secondary silicification of pre-existing limestones, in which, for instance, originally calcareous fossils have become completely silicified. It was formerly thought that this secondary silicification occurred by a diagenetic process during the later history of the limestones. But newer research has shown that, although silicification during diagenesis occurs, this is a rare event. The silicification of the silex nodules must have taken place hard upon the sedimentation of the primary limestone, even before the formation of the next higher limestone bed. Such immediate, upon-the-scene replacement is called *syn-sedimentary*, as opposed to any possible later diagenetic silicification.

The origin of the silica necessary for the replacement of the primary calcium carbonate of the silex nodules is thought to lie in a temporary explosion of silica secreting organisms, such as silica sponges. The latter possess a loosely built outer skeleton composed of numerous minute quartz needles, which strengthen the living tissue. Upon the death of the animal the silica needles fall apart, and, due to their minute size and needle shape, dissolve easily in sea water. A temporary bloom of silica sponges or other siliceous organisms is thought to lead to such an accumulation of free silica on the sea floor that a silica gel develops. This could then crystallize around scattered foci in the limestone and replace the original carbonate by a molecule-for-molecule process (RUTTEN, 1956).

It will be clear that neither of the two Phanerozoic types of cherts, the siliceous deep sea oozes or the replacement nodules, is comparable to the cherts of the banded iron formations of the Early and Middle Precambrian. The banded iron formations did originate either on the continents, in lakes (if we follow GOVETT, 1966), or in shallow seas bordering the continents (if we follow LEPP and GOLDICH, 1964), and not in the deep oceans. Nor were diatoms, Radiolaria, siliceous sponges or other advanced silica secreting organisms, such as held responsible for the formation of the silex nodules, available during these early times of the earth's history (RUTTEN, 1957). This means that the mode of formation of the laminated cherts, so typical for these older periods, must have been quite distinct from that of the cherts formed during the Phanerozoic. In a way this is no more than paraphrasing the conclusion arrived at already in the preceding chapter, namely that all banded iron formations, being restricted to the Early and Middle Precambrian, must have formed under conditions which are quite different from those prevailing at present.

We have no inkling as to the process or processes which formed the earlier cherts. The fact that in the younger history of the earth cherts are formed, if not biologically, still derived from material derived from living organisms, might induce us to consider whether in the older history of the earth the cherts could derive from living material too. But I doubt that life at that time was already dense enough to produce the enormous amount of silica represented by the older cherts. I would prefer to think in the first place of inorganic reactions made possible by the scarcity of oxygen in the primeval atmosphere.

One thing which the newer cherts tell us, and which complicates our quest, is that both types of younger cherts are not in equilibrium condition with normal sea water. On the one hand the deep-sea oozes form only if the rate of sedimentation is higher than the rate of destruction by solution, whereas the silex nodules form when a temporary—and no doubt localized—silica gel forms. As was the case with several of the processes discussed in the preceding chapter, it is the kinetics that count, not the equilibrium conditions. This may well have been the case with the formation of the cherts of the banded iron formations also.

4. GEOCHEMICAL INVENTORIES

As has been remarked in the introductory section of this chapter, a branch of the earth sciences which might be fruitful for our topic is the study of geochemical inventories. We mean the branch of geochemistry that draws up global estimates for the abundance of elements of the earth's crust.

But the trouble underlying all geochemical inventories is that they are based on considerable extrapolation from the data known to us from limited parts only of the upper layers of the crust. Through direct observation we know the general composition of the crust down to several kilometers only, thanks to the information gathered from mines and oil wells. Moreover, in the deeply eroded cores of former mountain chains the composition of the crust such as it was formerly buried down to perhaps 30 km can locally be studied too. From these meagre data geochemical inventories arrive at a global picture only by the use of far-stretched extrapolation. For instance, since both the structure and the composition of the crust is at present quite different under the continents and under the oceans, a geochemical inventory will be different, whether one postulates stable continents and oceans, or whether one accepts a certain mobility. Or, in other words, whether one is a "fixist" or a "drifter". Most of the older geology is fixist in its assumptions, but during the last decades the drifters have come to the forefront in a remarkable way. Another difference will be found in one's geochemical inventories, depending upon one's interpretation of igneous rocks such as granites. These were formerly thought to derive from the mantle underlying the crust, the latter thus recieving continuous accretion during its history. But against this *magmatic* assumption others interpret granites as ultrametamorphic parts of the older crust, which locally has become fluid: a *migma*. In their opinion granites do not indicate an accretion of crustal material, but rather a re-cycling of the same crustal material over and over again (Fig.104). The latter interpretation has won much adhesion in the last decades (NIEUWENKAMP, 1966).

Geochemistry is therefore not able to give unambiguous answers to the questions one would like it to solve. Moreover, most geochemical inventories are

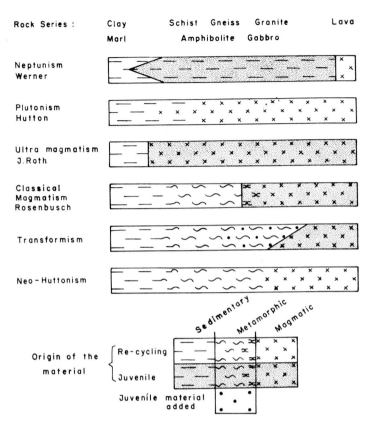

Fig.104. Review of the major petrological theories. The gray tint indicates that, according to the theory in question, the material is thought to be juvenile; the white tint that it is interpreted as re-cycled. At the top two continuous rock series are indicated, the upper line without, the lower line containing CaCO$_3$.

The major theory which held sway during the latter part of the former and the first half of the present century has been that of classical magmatism, mainly due to the German petrographer Rosenbusch, whereas modern petrologists tend more to hold transformistic, or even neo-Huttonian ideas. (From NIEUWENKAMP, 1966.)

based on fixist and magmatic interpretation of the data. There is, however, one notable exception, i.e. the elements making up the hydrosphere and the atmosphere, for which the expression "*excess volatiles*" has been coined by RUBEY (1951, 1955). When making geochemical inventories for the major rock-forming elements (Si, Al, Fe, Ca, Mg, Na, K etc.) and starting from fixist and magmatic assumptions, it seems evident to Rubey that the global quantities of these elements, such as they are at present found in sediments and as dissolved bases in sea water, can all be accounted for from the quantities of primary rocks weathered during geological time. Starting from a drifter's or migmatist's standpoint would hardly alter this conclusion. But neither of these assumptions can account for the quantity of a number of volatile materials (H_2O, CO_2, Cl, N, S and several others), which are much too abundant in the hydrosphere and the atmosphere to have been generated solely by the weathering of ancient rocks. The amount of these volatiles which are in excess over the amount that can be derived from rock weathering are called excess volatiles. Their global quantities are given in Table XXII.

TABLE XXII

GLOBAL QUANTITIES OF "EXCESS" VOLATILE MATERIAL IN PRESENT ATMOSPHERE AND HYDROSPHERE AND IN BURIED SEDIMENTARY ROCKS (10^{20} g)
(From RUBEY, 1951)

H_2O	16,600
Total C as CO_2*	910
Cl	300
N	42
S	22
H	10
B, Br, A, F, etc.	4

* The value for "total C" as is compounded from 248 "excess" units C and 506 "excess" units O.

Two possible explanations for the presence of excess volatiles have been proposed. Either the waters of the ocean plus all excess volatiles of ocean and atmosphere have been inherited from a primitive hot atmosphere which later cooled and separated into hydrosphere and atmosphere, or they have escaped from the interior of the earth through *outgassing* during geologic time. From the geochemical data it can be shown that an original hot atmosphere with a composition in accord with the quantities of excess volatiles would have been entirely unstable. Moreover, quite different minerals would have formed in the contemporary sediments. The second explanation, which is moreover in better accordance with the modern astronomical views referred to in Chapter 5, must therefore be preferred.

For our topic the most important aspect of the inventories of the volatiles lies in the global quantities of O_2 and CO_2 in relation to the amount of carbon fixed in sediments. If, as we have assumed, and as we will go into more detail in the next chapters, all present atmospheric free oxygen is biogenic and derived from the photodissociation of atmospheric CO_2, what will in this case have become of the carbon to which the CO_2 has been reduced? We know of course that it is fixed as plant material during photosynthesis, whereas upon the death of the organisms it can be fixed in sediments. Here it may give rise to the fossil fuels or caustobiolites, such as coal and oil. Their main constituent is carbon of biogenic origin.

If we now compare the global quantities, we find that there is at present in the combined atmosphere and hydrosphere much more O_2 than CO_2. The global quantity of free oxygen, mainly present in the atmosphere, is estimated at about $60 \cdot 10^{18}$ moles, whereas that of CO_2, mainly present in the hydrosphere, is estimated at only $3 \cdot 10^{18}$ moles (RUTTEN, 1966). To account for the present volume of oxygen, considerable outgassing of CO_2 must consequently have been going on all through geologic time. This is only to be expected, because CO_2 is the most common of volcanic gases.

The quantity of fossil caustobiolites, on the other hand, is estimated at $248 \cdot 10^{20}$ g C, or $20 \cdot 10^{20}$ g-at. C. Compared with the $60 \cdot 10^{10}$ moles O_2 of the present atmosphere, this seems ample. Most of the oxygen corresponding to the apparent surplus of carbon will have gone into the oxidation of the primary rock-forming minerals of the earth's crust. The difficulties encountered in drawing up more meaningful geochemical inventories, are, however, well illustrated by the sulfur cycle which will be reviewed in the next section.

5. THE SULFUR CYCLE

The sulfur cycle will serve as an example of a more detailed geochemical inventory, related to one element only. As seen from Fig.105, sulfur is found in large quantities on top of and in the crust of the earth, both in its oxidized form as sulphates and in its reduced form as sulphides. The data which have gone in establishing the cycle as presented in Fig.105 could be assembled thanks to the fact that sulfur has two stable isotopes, the common ^{32}S and the heavier isotope ^{34}S, which occurs only in an amount of about 4.2% of the lighter isotope. Fractionation and biogenic activities often preferably select either the lighter or the heavier isotope, and in this way quantities of "light" or "heavy" sulfur can be followed through the cycle.

In its oxidized form sulfur occurs mainly in solution in sea water and crystallized as gypsum in the so-called evaporites or salt layers. In fine-grained sediments of the black shale type sulphates are reduced biogenically, mainly by microbes of the *Desulphovibrio desulphuricans* type, to sulphides. Primary sulphides occur moreover in igneous and in metamorphic rocks and in ore veins. The amount of sulfur in rivers and fresh water lakes and in the atmosphere, finally, is negligible compared to the quantities contained in sea water and in the rocks just mentioned.

Although the amounts of oxidized and reduced sulfur can be well accounted in the sulfur cycle itself, a certain uneasiness creeps in when we compare the quantities involved in the sulfur cycle with those cited in the preceding section on free oxygen and carbon dioxide. The amount of sulfur present as sulphide is $9.7 \cdot 10^{15}$ tons, which is $9.7 \cdot 10^{21}$ g, a quantity roughly equivalent to $30,000 \cdot 10^{18}$ g-at. For

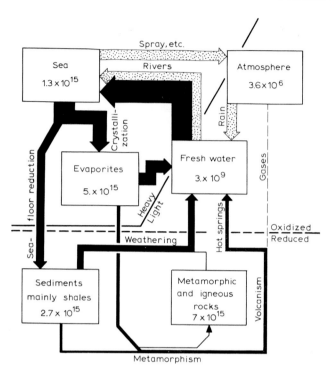

Fig.105. The geochemical cycle of sulfur. Masses are in metric tons. Most material above the heavy dashed line is oxidized to sulphate, whereas below this line it mostly occurs as sulphide. The heavy line separates the domains in which the isotopic composition of sulfur shows a slight domination of the ^{34}S isotope from those containing "light" sulfur, containing a slightly higher fraction of the ^{32}S isotope. (From Holser and Kaplan, 1966.)

the sulfur present as sulphate these figures are $6.3 \cdot 10^{15}$ tons, that is $6.3 \cdot 10^{21}$ g, which is roughly equivalent to $20,000 \cdot 10^{18}$ g-at. Assuming that all sulfur originally was present in its reduced form and assuming moreover that all oxidation of sulfur occurred by way of biologically liberated atmospheric oxygen, we ought to find at present fixed in the earth's crust $20,000 \cdot 10^{18}$ g-at. of carbon only as a corollary to the global amount of sulfur. This does not compare favourably with the total figure of $2,000 \cdot 10^{18}$ g-at. C cited in the preceding section.

Something, or even various things, seem to be wrong in the chemical inventories used. My first guess would be that there is of course much more carbon fixed in the crust, in a finely disseminated state, than is represented by the figure for the fossil fuels only. The global amount of this finely disseminated carbon, in sediments such as black shales, or the banded cherts of the Precambrian, is, however, difficult to assess. Another possibility, and one for which it seems entirely impossible to make a quantitative evaluation, is that part of the sulfur had already been oxidized during the very early history of the earth. As we will see in the next chapters, atmospheric oxygen was at that time produced by inorganic photodissociation of water and no fixed carbon was produced by this process. Although the actual level of atmospheric oxygen has remained very low during this early part of the earth's history, i.e. below the Urey level of 0.001 PAL O_2 (see Fig.127) oxidation would, as BERKNER and MARSHALL (1965, 1966) have repeatedly pointed out, be very active at that time because ozone would still form at or near the earth's surface.

Perhaps other processes have been at work too. I will not try to give a full evaluation of all possibilities, but I hope that the example given illustrates the difficulties inherent in the interpretation of geochemical inventories.

6. THE IMPORTANCE OF CLAYS

Clay is one of the most abundant rock types, both on the present surface of the earth and in sediments. From what we know of the oldest sediments laid down in Early Precambrian times, it seems to have been just about as abundant then. Clays are very fine-grained rocks. They consist of a group of related minerals, the *clay minerals*, whose study has only become possible through the development of Roentgen—or X-ray—analysis. All clay minerals are similar in composition and

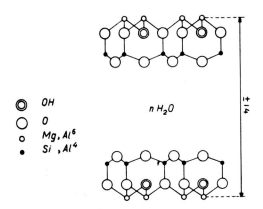

Fig.106. Schematic section of the crystal lattice of montmorillonite, one of the more simply built clay minerals. Each layer of silicate is separated by a loosely bound number of water molecules, through which clays swell when wetted and shrink when drying. Maximum distance between silicate layers is about 14 Å.

Clays may have been important for the origin of life in several ways. One is the extreme smallness of individual crystals and grains and their flaky form, which results in a very large surface-weight ratio. This would make possible the adsorption of great quantities of the various compounds of the "thin soup", a fact already emphasized by BERNAL in 1951.

Another important property of clays, which will be considered in the next section, is their ability to exchange cations when the composition of the surrounding fresh water or sea water changes, and by the buffering quality they consequently possess. (From BIJVOET et al., 1948.)

in structure. They are formed by silicates of aluminium and various other cations, and contain much water. They have a decidedly flaky structure, which resembles that of the micas, the main difference being that individual crystals are much smaller and that the chemical composition shows a far wider range of variation.

The crystal lattice of the clay minerals consists of well-defined, thin, parallel, wafer-like layers of strongly bonded ions, separated by voids in which only a small number of cations or of hydroxyl groups are found. The distance between the parallel layers varies both according to the number of cations and to that of the hydroxyl groups incorporated in the voids. In the case of montmorillonite, shown in Fig.106, water molecules can moreover be incorporated in the voids between parallel flakes. This enables clays to swell when wetted, and to shrink when drying out.

7. THE COMPOSITION OF OCEAN WATERS

Having talked so much about the "thin soup", which, in the mind of so many authors is equivalent to "oceanic thin soup", it seems appropriate to discuss here briefly the properties of sea water. The total volume of the present-day oceans has been calculated as $1.37 \cdot 10^9$ km^3, or $1.37 \cdot 10^{21}$ l. Nothing is known of the volumes during the earlier periods of the earth's history. Apart from local variations due to evaporation or to the influx of fresh river waters, the composition of ocean water is remarkably constant the world over. Its main constituents are given in Table XXIII. The salinity of the sea is 3.5%, its pH is 8.2 ± 0.2 and its redox potential is 12.5 ± 0.2.

TABLE XXIII

THE CONCENTRATION OF THE MAJOR COMPONENTS OF NORMAL SEA WATER
(From HOLLAND, 1965)

Components	Concentration (mmol./l)	Components	Concentration (mmol./l)
Na$^+$	480.80	Cl$^-$	560.70
Mg^{2+}	54.78	SO$_4^{2-}$	28.84
Ca^{2+}	10.46	HCO$_3^-$	2.38
K$^+$	10.18	CO$_3^{2-}$	0.27
Sr^{2+}	0.10	Br$^-$	0.85

The early interpretation of the salinity of sea water was that the oceans had originally started out as fresh water bodies, which were gradually salted by the influx of dissolved materials brought by the rivers. Parenthetically remarked.

this concept has served as one of the earliest "clocks" to arrive at an estimate of the age of the earth. Taking the present yearly global input of dissolved material by the rivers, and taking into account the present oceanic volume and salinity, it seemed easy to calculate a starting point. The method is, however, as insufficient as it seemed easy, because nothing is known about the input of dissolved material by former rivers, nor of the volume of the former oceans, nor also about amount of dissolved matter removed from the oceans by the deposition of limestone and evaporites, or of a possible re-cycling of this material through renewed erosion.

This view of an increasingly salty ocean was only seriously challenged in the 1950's, when, for a variety of reasons, it became generally accepted that the residence time of individual ions in the ocean is short in relation to the length of the history of the earth. Or, in other words, it was now thought that the salinity of ocean waters was the result of transitory, not of accumulative, processes. It was at that time admitted that the nonetheless remarkably constant value of the pH of ocean waters the world over was due to buffering in the carbonate system. This can, in its simplest form, be thought of as the relationships existing between $CaCO_3$, Ca^{2+}, CO_3^{2-}, HCO_3^- and H_2CO_3.

The latter view was thereupon challenged in 1961 by the Swedish chemist SILLÈN, who conceived a working model of sea water by using its main constituents and applying their reactivity constants and Gibb's phase rule. Sillèns' model, which is in excellent agreement with the actual composition of sea waters, has since been vigorously defended in a series of papers, often with aggressive titles, such as: "How has sea water got its present composition?". It has survived without a major change, so, apart from the initial 1961 publication, it is at this point enough to refer to SILLÈN (1967a, b).

The theory can be summarized as follows. The model contains only the eight components H_2O, HCl, SiO_2, $Al(OH)_3$, $NaOH$, KOH, MgO and CaO. Seven phases will be at equilibrium. These are (1) an aqueous solution of about the composition of sea water, (2) quartz, (3-6) the clay minerals kaolinite, hydromica or illite, chlorite and montmorillonite, and (7) phillipsite or some other zeolite. The composition, both of the aqueous solution and each of the solid states, will in this equilibrium be fixed, once the pressure, the temperature and the amount of Cl^- ions are given. Adding CO_2 will bring about the deposition of limestone, $CaCO_3$; adding FeO and Fe_2O_3 that of glauconite and goethite, but the composition of the solution, that is of sea water, will hardly change.

According to Sillèn, the stable composition of sea water is not the result of buffering by the carbonate system, but by what he, as a chemist, describes as "fine-grained alumino-silicates". That is what a geologist would describe als "clay minerals". Geochemists and clay mineralogists have since confirmed this view in various ways (HOLLAND, 1965; MACKENZIE and GARRELS, 1966; MAC KENZIE et al., 1967; MILLOT et al., 1966), and a simple example will be given to show how this process operates. In reality most clay minerals are much more complicated than the simplified version of a montmorillonite given in Fig.106. Moreover, there are not only two-layer, but also three-layer clay minerals, whereas the relative amount of cations in each of these layers is quite variable. But there is a general difference in composition between the clay minerals occurring in rivers and lakes on the continents and those found in the oceans. Most of the waters in rivers and lakes, and especially those of the larger river basins in which marls and limestones are found, contain an excess of calcium. These so-called *hard waters* tend to leach out the sodium and magnesium which was originally present in the crystal lattices of the clay minerals, which are consequently "degraded". Upon entering the ocean, where the water with its large amount of Na can be called *alkaline*, they are "regraded" (French: "agradation"). That is the sodium and the magnesium is again put in. It is this property of *degradation* and *regradation* dependent upon the composition of the surrounding water, which constitutes the buffering mechanism of clay minerals.

Schematically the reactions of degraded clay minerals upon entering the sea can be described as follows. A first reaction involves extensive base exchange. Calcium, which usually predominates over other cations in river-borne clay minerals, is exchanged for Na^+, Mg^{2+} and K^+ on contact with ocean water. The second reaction of river-borne clay minerals with ocean water consists of a reconstitution of degraded illite and chlorite, in which aluminum is the only cation left. This involves the uptake of potassium and/or magnesium atoms in the crystal lattices. A third reaction which possibly takes place involves the formation of altogether new—so-called *authigenic*—clay minerals from the combination of the dissolved material in river water and sea water (HOLLAND, 1965).

As an example, the reaction between a calcic river montmorillonite and a sodic ocean montmorillonite can be written as follows:

$$Ca_{0.17}Al_{2.33}Si_{3.67}O_{10}(OH)_2 + 0.33\ Na^+ =$$
$$Na_{0.33}Al_{2.33}Si_{3.67}O_{10}(OH)_2 + 0.17Ca^+$$

The reaction for the regradation—or reconstitution—of degraded alumino-silicates reads schematically:

$$3.5\ Al_2Si_{2.4}O_{5.8}(OH)_4 + Na^+ + SiO_2 + HCO_3^- =$$
$$3\ Na_{0.33}Al_{2.33}Si_{3.67}O_{10}(OH)_2 + CO_2 + 4.5\ H_2O$$

8. SILLÈN's "MYTH OF A PREBIOTIC SOUP"

Sillèn's model of the stable chemistry of ocean water suffers no direct effects of lower oxygen pressures (HOLLAND, 1965). Ocean water chemistry itself consequently can give us no clues to reconstruct the dependency on the oxygen pressure in the contemporary atmosphere. On the other hand, we might try to find from the reactivity values in Sillèn's sea-water model, at what level free oxygen would become prohibitive for the existence of free "organic" molecules. Or, in other words, above what atmospheric level of oxygen, molecules like CH_4 would be "burnt up" in the contemporary oceans.

The necessary calculations have been executed also by Sillèn, who expanded his considerations on the equilibrium between solid, liquid and vapour phases by incorporating the possible equilibria of carbon compounds—or "species with C", such as he, as a chemist, calls them—with ocean water. His formal results are that, for instance, CH_4 would only have "meaningful chemical activity"—which is a chemist's way of saying that it could exist—at extremely low pressures of oxygen (Fig.107). According to these calculations CH_4 could only be present as an impor- tant constituent below log $p(O_2) = -75$. This is, however, a purely formal result. For if we take the present amount of $6 \cdot 10^{23}$ moles of free oxygen, which leads to a partial pressure $p(O_2)$ of 0.21 atm, and multiply this with the Avogadro number, we find that there are at present only $3.6 \cdot 10^{43}$ molecules of free atmospheric oxygen present on the whole earth. A partial pressure of log $p(O_2) = -75$ would mean that the whole earth would only possess $6/10^{23}$ parts of one molecule of oxygen. Which is indeed an anoxygenic atmosphere.

These calculations stem from, an equilibrium chemist. They are based upon, and extrapolated from, the equilibrium and reactivity constants measured in solu- tions in which the ions in question occur in measurable concentrations. Such an

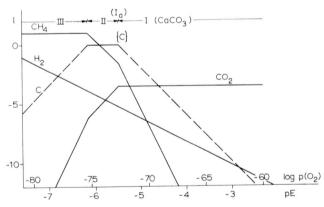

Fig.107. Activity logarithms l for $C_{(solid)}$, CH_{4gas}, CO_{2gas} and H_{2gas} as a function of log $p(O_2)$ and pE in Sillèn's model for ocean water.
        The ranges are: (I) $CaCO_{3(solid)}$ present. (Ia) which forms the boundary between I and II, both $CaCO_{3(solid)}$ and $C_{(solid)}$ and present. (II) $C_{(solid)}$ present, and (III) neither $CaCO_3$ nor C present. The diagram shows only the reducing part of range I. At present log $p(O_2) = 7 \cdot 10^{-1}$, whereas $pE = \pm 12,5$.
        CH_4 reaches above zero chemical activity, that is, its existence is limited to the field where the partial pressure of oxygen falls below log $p(O_2) = -75$. (From SILLÈN, 1965.)

extrapolation may be formally correct, but is of no importance for our problem. Of course, some of the equilibrium data are still insufficiently known in relation to the level of free oxygen. SILLÈN (1965) quotes, for instance, the buffering reactions between hematite and magnetite, which can be written as follows

$$12 \ FeOOH_{(solid)} \rightleftarrows 4Fe_3O_{4(solid)} + 6 \ H_2O + O_{2(gas)}$$

Dependent upon the reactivity values used, this equilibrium could be attained within the wide range of $p(O_2) = 10^{-1.1}$, which, incidentally is close to the present value of $p(O_2) = 10^{-0.7}$, as far down as $p(O_2) = 10^{-117}$. These are, however uncertainties which surely will be investigated in the future.

Leaving alone the uncertainties mentioned, we might still submit that such extrapolation of equilibrium reactions is entirely meaningless. Long before we arrive at such low concentrations, kinetics take over as the ruling factor, and no true equilibrium is ever attained. This paraphrases the conclusions reached already earlier that there was in general no true equilibrium between the lithosphere and the anoxygenic primeval atmosphere. As we saw, the type of sediments formed during the early history of the earth does not depend—or does not only depend— on the equilibria, but mainly on the kinetics of the relevant processes. That is, on the rate of oxidation versus those of weathering, transportation and sedimenta- tion. This does not invalidate Sillèn's calculations regarding the composition of ocean waters, which have been warmly welcomed and adopted by the geo- chemists concerned. It only refers to the unwarranted extrapolations cited above.

From the two next chapters it will be clear that we are most interested in this question of equilibria versus kinetics in the range of free oxygen of 0.001 PAL (= *Present Atmospheric Level*) to 0.1 PAL, which is equivalent to partial pressures of oxygen from around 0.0002 atm to 0.02 atm, that is for $p(O_2)$ to range from $7 \cdot 10^{-4}$ to $7 \cdot 10^{-3}$. For the understanding of the "myth of the prebiotic soup" the most important property would probably be the diffusion rate of molecules of oxygen through ocean water, which would determine whether "species with C" could exist in the primitive ocean. However, nothing is known as yet about the relative importance of kinetics versus equilibria at these levels of free oxygen.

As a way out of his absurd conclusion, Sillèn proposed a number of possi- bilities, two of which will be discussed here. The first is that life started from "organic" substances formed by equilibrium reactions, but in unusual surround- ings, such as highly acidic volcanic waters, or in layers adsorbed on clays in primeval lakes and lagoons, or in highly evaporated solutions. In such partly closed systems neither the amounts of the components, nor the other restrictions given for the total system, need to be valid. The second is that life started from "organic" substances formed by non-equilibrium reactions, for instance the in- organic photosynthesis produced by the shorter ultraviolet sunlight producing more of these materials than could be destroyed by other reactions.

The first mode of escaping the too stringent equilibrium constants has been cited already and will be further discussed in the next chapter. Its modern equivalent lies in lagoons, where, although oceanic salinity is stabilized at 3.5%, nevertheless gypsum and salt layers are deposited by evaporation in dry climates. In Chapter 10 we saw how this type of "unusual" environment of lakes and lagoons must have been much more common than at present in the base-levelled-continents and shallow shelf seas environment of the quiet geosynclinal periods of each successive orogenetic cycle. This is the reason why I as a geologist would prefer to think rather of this type of environment than of that of highly acidic volcanic waters, when trying to visualize the most probable of the "unusual" environments cited by Sillèn. In this type of environment we might also readily find the layers of clays in lakes or lagoons on which adsorption processes could have taken place. In this model BERNAL's (1951) "development of life on the shores of the oceans" (see section 4 of Chapter 10) will have been more probably a "development of life in lakes and lagoons". Moreover we have to take notice of the fact that, according to the equilibrium reactions cited above, the "thin soup" has most probably been a "lagoonal thin soup", or a "limnic thin soup", and not an oceanic one.

The second possibility indicated above, e.g. that life started from "organic" substances formed by non-equilibrium reactions, is to my mind just as important as the first one. It means that pre-life was maintained without being in equilibrium with its surroundings. Of course, life is not in equilibrium with its surroundings too, but this is accomplished by biochemical reactions within the cell, shielded from the exterior world by semipermeable membranes. Pre-life, without membranes, could only persist by a rate of production which was higher than the rate of its destruction. In view of the large influx of light quanta from the sun on the one hand and the difficulties of destruction by oxidation will have encounterd in anoxygenic hydrosphere, where oxygen was not as readily available as at present, this seems to be a good possibility, to say the least. But here again, we are diverted from equilibrium reactions proper to the rates governing synthesis and destruction, which is basically a problem of kinematics.

Our conclusion therefore must be that equilibrium conditions seem to exclude the formation of the "thin soup" in normal oceanic conditions, even under an anoxygenic atmosphere. It could, on the other hand, develop in unusual surroundings, such as lakes and lagoons. But even there it probably was under non-equilibrium conditions with its surroundings. Equilibrium calculations, it follows, are to be used with a certain reserve, when applied to the study of the origin of life.

9. STABLE ISOTOPES IN ORGANICALLY AND INORGANICALLY
FORMED COMPOUNDS

In Chapter 3 we have come across the fact that many elements consist of
a mixture of isotopes which are chemically identical and only differ in physical
properties, such as atomic weight. In that chapter we only discussed those isotopes
which are unstable and can be used as physical clocks in absolute dating. There
are, however, many more stable isotopes than unstable isotopes, and at this point
we will turn our attention to the former group.

Only one of the many aspects which have been studied in relation to stable
isotopes will be treated here, i.e. the question of biogenic versus abiogenic origin
of certain compounds. In this regard interest has centered on the elements carbon
(C), oxygen (O) and sulfur (S). Together with the literature on other elements
important for isotope geology, research on these elements up to 1959 has been
fully reviewed by RANKAMA (1963). For sulfur the newer publications by DECHOW
and JENSEN (1965) and HOLSER and KAPLAN (1966) may moreover be referred to.
I will limit myself to a discussion of the isotopes of carbon, since these are not
only the most important for our subject, but may also serve as an example of the
problems encountered in the study of the stable isotopes of other elements.

In the atmosphere four types of carbon dioxide are found, which differ in
their isotopic composition, $^{12}C^{16}O_2$, $^{13}C^{16}O_2$, $^{12}C^{16}O^{18}O$ and $^{12}C^{18}O_2$. The
first three are the most abundant, whilst ^{12}C predominates by far over ^{13}C
(Table XXIV).

TABLE XXIV

ABUNDANCE VALUES OF ISOTOPIC FRACTIONS OF ATMOSPHERIC CARBON DIOXIDE
(From RANKAMA, 1963)

Molecule	Centimetres (NTP) reduced to sea level
$^{12}C\ ^{16}O_2$	320
$^{13}C\ ^{16}O_2$	1.54
$^{12}C\ ^{16}O\ ^{18}O$	0.67

At present there is a definite fractionation of carbon isotopes during the
assimilation of CO_2 by photosynthesis in green plants. These show $^{12}C/^{13}C$ ratios
varying between 90.0 and 92.9%, as measured against the ratio found in the so-
called "Chicago standard", which is based on the composition of a fossil belemnite.
This same ratio for non-biogenic carbon varies from 88.0 to 90.2%. It is not known
in detail to what process or processes this difference in isotopic ratio of carbon is

due. Photosynthesis consists of a series of kinetic processes and not of equilibrium reactions. Consequently, the kinetics of each succeeding step are important and kinetic reaction will in general favour the lighter isotope. On the other hand there are many other processes operating during the metabolism of green plants, one of these being respiration. Several of these are equilibrium reactions and it is known that during association–dissociation reactions the heavier isotope is in general favoured for compounds with the strongest bonding. A reasonable explanation for both the deviation of the $^{12}C/^{13}C$ ratios and of their variation, as found in present-day green plants, is that the deviation is mainly caused by photosynthesis and that the variation is due to the influence of other metabolic processes.

Whether this is an acceptable explanation or not, the fact remains that when we go back in the geological record, a comparable difference in isotope ratios is found to exist. The relative amount of ^{12}C is always higher, if the carbon occurs as clearly biogenic coal and graphite, than when it is found in igneous rocks. Probably the fixation of carbon in the latter environment was more fully controlled by equilibrium processes, which would favour an enrichment of the ^{13}C isotope.

The important point for our topic now lies in the fact that analyses of coal and graphite found in the sedimentary series of Early and Middle Precambrian show isotope ratios comparable to those of present-day plants. The conclusion has therefore been drawn, first that all of this carbon is biogenic in origin, and second, that life was already present on earth during the time of deposition of the oldest sedimentary rocks known to science.

This conclusion, if not actually erroneous, certainly is unfounded at present. Similarity of isotope fractionation does not indicate more than similarity of processes of formation. In the case in question there can be no doubt that during inorganic photosynthesis under the primeval atmosphere kinetics has played as important a part as during organic photosynthesis under the present atmosphere. Our conclusion must therefore be in the negative, i.e. that ratios of stable isotopes of carbon comparable to those now occurring in green plants and in biogenic deposits do not prove a biogenic formation of ancient sedimentary coal or graphite. Similar considerations apply to the stable isotope ratios of other elements.

10. METABOLISM OF LIME-DEPOSITING ORGANISMS

In view of the antiquity of limestone-depositing organisms, as related in section 8 of Chapter 12, their metabolism must in the past have been different from the present limestone-depositing algae. 2.7 billion years ago the earth was surrounded by an atmosphere in which the O_2 concentration was at most only one percent of the present (Fig.127) and the metabolism of the living world must at that time have been quite distinct from what it is today. This does, however, not affect directly the faculty of lime deposition. MacGregor's biogenic limestones could equally well have been deposited by primitive photosynthetic organisms, similar in their metabolism to the present-day green-algae, as by organisms with a metabolism similar to the fermentations and other reactions at present encountered among various kinds of anaerobic organisms. Although it would lead me too far to cover this aspect in detail, a summary of the possibilities seems desirable. It is based on information kindly supplied by Professor C. B. van Niel (personal communication, 1967).

Lime deposition may be the result of inorganic as well as of organic processes. The former are illustrated by calcite deposits such as formed in stalactite caves. The main product of present-day biogenic limestone deposition is the *travertine* found around fresh water wells in sunny climates.

Calcite deposits in stalactite caves are found as layers covering the cave floors, as stalactites and as stalagmites. All are formed by the partial evaporation of CO_2 from calcium carbonate-rich groundwater, circulating in the rock surrounding the caves and dripping from their ceilings. The equation can be summarily written as follows:

$$Ca(HCO_3)_2 \longrightarrow CO_2 + H_2O + CaCO_3$$

In constrast to the situation in dark caves, sunlight will permit the mass development of algae, whose photosynthetic activity deprives the water of CO_2, with the consequent precipitation of $CaCO_3$, according to the formula:

$$Ca(HCO_3)_2 \xrightarrow{\text{light}} (CH_2O) + O_2 + CaCO_3$$

In this and subsequent equations pertaining to photosynthesis, the term (CH_2O) symbolizes the organic matter synthesized by the organisms. Travertine, widely used as a building or a finishing stone (e.g. "Stazione Termini" in Rome), is, for instance, widely developed in Mediterranean countries, wherever limestone is present in the subsoil. It also often occurs around hot springs, the Mammoth Hot Springs area in Yellowstone Park being a spectacular example. Travertine is associated with localities where calcium bicarbonate-rich groundwater emerges into a sunlit environment.

Inorganic deposition of limestone through evaporation of spring water will of course occur concurrently with the biogenic deposition through photosynthesis. However, in a sunlit environment the volume of inorganic deposition is so much less than that produced through photosynthesis by algae, so as to be hardly noticeable. It is only in a dark environment, where photosynthesis is absent, that we will find the inorganic limestone deposits so typical of limestone caves.

But biogenic limestone can, perhaps as surprisingly for other laymen as it was for myself, also be deposited by several kinds of bacteria able to grow in an O_2-free environment. They belong to the groups of the chemo- and photoorganotrophic microorganisms and the photolithotrophic green and purple sulfur bacteria (see section 7 of Chapter 7). These, or metabolically similar creatures, may have been important agents of limestone deposition under the anoxygenic primeval atmosphere. Moreover, because of the probability mentioned earlier that pre-life and early life have been contemporaneous over a long time span, it is probable that a group of photo"organo"trophic organisms has to be added.

Generally speaking, deposition of $CaCO_3$ can have been caused by such organisms as a result of metabolic activities which tend to increase the alkalinity of the milieu. This causes a shift towards CO_3^{2-} in the equilibria:

$$H_2O + CO_2 \rightleftarrows H_2CO_3$$
$$H_2CO_3 \rightleftarrows H^+ + HCO_3^-$$
$$HCO_3^- \rightleftarrows H^+ + CO_3^{2-}$$
$$H^+ + OH^- \rightleftarrows H_2O$$

And if the concentration of CO_3^{2-} in an environment containing Ca-ions exceeds the solubility product of $CaCO_3$ (about $1 \cdot 10^{-8}M$), the latter precipitates. Specific examples of such anaerobic processes known to occur today are the methane fermentation, the dissimilatory sulfate and nitrate reductions and the bacterial photosynthesis. It is reasonable to assume that these or biochemically similar metabolic processes were among the early manifestations of life on earth, and consequently could have been operative in the formation of Early and Middle Precambrian biogenic limestone deposits.

In the methane fermentation (SÖHNGEN, 1906; SCHNELLEN, 1947; BARKER, 1956) salts of organic compounds, particularly of fatty acids, are oxidized with the simultaneous reduction of CO_2 to CH_4. Söhngen established that the fermentation of Ca-acetate, Ca-butyrate and Ca-caproate can be expressed, respectively, by the following equations:

$$(C_2H_4O_2)_2Ca + H_2O \rightarrow CH_4 + CO_2 + CaCO_3$$
$$(C_4H_7O_2)_2Ca + 3 H_2O \rightarrow 5 CH_4 + 2 CO_2 + CaCO_3$$
$$(C_6H_{11}O_2)_2Ca + 5 H_2O \rightarrow 8 CH_4 + 3CO_2 + CaCO_3$$

Because CO_2 is produced in these fermentations, one might perhaps expect that the Ca-ions would remain in solution as bicarbonate. However, the simultaneous formation of the water-insoluble methane, which must escape into the atmosphere, tends to sweep out the CO_2 as well. Thus the inside of the bottles in which such fermentations take place soon becomes coated with a lime deposit.

During sulfate reduction, in which appropriate oxidizable substance is oxidized with the concomitant reduction of sulfate to sulfide (POSTGATE and CAMPBELL, 1967), the increase in alkalinity is due to the fact that sulfuric acid is a very much stronger acid than is hydrogen sulfide. Similarly, during nitrate reduction the strongly acidic nitric acid is converted to non-acidic reduction products such as N_2O, N_2, NH_3, along with the oxidation of a substrate (BEIJERINCK and MINKMAN, 1910; ALLEN and VAN NIEL, 1952).

The currently existing photosynthetic bacteria, finally, are represented by three groups: (*1*) the anaerobic green sulfur bacteria (*Chlorobacteria*), (*2*) the anaerobic purple or red sulfur bacteria (*Thiorhodacea*) and (*3*) the facultatively aerobic red and brown non-sulfur bacteria (*Athiorhodacea*). All three groups share two outstanding features. Firstly, in the absence of O_2 they are dependent for their development on a supply of radiant energy, with wavelengths in the region between 7,300 Å and 10,000 Å the most effective. And secondly, they do not produce O_2.

The first two groups are photolithotrophs (see section 7 of Chapter 8). They oxidize inorganic sulfur compounds, such as H_2S, elementary sulfur or thiosulfate, to sulfate; with the concomitant assimilation (i.e. reduction) of CO_2 (VAN NIEL, 1931; LARSEN, 1952). Their chemical activities are approximately expressed by the following equations:

$$2H_2S + CO_2 \xrightarrow{\text{light}} (CH_2O) + H_2O + 2S$$

$$H_2S + 2H_2O + 2CO_2 \xrightarrow{\text{light}} 2(CH_2O) + H_2SO_4$$

The *Athiorhodacea* are photoorganotrophs which normally use organic substances, principally fatty acids, and convert them into cell material. If the substrate is more reduced than (CH_2O), this is accompanied by the assimilation of CO_2 (VAN NIEL, 1944), The utilization of acetic and butyric acids, for example, may be expressed by the overall equations:

$$C_2H_4O_2 \xrightarrow{\text{light}} 2(CH_2O)$$

$$C_4H_8O_2 + H_2O + CO_2 \xrightarrow{\text{light}} 5(CH_2O)$$

All these processes lead to an increase in pH owing to the conversion of acidic substances, including CO_2, into neutral cell material. This, in turn, causes the precipitation of $CaCO_3$ in any Ca-ion containing milieu.

Summarizing this information on the metabolism of organisms at present engaged in lime deposition, we may distinguish between two groups. First are the green algae who live in an aerobic, oxygenic environment, and are eucaryotic. And second the anaerobic, procaryotic, fermenters and photosynthesizers. Although under the present atmosphere limestone deposition by algae is by far the most important, it is reasonable to postulate biogenic deposition of limestone by organisms similar to the present fermenters and anaerobic photosynthesizers in the earth's early history. Such early life was, of course, aradiatic for the shorter ultraviolet sunlight, but, being anoxygenic, it could exist in free contact with atmosphere and hydrosphere.

VAN NIEL (1956) has postulated on theoretical grounds that the earliest organisms participating in limestone deposition were non-photosynthetic, and that these have been followed by microbes displaying a bacterial type of photosynthesis as a logical intermediary stage in the evolution leading to green-plant photosynthesis. But we must, of course, in such reconstructions always be aware of the dangers of comparative biochemistry (see section 5 of Chapter 9).

11. PROCESSES OF FOSSILIZATION AND THE DISTINCTION BETWEEN PROCARYOTIC AND EUCARYOTIC FOSSIL ORGANISMS

The remarkable successes attained during recent years in the search for fossils in the Precambrian has in some cases resulted in too optimistic ideas about the possibility to distinguish between procaryotic and eucaryotic fossil organisms. Moreover, a serious misconception frequently found in the literature is that these fossils are "organically or biologically" preserved. Fossilization, in the overwhelming majority, is a replacement of organic material by rock-forming minerals. Such a replacement will have its least visible effect on hard parts, such as shells, skeletons and teeth, which the organisms may already possess during their life, and which already consist largely of mineral substances like calcite, silica or phosphate. But even here recrystallization, and often partial or wholesale restitution will take place. Soft parts have as such only been preserved in extremely uncommon circumstances, such as burial in an asphalt pit or under an ice cap. Even in coal swamps the original organic material is first broken down and only conserved as newly polymerized humic acids.

During *silicification*, which is the normal fossilization process for Precambrian fossils, wholesale replacement occurs. To give an impression of the havoc silicification entails, even in the Phanerozoic, Fig.108 and 109 are included. The first is a photograph of a living radiolaria, the second of a thin section of fossilized radiolaria in a chert of Mesozoic age. Only the hard shell is conserved during fossilization, nothing of the delicate spines, whereas the whole interior, and probably the wall too, is completely recrystallized.

The organisms of the Precambrian had no hard shells. The morphology of the original organism is, in exceptional cases only, nevertheless preserved during the molecule by molecule replacement of the silicification process, but only because remnants of the biological material have migrated towards the outer boundary of the former organism and effect a distinct coloration along this boundary. The original biological material has been transformed during this process. Most of it has been transformed to kerogen and only minimal amounts of degraded original

Fig.108. Photograph of a living radiolaria, showing the needle-like spines surrounding the globular test on all sides; × 250.

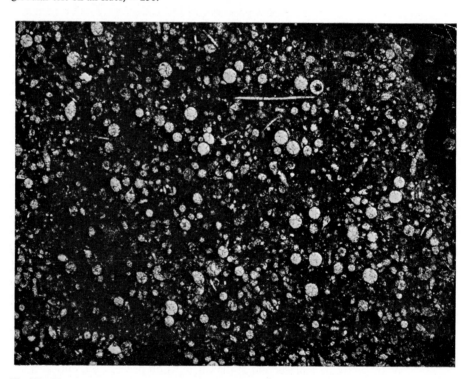

Fig.109. Photograph of a thin-section of fossil radiolaria of the Mesozoic of Curaçao. Nothing remains of the spines, only the wall of the test with its radial structure. During further silicification this will also be destroyed, and the—still recognizable—radiolaria will only be represented by clear blobs of silica. As can be seen from the picture taken under polarized light, the interior of the cell is made up by an irregular aggregate of newly formed quartz crystals, which show various shades between light and dark according to their optical orientation.

material remain. It is consequently quite true that "biological material" has been preserved, some way or other, in these fossils. But it is not true, on the other hand, that the organisms have been "biologically preserved" during fossilization.

It follows that during fossilization the internal structure of the cell is almost always destroyed. The inner part of the cell is homogeneized and replaced by new material. This process is so common, even for deposits of younger age, that it is indeed extremely rare to find a fossil where the inner structure of individual cells has been preserved. Normally, it is therefore impossible to distinguish in fossil material, whether the former organisms had procaryotic or eucaryotic cells.

The new wave of interest in Precambrian fossils may eventually shed more light on this problem. A first step has already been taken by SCHOPF (1968) in his study of the Bitter Springs Formation of Central Australia (Fig.110). In carbonaceous cherts of this formation a rich microflora has been found, which need not to interest us particularly here, because it is of the same general evolutionary level as that described by Pflug from the Belt Series of North America (see section 6 of Chapter 12). But the preservation of the Australian microorganisms is so good, that in several cases internal cellular structures seem to be preserved, which can

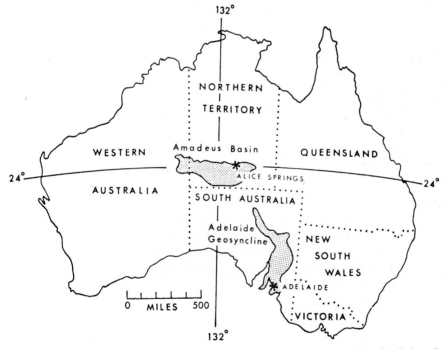

Fig.110. Map of Australia showing the two basins from which Late Precambrian fossils have been described. The Bitter Springs Formation is found in the Amadeus Basin, the fossils being found east of Alice Springs. The much younger Ediacara fauna, described in section 5 of Chapter 12, is found in the southern part of the Adelaide Basin. (From SCHOPF, 1968.)

be interpreted as the residue of a nucleus (Fig.111). If this interpretation is correct, these fossils represent remnants of eucaryotic cells. Moreover, in fossils of another organism found in the same deposit, it seems as if various stages of the development of a single species have been preserved. If this is so, these stages can be assembled in a suite which is similar to the mitotic division occurring at present only in

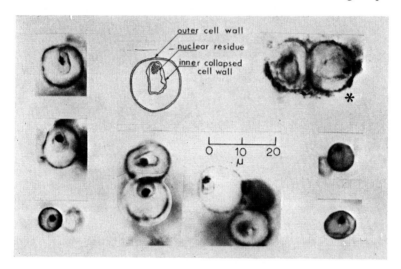

Fig.111. Fossilized microorganisms from the Bitter Springs Formation in Australia called *Caryosphaeroides pristina*.
Within the cells remnants of intracellular structure are clearly visible. As is shown in the drawing these have been interpreted to represent a nuclear residue and an inner cell wall. This indicates that the original organisms have been eucaryotic. (From Schopf, 1968.)

eucaryotic cells (Fig.112). Although the second interpretation is less convincing than the first one, based on direct observation of intracellular structures, both of them taken together seem to offer valid proof of the presence of eucaryotic organisms at the time of the sedimentation of the Bitter Springs Formation.

The age of the Bitter Springs Formation is not exactly known, but it must be around 1 billion years. It is not surprising to find eucaryotic organisms already in existence at that time, but it shows the way in which paleontological studies might in future indicate at what time the eucaryotic cells developed. It is only through detailed work on extremely well-preserved material that reliable data in this vein may be acquired.

It must at this point be stressed that a fossil cell in which no inner structure is visible does not prove that the organism was procaryotic. Accordingly, statements to the effect that "life was still procaryotic at that and that time" are not based on fact. And when such statements stem from paleontologists of renown, they are downright misleading. Such statements are made because in the Early and Middle Precambrian only fossils without inner structure have as yet been found. It is, however, not known whether this is because the former organisms really were procaryotic, or whether it is due to the destruction of intercellular organelles during fossilization.

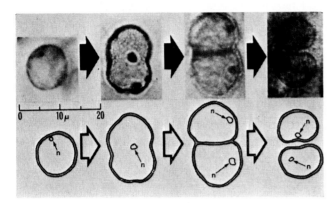

Fig.112. Cells of a fossilized microorganism from the Bitter Springs Formation of Australia called *Glenobotrydion aenigmaticus*, which show different forms and moreover a different position of the black dot found within the cells thought to represent the fossilized remnant of a nucleus.
    Arranged in a series these various stages of a single species of a microorganism can be interpreted to represent mitotic division characteristic for eucaryotic cells, as is indicated in the series of drawings. (From SCHOPF, 1968.)

Similarly, it is at present impossible to state when the eucaryotic organisms first arrived on earth. Only when we find fossilized organelles within a fossil cell, can we be sure that the organism in question has been eucaryotic. Unfortunately this happens to be such a rare event that statements such as "The first eucaryotic organisms date back to this or that point in the Precambrian" are misleading, too. As long as we have only one well-documented description of eucaryotic cells in the Precambrian—those found by SCHOPF (1968), as reported above—it is some- what early to speak about their stratigraphic distribution.

12. THE ALGAE AND (PALEO)BOTANICAL NOMENCLATURE

Another misleading situation exists in regard to the nomenclature of the algae. This rests upont a difference in the principles underlying the nomenclature of higher plants and animals and that used in paleobotany and in some groups– amongst them the algae–of neobotany. In higher plants and animals the nomen- clature is based, as far as possible, on genetic relationships. Two species of higher plants and animals, although closely similar at first sight, are classed in unrelated taxa, whenever it can be shown that they differ in their anatomical structure, or in certain aspects of their life cycle, or in some other way which indicates that they are not related genetically. This modern attitude in the classification of living forms goes back in its essence to John Ray (1627–1705), who, for instance in 1693, classed the dolphin and the bat with the mammals and not with the fishes and the birds.

The rules of botanical nomenclature differ from those in zoology in permitting in some cases a classification based only on outer resemblance. This applies to most of paleobotany and to both living and fossil representatives of the algae. Such organisms are classified in so-called *form groups*, solely on the basis of their morphology. The reason for this is that for these groups a genetic classification cannot yet be set up. So, one has to contend oneself with the next best, i.e. a classification based on outer form.

All this has, of course, been laid down quite strictly in the codices of the international rules on botanical nomenclature, but it remains confusing to the layman. Almost anyone amongst us has been brought up so exclusively on the genetically based classification of higher plants and animals, that wherever he finds the mention "related species" he understands this to mean species which, however different they may be in outer form, have a similar anatomy, and have the same ancestry. For him a bat is not a "species related to a bird", nor is a dolphin "related to a fish". Such a statement, however, would be acceptable according to the form group classification of paleobotany.

So, when an eminent paleobotanist like Professor E. S. Barghoorn tells us that the Gunflint fossil *Animikiea septata* (BARGHOORN and TYLER, 1965; see section 13 of Chapter 12) "in general appearance, size and form resembles species of the extant (= present-day) genera *Oscillaria* and *Lyngbya*", such a statement is undoubtedly correct in paleobotanical nomenclature. To the general public such a statement is misleading, because it is not aware that this means no more than what it says, i.e. that there is a resemblance of its morphology, of its outer form only. No genetic meaning is implied, according to paleobotanical nomenclature, but because it so happens that the two genera mentioned belong to the blue-green algae, a non-botanist will get the idea that the fossil forms must have belonged to the blue-green algae too. A similar misleading inference will forcibly be drawn by the non-paleobotanist, when he reads that a certain form of *Gunflintia minuta* BARGHOORN is said to be "reminiscent in structure of certain blue-green algae and of an iron bacterium". For he will not realize that "structure", as used in this context, applies to outer form only.

In the microbial world the inferences drawn from outer resemblance can be much farther off the mark than is the case in higher plants and animals, because of the great variety in the processes of metabolism shown by microbes (see section 8 of Chapter 8). Translated into the context of higher plants and animals we would, when applying the concept of form groups to their nomenclature, not only classify a dolphin with the fishes and a bat with the birds, but we would classify stick insects and leaf insects together with the trees on which they feed. And not only that, we would consider them to be extremely closely related species, because, as every one knows, they do show such a perfect resemblance in outer form.

For our quest into the circumstances surrounding the origin of life on earth we are, of course, interested in real—that is functional and genetic—similarity, not in resemblances in outer form only. Paleobotanical statements such as cited above, although correct on a nomenclatorial base, should, when addressed to a more general public, state clearly that they are meaningless when it comes to questions of kinship.

13. ORGANIC PROCESSES AND THE ENDOGENIC PROCESSES OF THE EARTH

The subject of this section is a controversial one and it cannot be given full treatment here. It is nevertheless important to realize that there is a problem, whether organic developments has been influenced by endogenic processes or not. The title chosen here is a little wider than the usual formulation of the problem, which is whether during the Phanerozoic so-called "faunal crises" are related to endogenic processes such as mountain building, or not. Embracing the total history of life, and taking into account both the evolution and the "crises", we might ask whether the main milestones in the evolution of life—such as, for instance, its first appearance, or the first appearance of eucaryotic cells, or of animals, etc.—are related in time to endogenic processes or not.

For the Phanerozoic the question is, as said above, controversial. There is some vague correlation between the first and last appearances of certain animal groups and preceding orogenetic periods, but there certainly is no strict time dependence. Personally I am always reminded of the farmer who obstinately states that a change of weather is correlated with the phases of the moon, but who refuses to tell in which way it is correlated, if good or bad weather will follow a certain phase, nor how many days will lapse between the appearance of a certain phase and the turn of the weather.

I believe that a good deal of the emphasis laid by some paleontologists on the existence of "faunal crises" is due to over-simplification and over-schematization. As to the over-simplification, it must be remembered that geologists usually cramp, if not the history of all of the Phanerozoic, at least the history of a complete system into a single-page diagram (see, for instance, NEWELL, 1967). Periods of accelerated evolution, and we know such periods occur both in the development and in the dying-out stages of the history of certain groups of animals, tend to be represented by a single line. In reality they will have been of the order of 10 million years or even longer. But notwithstanding their evident gradualness, when followed in detail, the faulty impression of a sudden "crisis" becomes anchored in our minds, as a result of the over-simplification representing such a period by a single line in the diagrams.

The over-schematization results from the fact that the major time boundaries in geology, the boundaries between systems or those dividing systems in two or three parts, have originally been based on the evolution of a relatively small group of fossils which were well documented in the area at the time when the main boundaries in geologic history were set up. The major boundaries were drawn on the evidence supplied by, for instance, corals, or brachiopods, or ammonites, without taking into account the evolution of other major groups of fossils. The time ranges of the fossils belonging to those other groups were—and still are—often but imperfectly known, mainly for two reasons. For one thing, they may not be well represented in the area in which the boundary in question was set up, whilst for another thing correlation with other areas in which those groups are well represented, is often difficult. Practically every boundary between the systems of the Phanerozoic, and a number of those dividing a system into a lower-, middle- and upper part, have over the last decades been the subject of one or more symposia of the paleontologists concerned, without their reaching a clear outcome.

A result of this uncertainty is that the ranges of such groups, to be on the safe side, are given as including all of the system or systems in which they are known to occur at some point. This leads to the faulty impression that all of these fossils, belonging to unrelated groups, have lived exactly during one or more geological systems, and thus to the equally unfounded impression that explosive evolution, over a larger part of the animal world, has taken place exactly at the base of some systems, whereas "faunal crises" have occurred at the top of the same or other systems.

The most striking example of such over-simplification, which might be treated in some detail here, is the Permian, where the two richest fossil localities are found on the island of Timor. The marine faunae found in these localities are not only extremely varied, but also mutually exclusive. That is, almost no species found in one locality are found in the other one. Since both localities represent an ecologically similar coral reef environment, this can only mean that they are not of the same age. The fauna present in one locality had died out and was almost completely replaced before the time of deposition of the other. Both faunae, it follows, have lived during a part of the Permian only. But because of their exclusiveness, and because they also show no relationships to other marine Permian faunas, such as found in Kansas or on the island of Sardinia, we have no way to find out which of the two faunae in either of the localities is the younger, and which the older. This can, moreover, not be established from the regional geology, because these Permian fossil localities are incorporated in different tectonic structures formed during the Alpine orogeny, and as such cannot be correlated in the field.

Although we thus know definitely that the animals found in these two fossil localities of the Permian lived only during a part of that system, all textbooks list the many hundreds of genera and species from Timor as "Permian", as extending exactly from the base to the top of this system. Now it so happens that the "faunal crisis" at the end of the Permian is the one most evident in the history of the Phanerozoic. Since it rests, in its major part, on the evidence from the fossils from Timor, it will be clear that the factual basis for the occurrence of "faunal crises" during the Phanerozoic stands in need of a critical review.

The whole question is, of course, still under discussion, and I know that I have many adversaries who are fervently in favour of a strict correlation between "faunal crises" and certain endogenic processes. Curiously enough, organic evolution, "faunal development", is not nearly as often referred to in this context. This is important for our subject, since we are more interested in the origin of life and in its deployment, than in its later history of one group of animals taking the lead, whilst other groups of animals die out.

In regard to this deployment of life, as opposed to the "faunal crises", we might, for instance, for the Phanerozoic cite the evolution at its very base, when "animals began to wear shells". This took place during a quiet geosynclinal period, without any orogenetic revolutions. So quiet was the history of the earth during that time that it is even in many places difficult to draw the boundary between the Precambrian and the Cambrian in the orderly succession of a concordant series of sedimentary rocks. For other major developments of the animal world during the Phanerozoic, such as those of the first amphibians, that of the first mammals, many of the main groups of ammonites and pelecypods, and many others, one also finds that they took place during the quieter periods of the earth's history and cannot be brought in any sensible correlation with endogenic processes such as the major phases of mountain building. So, as far as the evidence for the Phanerozoic goes, we may conclude that, even if a correlation between "faunal crises" and the major orogenetic periods is mistakenly sustained by many geologists, there is no such correlation in regard to the major developments of life.

"If this is so", the reader may well be tempted to ask, "why then has the development of life in the Precambrian been bound up so tightly with the boundary between Middle and Late Precambrian, which, as interpreted by GOLDICH (1968), and as followed in this text, is a division based on the dating of igneous rocks, and thus on a major orogeny". The answer is easy and simply reads: "For want of better data".

As we have seen, the stratigraphy of the Precambrian is still in its infancy. Also it is still beset by the difficulty of correlating the old and meaningless divisions based on relative dating with the newer classification based on absolute dating. As long as this is the case, we can, for a problem like ours, in which we have to have world-wide correlations, use only the simplest of subdivisions. The regional geologist, interested in a limited part of a single old shield, may well use a lithostratigraphic division which employs terms such as, say, the "Huronian". But he will not be sure whether his "Huronian" covers the same time span as that of another geologist's "Huronian" in another old shield, or even in another part of his own old shield. In fact the chances are that they do not.

To conclude, we have been forced to use the subdivision of the Precambrian based on absolute dating and thus on a succession of major orogenies, because this is at present the only global chronostratigraphy available to us. I am convinced that further studies will show how the main developments of life have taken place during the quiet geosynclinal periods. It is probable that, just as in the Phanerozoic, they will not have been related to the major orogenies, and have occurred more or less at random in relation to such a succession. But we need many more data to prove this, and at present such a refinement of the history of life is impossible.

14. UNIFORMITY OF SURFACE TEMPERATURE OF THE EARTH

There is another aspect to be mentioned in this chapter, that is the uniformity of surface temperature of the earth, which apparently has persisted over, say, the last three billion years. This is not so much a question of the origin of life, but it has made possible its conservation and further evolution.

For if, at one time or another, life had become exterminated, and no further biogenic oxygen would have been produced, the oxygen existing at that time would have been quickly used up in oxidation of rock minerals. A situation comparable to the primeval one would have become reinstated. From that point onward all early processes of inorganic photosynthesis and those of biopoesis would have had to be repeated. However, the various selective processes operating during biopoesis and the further evolution of life are influenced by so many independent variables, such as mutations and environmental factors, that the eventual outcome of such a repetition would certainly have been different from the former pattern. Even with a similar type of biochemistry, the morphological expression of such a second cycle of life on earth would have been quite different from the first one. Since up to now nothing is known about such a repetition of the evolution of life on earth, it seems safe to assume that over the last three billion years or so the temperature at the surface of the earth has always remained favourable for life.

Life has been protein-based over this period of approximately three billion years, and we know that protein is a subtle compound. It survives neither freezing nor heating over any geologic length of time. It follows that all during this period the average temperature of the earth cannot have varied more than a few centigrades. Or, in other words, the temperature of the earth has shown remarkable stability all during geologic history.

This statement is perhaps in need of qualification for the general reader, who, by his familiarity with the normal geological literature, may have become convinced of the strong variations in temperature in the geologic past, leading, for instance, to the Ice Ages. It is, of course, true that there have been Ice Ages in the geologic past. The most recent one even occurred in what, in geologic history, can be called a very short while ago, within the last fifty thousand years. It is even probable that we do live at the moment, not in a Postglacial, but in an Interglacial Age, although, parenthetically remarked, man may well prevent, willingly or unwillingly, the advent of the next glacial period by the effects of industrialization. Other Ice Ages have occurred earlier, such as during the Late Carboniferous and Permian Systems, between 250 and 200 m.y. ago, and during the Late Precambrian, estimated at 600 m.y. ago (SCHWARZBACH, 1961, 1968). Apart from Ice Ages there have probably occurred Heat Ages too, but these are far more difficult to detect. As an example the Permian may be cited, when, following upon the Ice Age, a great quantity of evaporite deposits formed all over the world.

But what does it mean that there have repeatedly been Ice Ages and Heat Ages in the geologic history of the earth? They indicate strong climatic variations, but these have occurred only regionally, in definite zones on earth, and have never affected the globe as a whole. For instance, neither the tropical rain forest, nor the coral reefs disappeared during the recent Ice Ages, although of course their northern and southern boundaries have shifted. But never did the whole surface of the earth freeze over, nor has it ever become unbearably hot. So, notwithstanding the occurrence of Ice Ages, and probably of Heat Ages too, in the geologic past, the globe has never become inhabitable.

A major factor in this surprisingly regular surface temperature of the earth over all of geologic history probably is the *glasshouse effect* of the atmosphere. This tends to counteract the variations of the independent variables which together determine the surface temperature of the earth. The most important of these variables are solar radiation, heat flow from the interior of the earth, and heat retention by the atmosphere.

Solar radiation is the main heat source for the surface of the earth. According to modern theories it must have remained constant over the last several billion years. The simple nuclear reactions generating heat in the sun apparently are reasonably well understood and can be extrapolated backwards into the geologic history with relative confidence. If we do not take into account smaller variations such as sun spot cycles and the like, there seems to be little doubt that the temperature of the sun has been constant during the few billion years of the geologic history of the earth.[1]

---

[1] For a more general discussion of this question in regard of life on other planets, see section 6 of Chapter 17.

The heat flow from the earth's interior, on the other hand, is still imperfectly known. It is often thought to have been much stronger during the earlier history of the earth, due to the larger amount of radioactive elements which were still present at that time and have since decayed. But we can no more than put on record that the difference evidently has not been large enough to be prohibitive for protein-based life.

The heat retention by our atmosphere is an extremely complex process, possessing various feedback mechanisms counteracting either cooling or heating by the primary variations in heat flow. For example, any initial warming up of the earth's surface, either by stronger solar radiation, or by stronger heat flow from the interior of the earth, would be counteracted by increased evaporation and hence by stronger cloud formation. This would in its turn affect the reflectivity of the atmosphere, the so-called *albedo* of the earth, resulting in a smaller amount of solar radiation reaching its surface. Other, more complex, reactions, such as a shift in the so-called *tropopause*, the boundary between the lower part of the atmosphere or troposphere and the next higher part, the stratosphere, would also enter into the process, all tending to buffer the effects of primary variations in solar radiation and terrestrial heat flow.

It follows that the heat retention by the atmosphere, with its various feedback mechanisms, seems able to buffer effectively rather large variations in primary heat flow. Our atmosphere consequently is a very good glasshouse indeed, and this fact must have played a major part in the origin and evolution of life on earth. As this quality mainly rests upon the presence of water vapour, which is assumed to have been present in the primeval atmosphere also, we may assume that this glasshouse effect was already in operation during the earliest geological history.

REFERENCES

ALLEN, D. E. and GILLARD, R. D., 1967. Stereoselective effects in peptide complexes. *Chem Commun.*, 1967:1091–1092.
ALLEN, M. B. and VAN NIEL, C. B., 1952. Experiments on bacterial denitrification. *J. Bacteriol.*, 64:397–413.
BARGHOORN, E. S. and TYLER, S. A., 1965. Microorganisms from the Gunflint chert. *Science*, 147:563–577.
BARKER, H. A., 1956. *Bacterial Fermentations*. CIBA Lectures in Microbial Biochemistry. Wiley, New York, N.Y., 95 pp.
BERKNER, L. V. and MARSHALL, L. C., 1965. On the origin and rise of oxygen concentration in the earth's atmosphere. *J. Atmospheric Sci.*, 22:225–261.
BERKNER, L. V. and MARSHALL, L. C., 1966. Limitation on oxygen concentration in a primitive planetary atmosphere. *J. Atmospheric Sci.*, 23:133–143.
BERNAL, J. D., 1951. *The Physical Basis of Life*. Routledge and Paul, London, 80 pp.
BEIJERINCK, M. W. and MINKMAN, D. C. J., 1910. Bildung und Verbrauch von Stickstoffoxydul durch Bakterien. *Zentr. Bakteriol. Parasitenk.*, 25:30–63.
BIJVOET, J. M., KOLKMEIJER, N. H. and MAC GILLAVRY, C. H., 1948. *Röntgenanalyse van Kristallen*. Centen, Amsterdam, 300 pp.

CALVIN, M., 1969. *Chemical Evolution.* Clarendon, Oxford, 278 pp.

CLOUD, P. E., 1968. Pre-Metazoans and the origin of Metazoa In: E. T. DRAKE (Editor), *Evolution and Environment.* Yale Univ. Press, New Haven, Conn., pp.1–72.

DECHOW, E. and JENSEN, M. L., 1965. Sulfur isotopes of some Central African sulphide deposits. *Econ. Geol.*, 60:894–941.

GARAY, A. S., 1968. Origin and role of optical isomery in life. *Nature*, 219:338–340.

GOLDICH, S. S., 1968. Geochronology in the Lake Superior region. *Can. J. Earth Sci.*, 5:715–724.

GOLDICH, S. S., NIER, A. O., BAADSGAARD, H., HOFFMAN, J. H., and KRUEGER H., 1961. The Precambrian geology and geochronology of Minnesota. *Minnesota Geol. Surv., Bull.*, 41: 193 pp.

GOVETT, G. J. S., 1966. Origin of banded iron formations. *Geol. Soc. Am. Bull.*, 77:1191–1212.

GRATZER, W. B. and COWBURN, D. A., 1969. Optical activity of biopolymers. *Nature*, 222: 426–431.

HAVINGA, E., 1954. Spontaneous generation of optically active substances. *Biochim. Biophys. Acta*, 13:171–174.

HOLLAND, H. D., 1965. The history of ocean water and its effect on the chemistry of the atmosphere. *Proc. Natl. Acad. Sci. U.S.*, 53: 1173–1183.

HOLSER, W. T. and KAPLAN, I. R., 1966. Isotope geochemistry of sedimentary sulphates. *Chem. Geol.*, 1:93–135.

KLABUNOVSKI, E. I., 1959. Absolute asymmetric synthesis and asymmetric catalysis. In: A. I. OPARIN (Editor), *Origin of Life on Earth.* Pergamon, London, pp. 158–168.

KROEPELIN, H., 1969. Racemisation of amino acids on silicates. In: P. A. SCHENK and I. HAVENAAR (Editors), *Advances in Organic Geochemistry 1968.* Pergamon, London, pp.535–542.

KVENVOLDEN, K. A., PETERSON, E. and POLLOCK, G. E., 1969. Optical configuration of amino-acids in Precambrian Fig Tree chert. *Nature*, 221: 141–143.

LARSEN, H., 1952. On the microbiology and biochemistry of the photosynthetic green sulfur bacteria. *Kgl. Norske Videnskab. Selskab., Skrifter.* 1:1–205.

LEPP, H. and GOLDICH, S. S., 1964. Origin of Precambrian iron formations. *Econ. Geol.*, 59: 1025–1060.

MACKENZIE, F. T. and GARRELS, R. M., 1966. Chemical mass balance between oceans and rivers. *Am. J. Sci.*, 264:507–525.

MACKENZIE, F. T., GARRELS, R. M., BRICKER, O. W. and BICKLEY, F., 1967. Silica in sea water: Control by silica minerals. *Science*, 155:1404–1405.

MACLEAN, I., EGLINTON, G., DOURACHI-ZADEH, K., ACKMAN, R. G. and HOOPER, S. N., 1968. Correlation of stereoisomerism in present day and geologically ancient fatty acids. *Nature*, 218:1019–1023.

MILLOT, G., LUCAS, J. and PAQUET, H., 1966. Évolution géochimique par dégradation et agradation des minéraux argileux dans l'hydrosphère. *Geol. Rundschau*, 55:1–20.

NEWELL, N., 1967. Revolutions in the history of life. In: C. C. ALBRITTON (Editor), *Uniformity and Simplicity—Geol. Soc. Am., Spec. Papers*, 89:63–92.

NIEUWENKAMP, W., 1966. Geschichtliche Entwicklung der heutigen petrogenetischen Vorstellungen. *Geol. Rundschau*, 55: 460–477.

POSTGATE, J. R. and CAMPBELL, L. L., 1966. Classification of *Desulphovibrio* species, the non-sporulating sulfate-reducing bacteria. *Bacteriol. Rev.*, 30:732–738.

RANKAMA, K., 1963. *Progress in Isotope Geology.* Interscience, New York, N.Y., 705 pp.

RUBEY, W. W., 1951. Geologic history of sea water. *Bull. Geol. Soc. Am.*, 62:1111–1147.

RUBEY, W. W., 1955. Development of the hydrosphere and the atmosphere, with special reference to probable composition of the early atmosphere. *Geol. Soc. Am., Spec. Papers*, 62: 631–650.

RUTTEN, M. G., 1956. Remarks on the genesis of flints. *Am. J. Sci.*, 255:432–439.

RUTTEN, M. G., 1957. Origin of life on earth, its evolution and actualism. *Evolution*, 11:56–59.

RUTTEN, M. G., 1966. Geologic data on atmospheric history. *Palaeogeography, Paleoclimatol., Palaeoecol.*, 2: 47–57.

SCHELMAN, J. A., 1968. Symmetry rules for optical rotation. *Account Chem. Res.*, 1:144–151.

SCHNELLEN, C. G. T. P., 1947. *Onderzoekingen over de Methaangisting.* Thesis, Tech. Univ., Delft, 137 pp.

SCHOPF, J. W., 1968. Microflora of the Bitter Springs Formation, Late Precambrian, Central Australia. *J. Paleontol.*, 42:651–688.

SCHWARZBACH, M., 1961. *Das Klima der Vorzeit*, 2nd ed. Enke, Stuttgart, 275 pp.

SCHWARZBACH, M., 1963. *Climates of the Past*. Pergamon, London, 328 pp.

SCHWARZBACH, M., 1968. Neuere Eiszeithypothesen. *Eiszeitalter Gegenwart*, 19:250–261.

SILLÈN, L. G., 1961. The physical chemistry of sea water. Oceanography. *Am. Assoc. Advan. Sci.*, 549–581.

SILLÈN, L. G., 1963. How has sea water got its present composition? *Svensk Kemisk Tidsk.*, 75:161–177.

SILLÈN, L. G., 1965. Oxidation state of the earth's ocean and atmosphere. I. A model calculation on earlier states. The myth of the "probiotic soup". *Arkiv. Kemi*, 24:431–456.

SILLÈN, L. G., 1966a. Oxidation state of the earth's ocean and atmosphere. II. The behaviour of Fe, S and Mn in earlier states. Regulating mechanisms for O_2 and N_2. *Arkiv Kemi*, 25: 159–167.

SILLÈN, L. G., 1966b. Regulation of O_2, N_2 and CO_2 in the atmosphere: thoughts of a laboratory chemist. *Tellus*, 18: 198–206.

SILLÈN, L. G., 1967a. Gibbs phase rule and marine sediments. *Advan. Chem. Ser.*, 67:57–69.

SILLÈN, L. G., 1967b. The ocean as a chemical system. *Science*, 156:1189–1197.

SÖHNGEN, N. L., 1906. *Het ontstaan en het verdwijnen van waterstof onder den invloed van het organische leven.* Thesis, Tech. Univ., Delft, 137 pp.

TERENT'EV, A. P. and KLABUNOVSKII, E. I., 1959. The role of dissimetry in the origin of living material. In: A. I. OPARIN (Editor), *Origin of Life on Earth*. Pergamon, London, pp.95–105.

VAN NIEL, C. B., 1931. On the morphology and the physiology of the purple and green sulfur bacteria. *Arch. Mikrobiol.*, 3:1–112.

VAN NIEL, C. B. 1944. The culture, general physiology, morphology and classification of the non-sulphur purple and brown bacteria. *Bacteriol. Rev.*, 8:1–118.

VAN NIEL, C. B., 1956. Evolution as viewed by the microbiologist. In: A. J. KLUYVER and C. B. VAN NIEL (Editors), *The microbe's contribution to biology*. Harvard Univ. Press, Cambridge, Mass., pp.155–176.

VAN NIEL, C. B., 1963. A brief survey of the photosynthetic bacteria. In: H. GEST, A. SAN PIETRO and L. VERNON (Editors), *Bacterial photosynthesis*. Antioch Press, Yellow Springs, Ohio, pp.459–475.

WALD, G., 1957. The origin of optical activity. *Ann. N.Y. Acad. Sci.*, 69:352–368.

Chapter 15 | The Two Atmospheres: Anoxygenic versus Oxygenic; Pre-actualistic versus Actualistic

1. THE ANOXYGENIC PRIMEVAL VERSUS THE OXYGENIC PRESENT ATMOSPHERE

We have seen in the preceding chapters how biologists have postulated the existence of a primeval anoxygenic atmosphere as a conditio sine qua non for the origin of life through natural causes. Moreover, the existence of such an early atmosphere of reducing character could be deduced from astronomical data. In Chapter 13 we have subsequently seen that it is also evident from geological data that an anoxygenic atmosphere has existed.

We have arrived at a tentative estimate of the dates when these atmospheres were present on earth. The anoxygenic primeval atmosphere has existed up to 1.8 billion years ago, the oxygenic atmosphere from 1.4 billion years ago to the present day. Between those dates the transition from the anoxygenic to the oxygenic atmosphere must have taken place. We have also seen that the earth is exceptional as a planet in having an appreciable amount of free oxygen in its atmosphere. We have postulated that this oxygen is biological in origin, having been, and still being, produced by organisms capable of organic photosynthesis, such as green plants. We have, moreover, seen that life has been present on earth long before 1.8 billion years ago, which means that life has existed under the anoxygenic atmosphere. The molecular fossils, moreover, indicate that life was already capable at that time of organic photosynthesis.

In this chapter we will go into some more detail concerning the properties of the two atmospheres. But I thought it well to have given this rather lengthy recapitulation first. Also, we have to stress the fact how different the primeval atmosphere has been from that of the present. This difference is, on many counts, as extreme as that between black and white. Much of what was possible then, is not possible now, and vice versa.

As has already been indicated, this contrast between the two atmospheres, although postulated by all scientists proposing an origin of life through natural causes, has not always been well expressed in their texts. Often technical terms pertaining to the present atmosphere have been incorrectly used in describing processes under the primeval atmosphere. In Chapter 7 we have seen this in relation to the terms aerobic and anaerobic, which, when applied to the primeval atmosphere, often mean oxygenic and anoxygenic. That chapter also emphasized the importance of materials radiatic or aradiatic for ultraviolet light.

2. PRE-ACTUALISTIC AND ACTUALISTIC

In order to press home this antithesis between the primeval and the present atmosphere I think I am justified in designating the primeval atmosphere as *pre-actualistic*, in contrast to the present-day *actualistic* one.

I realize this is an equivocal proposition. For under the primeval atmosphere natural processes obeyed the same laws of nature as they do now. So there is, if one likes to raise this point, still uniformitarianism. Uniformitarianism at this level is even essential to the validity of all biochemical theory and experiment reported in the earlier chapters. Chemical bonds in the "thin soup" are postulated as having been exactly the same as chemical bonds at present, whether formed inorganically, or by life or in the laboratory. However, though there was uniformitarianism at this basic level, the results were not only different, but often exactly the opposite. So, because so many actual processes were not possible at that time, and vice versa, the designation of pre-actualistic seems justified.

As we have seen, the antithesis between the primeval and the present atmosphere also affected part of the exogenic geological processes. That is, as far as they were in actual contact with the contemporary atmosphere or hydrosphere. We are therefore also justified in speaking of pre-actualistic processes leading to the formation of Precambrian sediments of an aberrant nature, such as the gold-uranium reefs and the banded iron formations described in Chapter 13.

It follows that exogenic processes older than 1.8 billion years were pre-actualistic, as far as they were in direct contact with the atmosphere or the hydrosphere. This is not the case for the endogenic processes, which are not influenced by any atmosphere. As far as we know, endogenic processes, and notably the orogenetic cycle, have shown no fundamental differences since the time of the oldest known crustal rocks, 3.3 billion years old. It must consequently be realized that for all of the earlier part of the history of the crust of the earth, from at least 3.3 billion years till 1.8 billion years ago, pre-actualistic circumstances prevailed at the surface of the crust; actualistic circumstances within the crust.

3. ULTRAVIOLET SUNLIGHT

In Chapter 4 we have stated in a rather general way how ultraviolet sunlight is able to penetrate a primeval anoxygenic atmosphere and reach the surface of the earth. Because of the relatively high energy of ultraviolet rays all kinds of chemical reactions may then take place, which are now no longer possible. A schematic illustration is presented in Fig.12. A more precise evaluation has since been given by BERKNER and MARSHALL (1965, 1966). This work, which gives us a much better understanding of the processes in a primeval atmosphere, will form the basis of this and following sections. The authors mentioned have assembled all the newest data on solar radiation, on atmospheric composition and on the absorption of specific atmospheric gases, as supplied by a wide range of disciplines, ranging from nuclear physics to satellite recordings. The studies of BERKNER and MARSHALL center on the influence of atmospheres of varying composition on the shorter-waved part of the ultraviolet sunlight.

As is well known, sunlight is made up by light waves of strongly varying length. Approximately one octave of wave lengths, from slightly below 4,000 Å[1] for the violet, to slightly above 8,000 Å for the red, is that part of the spectrum we ourselves can percieve, and which, in our arrogance, we call the visible part of the spectrum. However, because of the absorption by the present atmosphere, there would be not so very much more visible, even if we had eyes which could percieve a wider spectrum of wave lengths. There is in fact only a narrow band beyond the violet—called the ultraviolet—and a somewhat broader band beyond the red—called the infrared—which are not absorbed by the atmosphere. To all intents, the present atmosphere is opaque to most of the wave lengths of the sunlight and absorbs all energy of the sunlight except for a window in the general region of the visible part of the spectrum. As we will see, this opaqueness of our present atmosphere depends for a very large part, at least for the shorter ultraviolet light, on free oxygen, O_2, and its accompanying ozone, O_3.

The reason why in the studies by BERKNER and MARSHALL (1965, 1966) special attention is paid to the ultraviolet and not to the infrared lies in the fact, already indicated in Fig.12, that the shorter the wavelength, the higher is the energy. Whereas, the higher the energy of the sunlight, the greater the possibility of inorganic photochemical reactions, either of dissociation, or of synthesis, taking place. In fact, the upper limit of wave lengths required for such reactions lies

---

[1] Å, the Ångström unit of length—named in honour of the Swedish physicist Anders J. Ångström, who was one of the foremost researchers of the solar spectrum—is 10^{-7} mm, that is one tenth of a millimicron. The reason why it is (still?) used in most spectrographic studies, though it is one order off from the metric system, is that it expresses the wave length of visible light in a convenient unit. Spectrometers were generally constructed so that they could detect about 1 Å difference in wave length.

considerably below that of the present ultraviolet light, so that it is the "shorter ultraviolet sunlight" which is of special interest to us. Its upper level lies at about 2,500 Å, and, as we will see, the calculations of Berkner and Marshall are indeed mainly concerned with this shorter ultraviolet sunlight, below wave lengths of 2,500 Å.

The energy of the sunlight which eventually reaches the surface of the earth is, of course, not only dependent upon the transmitting capabilities of the atmosphere, but also upon the energy of the sunlight itself. This does strongly decrease below 1,000 Å (Fig.113). Moreover, the absorption coefficients of the various

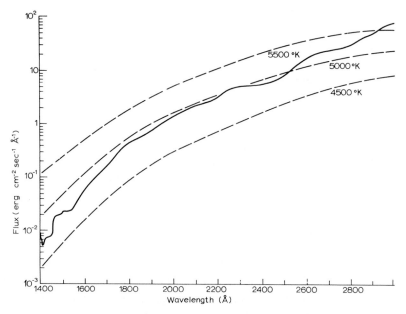

Fig.113. Solar intensity, that is the energy of the sunlight at the outer border of the earth' atmosphere for wave lengths of the ultraviolet between 1,400 Å and 3,000 Å.

For shorter wave lengths the solar intensity is even lower and fluctuates between 10^{-2} and 10^{-3} erg cm^{-2} sec^{-1} $Å^{-1}$. Or in words, between one hundredth and one thousandth of an erg per square centimeter, per second, for a wavelength of sunlight one Ångström wide. (From BERKNER and MARSHALL, 1965.)

gases rise sharply below wave lengths of 1,800 Å. Little solar energy consequently reached the surface of the earth in light of wave lengths below 1,600 Å during any time of its geologic history. It follows that the sunlight specially important for our study, i.e. for the possibility of inorganic processes of photosynthesis in a primeval anoxygenic atmosphere, is that portion of the sunlight between 1,600 Å and 2,500 Å.

4. ABSORPTION OF ULTRAVIOLET SUNLIGHT

In studying the penetration of sunlight in primitive atmospheres one must first decide which models will be studied. Based on the astronomical relations reviewed in Chapter 5, one might assume a primitive atmosphere to have contained H_2, N_2, H_2O, CO_2, A, CH_4 and traces of other gases, together with small amounts of O_2 and O_3. For all practical purposes the absorption of H_2, N_2, A and CH_4 is negligible in the region of wave lengths between 1,600 Å and 2,500 Å. Consequently, our study can be limited to the effects of water vapour, H_2O, carbon dioxide, CO_2, oxygen, O_2 and ozone, O_3. To be able to calculate the penetration of sunlight through an atmosphere of a certain composition, we must first know the energy of a particular wave length, which has already been given in Fig.113. Secondly, we have to know the absorption constant for that atmospheric wave length, and thirdly the amount of the various gases the sunlight has to pass through in traversing the atmosphere. In practical applications the last two values are always calculated separately for the respective atmospheric gases and the value for the total atmosphere is then found by addition.

The absorption coefficient of a gas, k, is the index of attenuation, I, of light passing through 1 cm of that gas at 0°C and at 1 atm (that is at STP conditions, see below). The intensity of light passing over a distance of x cm at STP is given by the formula:

$$I_x = I_0 e^{-kx},$$

in which I_0 is the initial intensity, and I_x the remnant intensity of the light after it has passed through the gas.

The amount of a certain gas present in the atmosphere, and which must be traversed by the sunlight in order to reach the surface of the earth, is normally given in two ways, one on an absolute scale, the other on a relative one. In the absolute scale the entire column of a specific gas is thought to be compressed to one atmosphere at 0°C. This is called the *Standard Temperature and Pressure* or STP. The amount of gas the sunlight has to penetrate is then equivalent to the length of the path through the thickness of this imaginary layer of gas. It is accordingly called *pathlength STP*, which in atmospheric physics is expressed by x and measured in cm. In the relative scale the amount of gas is compared to that found at present in the atmosphere. The latter is indicated as *Present Atmospheric Level*, or PAL, and in the relative scale amounts of atmospheric gases are given as a quotient of *PAL*.

The absorption coefficients of the four gases that are important in the absorption of sunlight with wave lengths between 1,600 Å and 2,500 Å are given in Fig.114. From this figure it follows that the absorption coefficients for water vapour, for oxygen and for carbon dioxide drop sharply between 1,600 Å and

2,000 Å. These gases consequently are easily penetrated by wave lengths longer than 2,000 Å. The opposite is true for ozone, which has an absorption coefficient which fluctuates between $k = 1$ and $k = 100$ all the way up to the wave length of 3,000 Å. The importance of atmospheric ozone in the absorption of the ultraviolet part of the sunlight is apparent from the Figs.114 and 115.

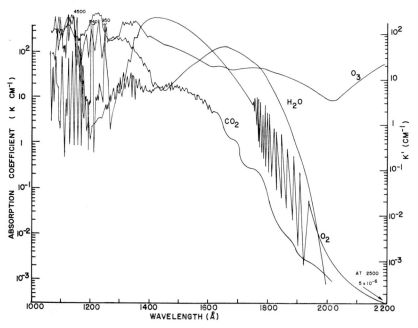

Fig.114. Absorption coefficients for the four atmospheric gases important for the attenuation of ultraviolet sunlight for wavelengths between 1,000 Å and 2,200 Å.
    The absorption coefficient k has been defined on p.341, 2. The righthand ordinate gives the values of k^1, a coefficient derived from k in calculating not total extinction, but extinction to 1 erg cm^{-2} sec^{-1} (50 Å)$^{-1}$, (see next section). (From BERKNER and MARSHALL, 1965.)

Taking up now the second variable important for the penetration of light through the atmosphere, we must take note of the fact that there is a marked difference in the vertical distribution of the various atmospheric gases. Normally a gas will be distributed exponentially with altitude, having its highest concentration at the earth's surface and thinning out upwards. However, out of the four gases important for our study, only two, i.e. O_2 and CO_2, follow this normal exponential distribution. Both H_2O and O_3 are aberrant in their occurrence throughout the atmosphere.

    To take water vapour first, the distribution of this gas is primarily controlled by the so-called *cold trap* in the troposphere, above 10 km altitude. There, at temperatures below $-40°C$, all water vapour will freeze and practically no water vapour can therefore reach the higher atmosphere. From Fig.120 we may see that of the total atmospheric water vapour, an amount which corresponds to over 10 m pathlength STP, only 1 cm is found above an altitude of 12 km. In section 6 of this chapter we will learn how important this feature is in keeping down the level of the free oxygen produced inorganically through photodissociation of water.

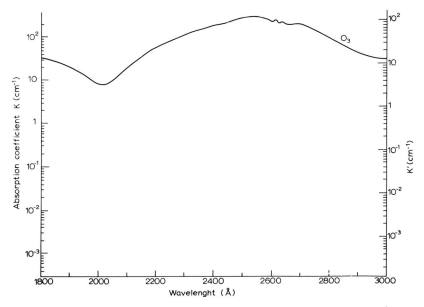

Fig.115. Absorption coefficients for ozone for wave lengths between 1,800 Å and 3,000 Å. (Cf. Fig.114. From BERKNER and MARSHALL, 1965.)

Ozone, on the other hand, shows an entirely different aberrant distribution with altitude. It is at present formed in the higher atmosphere as a result of irradiation of oxygen. This breaks up the molecules of oxygen into their chemically more active atoms, which in part recombine to ozone. This gas is thereupon carried downward by convection in the atmosphere. It has, however, no time to achieve an exponential distribution, for as soon as it reaches the surface of the earth it combines with minerals and organic matter in oxidation processes. The oxidation potential of ozone is so much greater than that of oxygen that, for our purposes, the disappearance of ozone at the earth's surface can be considered as instantaneous.

At present O_3 is therefore distributed about evenly throughout the atmosphere. In an anoxygenic atmosphere this would be quite different. For the radiation which now produces ozone in the higher atmosphere will then penetrate through the atmosphere down to the surface of hydrosphere and lithosphere. And what ozone there will be formed, will consequently arise right there. It will also be used up right away for oxidation of surface materials, and thus not have time to rise in the air. In an anoxygenic atmosphere most of the ozone will therefore be found in a relatively thin layer close to the earth's surface. The amount of ozone actually formed will, of course, depend upon the available amount of free oxygen, a fact well taken care of in Berkner and Marshall's calculations. Their result is expressed in the graphs of Fig.122.

5. ABSORPTION OF ULTRAVIOLET SUNLIGHT IN THE ATMOSPHERE

We may now tackle the question of how much sunlight will be absorbed in atmospheres of varying compositions. The curves we will work with indicate the amount of a certain atmospheric gas—expressed either in "path length STP", or in "PAL", or in both scales—which must be present in the atmosphere, to absorb light of a certain wave length "to extinction".

As in all absorption processes, sunlight in theory is never absorbed totally by either the atmosphere or the hydrosphere. There will always be some light transmitted. But, because this is an exponential process, this soon becomes an immaterially small amount. A conventional "limit of extinction" has therefore to be defined. In all of Berkner and Marshall's calculations this is taken at $1 \text{ erg cm}^{-2} \text{ sec}^{-1} (50 \text{ Å})^{-1}$. Or, in words, when the energy of a band of sunlight 50 Å wide is attenuated so far that it only transmits one erg per square centimeter per second. At this level of extinction the sun would appear about 50 times as bright as the full moon. No inorganic photochemical reactions of any importance would take place below this level.

Apart from the influence of the rather regular increase in the solar flux with sunlight of longer wave lengths, as figured in Fig.113, the curves representing the amount of gas needed to absorb a certain wave length to extinction, are the mirror images of the curves representing the absorption coefficients. For the smaller the absorption coefficient is, the more transparent a given gas is for light of a given wave length, and hence the thicker the layer of that gas must be, before extinction is ensured.

Of four atmospheric gases important for our subject, variations in the amount of water vapour and of carbon dioxide are found to have only minor influence in regard to the shielding of the ultraviolet. Oxygen, and its accompanying ozone are, on the other hand, found to have a decisive influence.

In the case of water vapour we learn from Fig.116 that it effectively shields ultraviolet sunlight below 2,000 Å at its present atmospheric level. Its influence does not, however, extend very much further, even at the 10 PAL level. As water vapour is in equilibrium with liquid water at the ocean's surfaces, whereas at its upper limit it is bounded by the cold trap in the troposphere, it is probable that its amount in the atmosphere will have shown only minor variations with time. It cannot at any time have had a major influence in shielding ultraviolet sunlight of wave lengths above 2,000 Å.

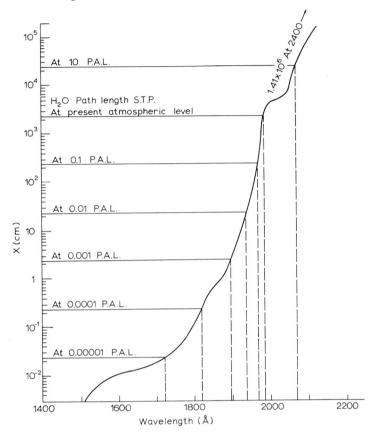

Fig.116. Thickness of H_2O required to absorb ultraviolet sunlight of different wavelengths to extinction [1 erg cm^{-2} sec^{-1} (50 Å)$^{-1}$]. The amount of H_2O is expressed in cm of x, or pathlength STP (see p.341). (From BERKNER and MARSHALL, 1965.)

A similar story holds good for carbon dioxide. From Fig.117 we may learn that CO_2 at its present atmospheric level only filters out wave lengths below 1,900 Å. At 30 times the present level this will reach up to 2,050 Å. It is, however, highly improbable that carbon dioxide could ever have reached such high atmospheric levels. In contrast to oxygen the bulk of the CO_2 is contained in the ocean waters and not in the atmosphere. Accordingly, the ocean provides an excellent buffer for the carbon dioxide content of the atmosphere, marine carbonate

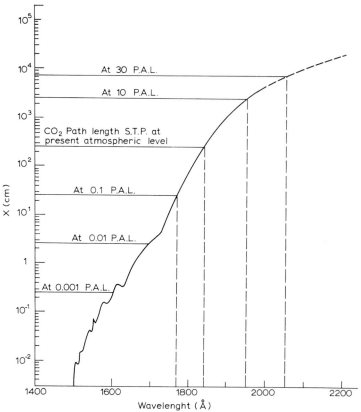

Fig.117. Thickness of CO_2 required to absorb ultraviolet sunlight to extinction. (Cf. Fig.116; From BERKNER and MARSHALL, 1965.)

precipitation acting as a "scavenger" mechanism with regard to excessive CO_2- levels in the atmosphere.

As indicated above, an entirely different picture is found for oxygen and its accompanying ozone. We have, throughout our narrative, stressed the important effects of variations in the amount of atmospheric oxygen for the penetration of ultraviolet sunlight, but we have had to do it in a rather general, qualitative way. In the publications of BERKNER and MARSHALL (1965, 1966) we have now a solid body of facts, from which numerical values have been derived which enable us to base our considerations on a quantitative basis. This is of so great importance that I have thought it worthwhile to expound stepwise the physical basis of our present knowledge on atmospheric absorption.

From Fig.118 we learn that, apart from the Schumann-Runge bands around 1,800 Å and 1,900 Å, which are due to atomic resonance, an increase in the amount of oxygen in the atmosphere entails a gradual and regular extinction of the shorter wavelengths of the ultraviolet. This goes up to a wave length of 2,500 Å at the present atmospheric level, when the absorption is taken over by ozone. So here we have an atmospheric gas with a marked influence on the composition of the sunlight which will reach the surface of the earth.

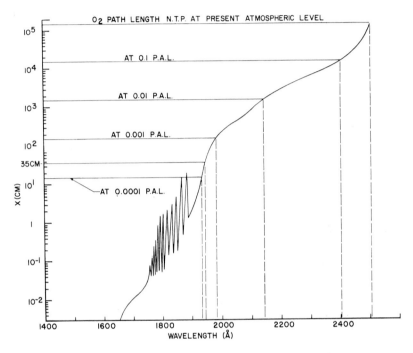

Fig.118. Thickness of O_2 required to absorb ultraviolet sunlight to extinction [1 erg cm^{-2} sec^{-1} (50 Å)$^{-1}$]. (Cf. Fig.116. From BERKNER and MARSHALL, 1965.)

The influence of ozone is complementary to that of oxygen, as follows from Fig.119. The completely different curve of its absorption coefficients is reflected in the different character of its "thickness required for extinction" curve.

We must at this point remember that the amount of ozone is directly dependent on the amount of oxygen. The possibility of an early atmosphere poor in oxygen, but rich in ozone, may be dismissed. Although the constants regulating the formation of ozone are not too well known, an estimate of the amount of ozone which will accompany a certain level of oxygen in a primeval atmosphere can, nevertheless, be given. This estimate is presented in Table XXV.

Comparing Fig.119 with Table XXV, we note that the level of ozone will already be but slightly lower than the present one, when the oxygen level is still at one tenth of the present. As ozone shields the ultraviolet light around 2,500 Å already at one tenth of its present level, oxygen and ozone together effectively extinguish all of the shorter ultraviolet light at one tenth of the present atmospheric level of oxygen (Cf. Fig.122).

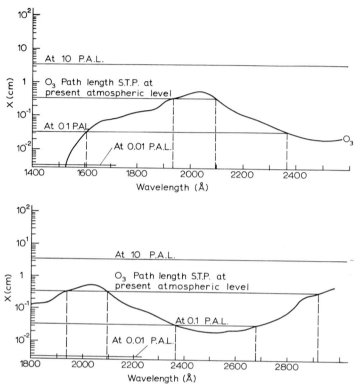

Fig.119. Thickness of O_3 required to absorb ultraviolet sunlight to extinction. (Cf. Fig.116. From BERKNER and MARSHALL, 1965.)

TABLE XXV

ESTIMATED PATH LENGTHS OF O_3 FOR VARIOUS AMOUNTS OF O_2
(From BERKNER and MARSHALL, 1965)

O_2 (STP) proportion of PAL	Approx. height of O_2 column (cm)	Average % O_2 (STP) in atmos. column	Approx. path length of O_2 (cm) (STP)
10.0	$65 \cdot 10^5$	$7 \cdot 10^{-8}$	0.5
1.0	$47 \cdot 10^5$	$7 \cdot 10^{-8}$	0.33
0.1	$28 \cdot 10^5$	$7 \cdot 10^{-8}$	0.2
0.01	$12 \cdot 10^5$	$4 \cdot 10^{-8}$	0.05
0.005	$12 \cdot 10^5$	$1.6 \cdot 10^{-8}$	0.02
0.001	$12 \cdot 10^5$	$4 \cdot 10^{-9}$	0.005

6. SELF-REGULATING MECHANISM LIMITING THE AMOUNT OF OXYGEN PRODUCED BY INORGANIC PHOTODISSOCIATION OF WATER AT ONE THOUSANDTH OF ITS PRESENT ATMOSPHERIC LEVEL

We are now in the position to decide upon a crucial question in atmospheric development. That is, whether the free oxygen of the present atmosphere could not have been formed inorganically, by photodissociation of water vapour through irradiation by ultraviolet sunlight. This has, in fact, been the earlier explanation of the genesis of the free oxygen of our atmosphere, and it is a theory which seems at first quite plausible. Since it cooled, there has, in all probability, always been sufficient water on earth, whilst water vapour is known to be dissociated by the shorter ultraviolet.

However, because of a mechanism which has first been suggested by H. C. Urey, the amount of oxygen which can be so produced is found to be sharply limited. The mechanism depends on the difference in vertical distribution between water vapour and oxygen. As we saw, water vapour is caught in the cold trap, whereas oxygen is distributed exponentially. The oxygen will therefore rise much higher in the atmosphere than water. Moreover, the same wave lengths of ultraviolet capable of dissociation of water, are also absorbed by oxygen. At a certain level of oxygen, this will consequently rise above the water caught in the cold trap. It will then effectively shield the water vapour from further irradiation by ultraviolet sunlight, and hence from further dissociation.

Calculations have shown that the level of the oxygen produced by inorganic photodissociation of water cannot rise higher than one thousandth of the Present Atmospheric Level (BERKNER and MARSHALL, 1966).

Photodissociation of water depends upon the energy acquired by the absorption of light quanta. It follows that this process will take place at those wave lengths where the absorption coefficient is high. In Fig.114 this is seen to be between 1,500 Å and 2,000 Å. But we also see in this figure that both CO_2 and O_2 have high absorption coefficients over these same wave lengths. So the question now resolves as to which gas can get at the ultraviolet sunlight first. As shown in Fig.120, oxygen—and carbon dioxide—are at a decisive advantage in this respect, once they are present in an amount which permits the exponential buildup of these gases into the higher atmosphere. They are then able to form a shield above the water vapour caught in the cold trap.

The absorption of the ultraviolet sunlight, summarized over the range of wave lengths from 1,500–1,800 Å for two models of the early atmosphere with variable contents of oxygen is given in Fig.121. Both models have water vapour at Present Atmosphere Level, based on the reasoning cited in the preceding section on the availability of oceans throughout geologic history. One model has CO_2 at PAL also, the other has CO_2 at 10 PAL. Not much influence is exerted by the

CO_2 level, due to the small amount present of this gas. It varies somewhat with the level of CO_2, but the main influence comes from the amount of O_2. It is quite safe to place the self-regulating mechanism at one thousandth of the Present Atmosphere Level of gas.

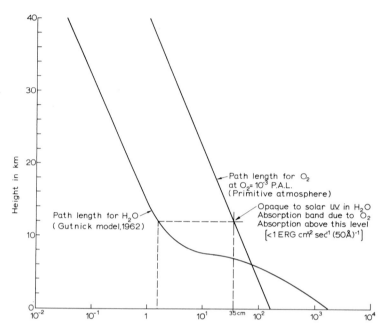

Fig.120. Integrated path lengths of water vapour at PAL and of oxygen at 10^{-3} PAL.
At this level of oxygen the water vapour is shielded from further dissociation by ultraviolet sunlight at a path length STP of 35 cm. This is already reached at an altitude of 12 km, that is higher than the cold trap in which the bulk of the water vapour is caught. Note that the scale of the path length is logarithmic, so the actual amount present diminishes very rapidly towards the left hand side of the picture.
The influence of carbon dioxide is similar to that of oxygen, but, as we see in Fig.121, its influence is small, due to the small amount of this gas in the atmosphere. (From BERKNER and MARSHALL, 1965.)

BERKNER and MARSHALL (1966) rightly insist that this is a very important mechanism, which cannot be broken by any inorganic process. The only ways in which it could be broken consequently would be (*1*) by some unkown extraterrestrial influence, (*2*) by supernatural intervention, and (*3*) by biogenic production of free oxygen. In the next paragraphs we will see how our atmosphere may have developed by the biogenic production of oxygen.

One further comment must be made at this point. That is that the self-regulating mechanism described above does only depend, in any atmosphere, on the presence of water vapour. The level at which the mechanism operates will shift somewhat with the amount of carbon dioxide, but this is not important. Consequently, it is a mechanism that must not only have operated in the primitive atmosphere of the earth, but in any other primitive atmosphere. It is a general mechanism for any primitive planetary atmosphere.

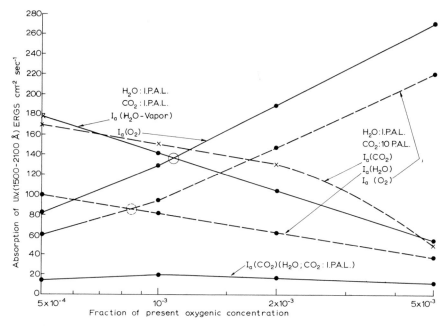

Fig.121. Self-regulating mechanism of inorganic production of oxygen in a primitive atmosphere.

Two atmospheric models are presented, both with $H_2O = 1$ PAL. One has $CO_2 = 1$ PAL (solid lines), the other has $CO_2 = 10$ PAL (dashed lines).

The lines indicate absorption to extinction of all ultraviolet sunlight between wave lengths of 1,500 Å and 1,800 Å. Below the level of $O_2 = 10^{-3}$ PAL absorption by H_2O overrides that of O_2. However, at higher levels of atmospheric oxygen the absorption by O_2 is the stronger. The water vapour is then shielded from the ultraviolet sunlight and no further dissociation can occur.

At the point at which the absorption lines of H_2O and O_2 cross lies the level of the self-regulating mechanism limiting inorganic formation of O_2. (From BERKNER and MARSHALL, 1966.)

The importance of the Urey self-regulating mechanism of atmospheric oxygen in relation to early life lies in the fact that any oxygen, of whatever formation, will tend to rise above the water vapour caught in the cold trap. The production of oxygen through organic photosynthesis, that is through dissociation of CO_2, instead of the inorganic photodissociation of H_2O, will consequently at first not lead to a higher level of O_2. The two processes are not additive, and any oxygen produced through organic photosynthesis will only lead to a decreased production of oxygen through inorganic photodissociation of water.

Only when photosynthesis has become so abundant that it can of its own accord produce a balance of atmospheric oxygen higher than 0.001 PAL, will the Urey self-regulating level be broken.

7. A SECOND SELF-REGULATING MECHANISM, LIMITING THE AMOUNT OF FREE ATMOSPHERIC OXYGEN AT ONE HUNDREDTH OF ITS PRESENT ATMOSPHERIC LEVEL. THE TRANSITION FROM FERMENTATION TO RESPIRATION AT THE PASTEUR POINT

We have discussed the transition from fermentation to respiration in the metabolism of certain groups of present-day microbes in sections 5 and 6 of Chapter 8. At this point we need only take into consideration the influence it must have had in the transition from the anoxygenic to the oxygenic atmosphere. The important fact is that the Pasteur mechanism seems to operate at present at about the same level of free atmospheric oxygen for various non-related microbes, classed together as the *facultative anaerobes*. As it is not confined to a single group of microbes, we may assume that this mechanism is controlled by some—as yet unkown—physicochemical threshold(s), and further that it has been operative in the past also. In that case we had better speak of *facultative respirators* than of facultative anaerobes.

This has already led us to postulate a feedback mechanism at 0.01 PAL free oxygen, because free oxygen produced above this level will have been consumed by the facultative respirators, which could switch again to fermentation every time the amount of free oxygen dropped beneath this level. We may now provisionally define an anoxygenic atmosphere as an atmosphere with a maximum of free oxygen of one percent of the present one.

It is, perhaps, wise to restate at this point our former conclusions that the amount of free oxygen in the primeval atmosphere could, as yet, not be decided upon from the evidence derived from former sediments. Only the extrapolation from the Pasteur mechanism, as known from the present-day microbial world offers a possibility in this direction. So if further microbial research should find that the Pasteur mechanism is not situated exactly at 0.01 PAL of free oxygen, our definition will have to be changed accordingly. A fool-proof definition of the anoxygenic atmosphere would therefore be: "An atmosphere with an amount less than or equal to the level at which the Pasteur mechanism is operative." But if we wish to compare the implications of microbiology with those of atmospheric physics, the needs must have a definition using numerical values, and thus the first, provisional, definition is preferred here.

The Pasteur Level has eventually been broken when organic photosynthesis was able to produce oxygen at such a rate that respiration (+oxidation of surface minerals + other oxygen losses) could no longer consume all the oxygen released. Presumably this has been due to the development of a better way of photosynthesis. For instance, by the ascendance of photolithotrophy over the earlier photo"organo"trophy. Or, perhaps, by the development of the eucaryotic cell (see section 8 of Chapter 8; SAGAN, 1967), in which photosynthesis is located in specialized organelles and can be much more effective than in procaryotic cells. At present such speculations can, however, be no more than considered guesses, since we have no data from which to draw any conclusion on this point.

8. PENETRATION OF ULTRAVIOLET SUNLIGHT IN WATER

After having thus gained information about the penetration of ultraviolet sunlight in atmospheres of different composition, our next move must be to look into the penetration of the ultraviolet in water. For early life cannot have subsisted on land, in direct contact with an atmosphere through which the lethal ultraviolet radiation penetrated freely. Early life must have been shielded either by rocks or soil, or by water in lakes and oceans.

A thin layer of rock or soil would have been sufficient to shield off the shorter ultraviolet, but communications, either through the pores of a layer of sand or clay, or from one natural cave to the other, will have been difficult. In larger water bodies, on the other hand, communication will have been easy, but a much thicker layer of water was needed, at least in the early days, to shield life from the ultraviolet sunlight. Since evolution depends, at least periodically, on easy communication, this must have been an important prerequisite for the development of early life and it seems probable that the bulk of life developed in larger water bodies. We will therefore in this section review the restrictions posed upon early life by the penetration of ultraviolet sunlight through the atmosphere and the upper layers of the hydrosphere. The relevant data have been summarized in Fig.122, figuring the penetration of ultraviolet sunlight of different wave lengths into liquid water under atmospheres with different amounts of free oxygen.

Comparing Fig.122 with the illustrations in the preceding sections, it will be seen that the present figure has been extended to cover longer wave lengths. The reason for this lies in the circumstance that in the earlier sections of this chapter we were interested mainly in the inorganic photochemical reactions leading to the synthesis of "organic" compounds. That is, in the spectrum of wave lengths up to 2,100 Å. At present our special interest lies in the lethal aspect of ultraviolet sunlight for living matter. It is not the possibility of synthesis, but the avoidance of destruction we are interested in now. The maximum absorption of ultraviolet light by living cells lies in the spectral range of 2,400–2,800 Å. Irradiation by these wave lengths may be lethal, even for energies lower than one erg per square centimeter per wave band 50 Å wide. Our interest has consequently shifted to somewhat longer wave lengths of the ultraviolet part of the sunlight.

In Fig.122 the combined influence of the absorption of the ultraviolet by both water and the oxygen with its accompanying ozone of the atmosphere has been represented. In the—imaginary—case of a water body without an accompanying atmosphere the distribution of the penetration of the ultraviolet sunlight would follow a smooth curve, the solid line in Fig.122. Water is practically opaque for the shorter wave lengths, as light of 1,800 Å would penetrate less than 1 cm, before being absorbed. Light of wave lengths around 2,800 Å would already be able to penetrate around 10 m before reaching extinction, whereas the red rays of the visible spectrum are able to penetrate down to 100 m.

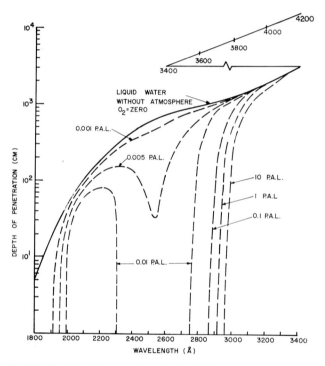

Fig.122. Depths of penetration of ultraviolet sunlight in liquid water for atmospheres with an amount of free oxygen of 0.001, 0.01, 0.1, 1 and 10 PAL.

Without absorption in the atmosphere ultraviolet sunlight would penetrate water less than 1 cm for the wave length of 1,800 Å, around 10 m for wave lengths around 2,800 Å, and around 100 m at the red end of the visible spectrum.

A primitive atmosphere, with O_2 stabilized at 0.001 PAL, through the Urey self-regulating mechanism, would not change anything in this situation. But at 0.01 PAL great differences would occur. The lethal radiation between 2,300 Å and 2,750 Å is then filtered out already in the atmosphere, whereas the shorter ultraviolet does not penetrate any more than 1 m into the water. At O_2 = 0.1 PAL all lethal radiation below 2,900 Å is already absorbed by the atmosphere, and life will be able to subsist on land. (From BERKNER and MARSHALL, 1965.)

This situation changes very little if one takes into consideration the effect of a primeval atmosphere in which the amount of oxygen is limited to 0.001 PAL.

At 0.01 PAL of O_2, a great difference has, on the other hand, set in. This does not result from large variations in the absorption by the water itself, or by the oxygen in the atmosphere, as we may deduct from Fig.118, but from the fact that at this level of free atmospheric oxygen the accompanying amount of ozone starts to become important.

The strong absorption by ozone of ultraviolet light with wave lengths between 2,400 and 2,700 Å (see Fig.119) leads to extinction of this part of the ultraviolet light already in the atmosphere. For the shorter wave lengths too, the absorption in the atmosphere has become important, and all ultraviolet sunlight is now absorbed by a water layer only 1 m thick.

At 0.1 PAL for O_2 the absorption within the atmosphere by the combined efforts of oxygen and its accompanying ozone reaches up to wave lengths of about 2,900 Å. This means that all lethal ultraviolet is now shielded by the atmosphere itself. Life consequently is no more in need of a protecting layer of water, and can move to the land.

9. LIMITED ENVIRONMENT OF EARLY LIFE

The facts gathered in the preceding sections can now be put together to enable us to form an opinion on the possible environment of early life. We will see that early life will have been severely limited in its environment by two factors. One is the lethal ultraviolet radiation penetrating the atmosphere and making the land inhabitable. The other is the low stage of organization of life itself, which prevented it from taking as full an advantage of its environment as would have been possible by the far better organized life of the present.

To take up the latter factor first, even if continental life were impossible, aquatic life of the present could still find a wide variation of possibilities. In a very general way it can be subdivided into three ecological groups: (1) the *nektonic* or free-swimming animals, (2) the *planktonic* or drifting, and (3) the *benthonic* or bottom dwelling organisms.

In early life, during the Early and Middle Precambrian, life had not yet so far progressed as to comprise free-swimming animals, so the nektonic group is ruled out altogether. Moreover, aquatic life had at that time to remain below 10 m of water, in order to be shielded from lethal ultraviolet sunlight. In present life organisms can adjust themselves to a specific water depth by regulating their specific gravity. So, in present-day life planktonic organisms always remaining below a depth of 10 m are quite conceivable. But this regulation is attained by a change in the specific weight of some or several of the organelles in the cell, which results in an overal change of the specific weight of the cell as a whole. This regulatory device consequently is only open to eucaryotic cells, which contain organelles, and not to procaryotic cells, which do not contain organelles within their cells. As early life has been procaryotic it did not have this faculty of regulating the water depth at which it floated. Planktonic early life would therefore forever stand in danger of being tossed upwards into the lethal upper ten meters of water by waves or currents. It follows that planktonic life at that early time will not have been possible either.

Early aquatic life had therefore to be bottom dwelling or benthonic. It could not occupy any bottom which was less than 10 m below the water surface. On the other hand, it could not expand profusely below 50 m, since the visible sunlight is so far absorbed below that level that no active organic photosynthesis can take place. As is schematically illustrated in Fig.123, life in these early days was confined to the bottom of lakes and seas in a zone roughly between 10 m and 50 m depth. For the ocean this means that life was only possible in a narrow zone along the shores. In big lakes, on the other hand, as long as they did not reach greater depths than 50 m, life could spread all over the bottom. BERKNER and MARSHALL (1965) who stressed these limiting environmental factors of early life, therefore are of the opinion that early life must have developed in large lakes.

10. EXTENSION OF THE ENVIRONMENT OF EARLY LIFE AT HIGHER LEVELS OF ATMOSPHERIC OXYGEN

From Fig.122 we see how with higher levels of atmospheric oxygen more and more environments were to be opened for early life. The most important changes in this respect seem to have taken place around the levels of $O_2 = 0.01$ PAL and $O_2 = 0.1$ PAL.

At around 0.01 PAL O_2 a considerable portion, and in effect the most lethal part, of the ultraviolet sunlight was already absorbed in the atmosphere. Moreover the ultraviolet of other wave lengths did not penetrate for more than 1 m into a water body. BERKNER and MARSHALL (1965) are of the opinion that around this level of atmospheric oxygen life could have developed a planktonic way of living. Even if it was thrown up into the upper meter of water, this presumably entailed only a temporary exposure to the ultraviolet sunlight, because waves and currents would tend to transport it downwards again in a relatively short time. Considering that such short exposures cannot be considered lethal, this means that life at that time could "conquer the sea". It still remains a moot point how much ultraviolet sunlight life could exactly stand, and whether this expansion to the sea could have occurred precisely at 0.01 PAL O_2, or perhaps at a little higher or a little lower level.

BERKNER and MARSHALL (1965) are so impressed with what they call the "conquering of the sea at the critical level of O_2 at 0.01 PAL", that they equate it with the beginning of the Cambrian. The first common occurrence of fossils in the geological record, cited in a spectacular way in most textbooks of geology, leads them to speak of the "explosion of life", which is thought to be inherent upon the acquisition of life of a new environmental habitat. We will discuss these implications in more detail in section 4 of the next chapter.

Still another aspect of the rise of the oxygen level in the atmosphere must now be mentioned, i.e. the fact that when water and atmosphere are in equilibrium the solubility of O_2 in water is the lower, the higher the temperature and the higher the salinity of the water is (L. V. Berkner and L. C. Marshall, personal communication, 1966). Under the same conditions fresh water contains about one fourth more oxygen than does normal oceanwater. Consequently the Pasteur Point (see section 5 of Chapter 8) would be reached in fresh water lakes at an earlier date than in the ocean at the time when the atmospheric level of oxygen gradually reached the critical level of 0.01 PAL.

Another critical level was then passed when atmospheric oxygen reached about 0.1 PAL. As we learnt from Fig.122, the lethal ultraviolet sunlight will be absorbed in the atmosphere at that level. Life could at that time dispense with a sheltering layer of water and expand to the land, in direct contact with the atmosphere. This "conquering of the land" at the "critical level of 0.1 PAL O_2" is correlated by BERKNER and MARSHALL (1965) with the beginning of the Silurian when the first major continental floras make their appearance in the fossil record.

Based upon all these considerations a series of generalized pictures of the environment of early life at the various critical levels of atmospheric oxygen is presented in Fig.123–125. These figures have in part been adapted from a personal communication from L. V. Berkner and L. C. Marshall (1966), but they have been redrawn and reinterpreted, so that these authors bear no responsibility for any mistaken ideas presented in them. The figures only try to visualize in a very general way the mode of development of the environment of early life, and no dating is proposed as yet. This will be pursued in the next chapter.

Fig.123 presents the early stages of atmospheric evolution. Just as in the two following figures the environments depicted are those of a continent with a large, shallow lake and the adjoining sea. In the latter a distinction is made between a shallow shelf sea, built up to a wave base assumed to be situated at 30 m depth, and the much deeper open ocean. In this primitive atmosphere, with about 0.001 PAL O_2, lethal sunlight will penetrate not only the atmosphere here, but also the upper ten meters of any water body. In these regions no life, unless otherwise sheltered, is possible. Planktonic life was at that time impossible too, because the primitive procaryotic cells did not yet have ways and means to regulate their floating depth to a safe distance below the lethal upper ten meters of water.

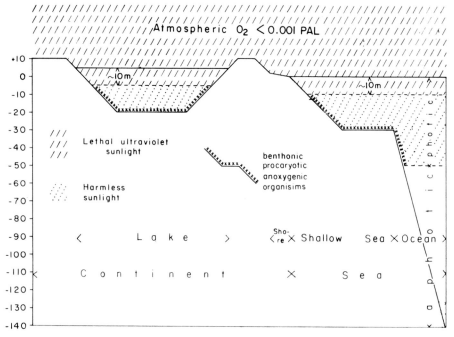

Fig.123. Sketch of the environment of life at the time of a primeval atmosphere with an oxygen level of 0.001 PAL or lower.

Consequently, only bottom life, procaryotic and anoxygenic, could develop. Just as at present this benthos can colonize the bottom of lakes and seas only so far as sufficient harmless sunlight penetrates to supply the energy for life. Although, according to the Berkner and Marshall definition of extinction [1 erg cm^{-2} sec^{-1} (50 Å)$^{-1}$] red sunlight penetrates water down to 100 m depth, there is in the present oceans already a marked attenuation of biomass around a depth of 50 m. This figure has therefore been adopted here as the boundary between the *photic* and the *aphotic* zones. Adoption of a figure of 100 m water depth for this boundary would not materially change the picture.

In Fig.124 an atmospheric level of O_2 of about 0.01 PAL has been reached. Most lethal ultraviolet radiation still penetrates the atmosphere (see Fig.122). But it now penetrates for only 1 meter into any water body, which makes the development of planktonic life possible, both in lakes and in the seas.

In another respect there must, however, have existed a sharp contrast between the development of lakes and oceans at about this level of atmospheric oxygen. Because of the fact that the fresh water of the lakes will take up O_2 more readily than the salt water of the sea, there is a relative supersaturation of oxygen in lakes. As we saw, the Pasteur Point is consequently reached at a lower level of

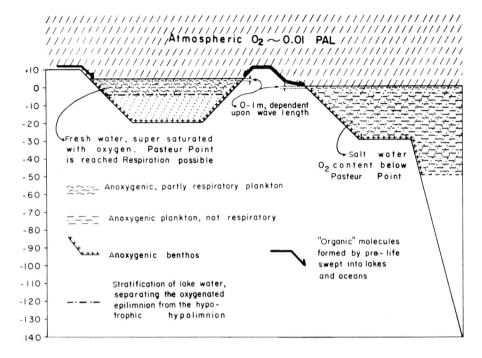

Fig.124. Sketch of the environment of life at the time of a primeval atmosphere with an oxygen level of about 0.01 PAL. (For additional legend, see Fig.123.)

atmospheric oxygen in lakes than in the ocean. It is not known exactly at what level of atmospheric oxygen this antithesis between lakes and oceans will develop, but it must be situated at about 0.01 PAL O_2. It is this stage in the development of the atmosphere and hydrosphere that has been figured here.

In the upper layers of lakes plankton which is still anoxygenic, but aerobic and, at least in part, respiratory, may now develop, whilst metabolically comparable benthos can develop on the shallower bottoms. But, due to the stratification occurring in most lakes (see section 13 of Chapter 13), the lower parts of such water bodies will not receive a constant supply of oxygen. Hypertrophic conditions will develop as a result of this—local—oxygen deficiency. Just as is the case in the lakes of our present time, a truly anaerobic bottom life will consequently develop at this point of atmospheric level on the bottoms of deeper lakes. As at present, this might of course be formed in part by facultative aerobes which could take up respiration when the seasonal turnover of the entire water body would destroy the stratification and permit the whole water body to become in equilibrium with the atmosphere of 0.01 PAL O_2.

In the sea, on the other hand, the Pasteur Point would not yet have been reached. The plankton which can now develop will have been anoxygenic and non-respiratory, whereas primitive anoxygenic benthos, similar to that figured in Fig.123, will continue, the only difference being that the benthos can now reach upwards to waterdepths of about 1 m. Although in the seas stagnant water bodies with hypertrophic bottom conditions may also develop, this is of much rarer occurrence than the normal stratification in lakes. It depends on special features of the relief of the sea floor. So, in this schematic picture, this possibility has been neglected and the sea is shown as well aerated down to the base of the photic zone. The emptiness of the ocean below the photic zone is also an over-simplification. For it is well known that life, both benthonic and nektonic, exists at greater depth. But the volume of life—more correctly the biomass—is so much smaller when compared with that of life above the photic zone, that this simplification seems warranted.

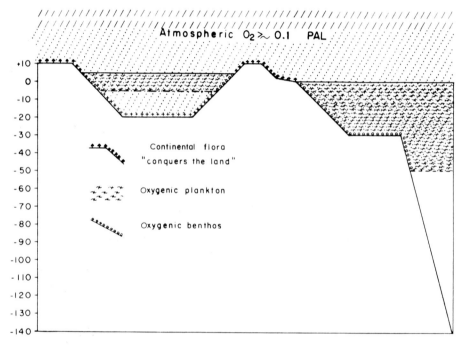

Fig.125. Sketch of the environment of life at the time of a primeval atmosphere with oxygen levels of 0.1 PAL or over. (For additional legend, see Fig.123 and Fig.124.)

In Fig.125, at atmospheric levels of 0.1 PAL O_2 and higher, the situation has become comparable to that of the present. Life has conquered the land, and oxygenic, respiratory plankton occurs in lakes and oceans.

11. OXYGEN AND CARBON DIOXIDE

So far we have been concerned almost exclusively with the relation between primeval and present-day atmospheres in regard to their oxygen content. We have arrived at the conclusion that all of the oxygen contained in our present atmosphere must be biogenic, that is produced by life. The mechanism used is organic photosynthesis, which results in the dissociation of atmospheric carbon dioxide. The time has come to look at that other side and to concern ourselves with atmospheric CO_2.

The geochemical properties of oxygen and carbon dioxide are quite different, so we have to realize that what can be applied to oxygen cannot be applied to carbon dioxide and vice versa. A short recapitulation of what has already been presented earlier seems worth while. A first point to note is that at present the larger part of the free oxygen is found in the atmosphere, whereas most of the carbon dioxide is found in the hydrosphere, dissolved in ocean water. Oxygen, as an atmospheric gas, is relatively free from interference by other atmospheric gases. It will, of course, influence life and exogenic geological processes, but variations in the level of atmospheric oxygen will have no major consequences for other components of the atmosphere. Carbon dioxide, on the other hand, shows a more complicated chain of reactions. A decrease in atmospheric carbon dioxide will first be counteracted by release of carbon dioxide from the oceans.

These equilibrium reactions between atmosphere and ocean are sluggish, when seen from a human standpoint, their duration being of the order of a thousand years. But at this rate they are of course practically instantaneous when seen from the angle of geologic history. The amount of carbon dioxide dissolved in ocean water is so much greater than that present in the atmosphere, that a short-lived depletion of atmospheric CO_2 cannot have had lasting geologic influence. It will be supplemented from the reservoir of CO_2 held by the oceans, which will not perceptibly change by such small depletion.

It is only when there is a continued addition or depletion of atmospheric carbon dioxide that the oceanic level of carbon dioxide will become involved. At that point another geochemical difference with oxygen will make itself felt, that is that carbon dioxide in ocean water is part of a rather complicated reaction system, in which other compounds play their role (see section 7 of Chapter 14). Nevertheless the amount of oxygen existing at present in the atmosphere and that bound to oxidized sulfur, and the amount of coal present as fossil caustobiolith (see sections 4 and 5 of Chapter 14) is so much larger than all carbon dioxide which could, by these geochemical reaction chains have been released, that we must postulate a more or less continuous supply of CO_2 over the geologic history. Geochemical investigations teach us that the oceanic level of carbon dioxide cannot have been more than one order of magnitude higher than it is at present and several geochemists think this is already to high a figure. So we cannot start out from a primeval ocean and atmosphere having a very high content of carbon dioxide, which over the geologic history has been used up by organic photosynthesis to produce the present oxygenic atmosphere.

A supply of CO_2 over the geologic history is not difficult to visualize, because it is by far the most common gas in volcanic exhalations (WHITE and WARING, 1963). So a ready source of CO_2 can be found in the continuous outgassing of the earth over billions of years. The rate of CO_2 production, however, will have probably changed with time, a subject taken up in the next section.

12. PRODUCTION AND CONSUMPTION OF OXYGEN AND OF CARBON DIOXIDE DURING SUCCESSIVE OROGENETIC CYCLES

In the preceding section we have seen how different oxygen and carbon dioxide are in relation to atmospheric history and in their importance to the origin and evolution of life. Basically, oxygen is produced by the consumption of carbon dioxide, so the history of the two gases are more or less complementary. The atmospheric level of a gas at any one time is the balance between production and consumption. In oxygen the production is almost exclusively due to organic photosynthesis. Its consumption is mainly the effect of two processes, i.e. those of organic respiration and of inorganic oxidation of surface minerals. In carbon dioxide, on the other hand, the main production is by an inorganic process, the outgassing of the earth, whereas the main consumption is by organic photodissociation.

One of the factors influencing the rate of the above-named processes is the variation in the rate of crustal movements during individual periods of successive orogenetic cycles (see section 2 of Chapter 10). Although the lack of data prescribes only a vague, qualitative approach, this influence is important enough to merit special consideration.

Taking oxygen first, the consumption of oxygen by the oxidation of surface materials will be relatively low during the quiet geosynclinal periods of any orogenetic cycle. The slow rate of the crustal movements during such periods will bring up a relatively small amount of crustal material anyhow. Moreover, most of this material will consist of rocks of the higher levels of the crust, that is predominantly of sedimentary rocks, already oxidized during an earlier cycle of weathering, transportation and sedimentation. Things are different during the orogenetic and post-orogenetic periods of an orogenetic cycle. Volcanic activity and stronger crustal movements do not only expose a much larger amount of crustal material, but these are moreover, at least in part, derived from the deeper, non-oxidized parts of the crust.

For oxygen there will consequently be less consumption during a geosynclinal period, then during the orogenetic and post-orogenetic periods. Because of the lack of data cited above, we do not know if these variations are large enough to influence the oxygen balance, but if this is the case, one could expect a gradual rise of atmospheric oxygen during geosynclinal periods and a sharp drop during orogenetic and post-orogenetic periods. This possible influence on the oxygen levels through the variations in the rate of crustal movements during orogenetic cycles is represented in Fig.126. Of course, such fluctuations could be quenched by the much stronger influence of a self-regulatory mechanism, such as that operating at the Pasteur Level as treated in section 7 of this chapter.

For carbon dioxide the influence of the differences in rate of crustal movements during successive orogenetic cycles is opposite from that for oxygen.

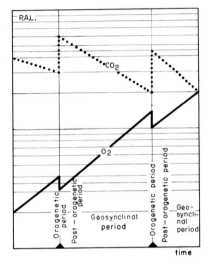

Fig.126. Schematic theoretical model of the influence on the rates of production and consumption of atmospheric oxygen and carbon dioxide by two successive orogenetic cycles.

Oxygen production is thought to increase regularly during geosynclinal periods with exponential expansion of life. Consumption of oxygen during orogenetic and post-orogenetic periods is, however, thought to be so severe, as to offset oxygen production and to cut back the atmospheric oxygen level. Of course, both the orogenetic and the post-orogenetic periods take up time. But as we have no precise idea how much time is actually represented by these periods, so for want of relevant data, this cut-back in atmospheric oxygen is indicated here by a vertical line. This is in keeping with the theoretical and schematic nature of this figure.

Carbon dioxide is consumed regularly by organic photosynthesis. During the orogenetic and post-orogenetic periods CO_2 production will, on the other hand, be strongly enlarged through greater volcanic activity, which will result in a net gain of CO_2. Again for want of data, both depletion and accretion have been represented by straight lines.

This model does not take into consideration other possible factors in production and in consumption of atmospheric gases. At least during part of atmospheric history, these may have been strong enough to override completely the variations in production and consumption of these gases due to variations in the rate of crustal movements in successive orogenetic cycles.

As a first consideration, we may assume that life, all other things being equal, has developed regularly and was not disturbed by the phenomenon of the orogenetic cycles. For even the "stronger crustal movements" of an orogenetic period proceed at a pace imperceptible to living things. And even when such a thing as "faunal crisis" has occurred (see section 13 of Chapter 14) this probably has not influenced the global biomass of life. Hence variations in the rates of crustal movements will not directly influence the rate of development of life. Depletion of CO_2 through organic photosynthesis therefore is a continuous process, and is not affected by any orogenetic cycle.

The production of CO_2 through the outgassing of the earth is, on the other hand, strongly dependent on orogeny. Volcanic activity is many times greater during the orogenetic and post-orogenetic, then during the geosynclinal periods of every orogenetic cycle. Again, we suffer from a lack of data, and we do not know how many times greater. But from the qualitative inferences drawn from the geological record it seems to be a strikingly big difference. It seems therefore reasonable to suppose that the major part of the production of CO_2 through the outgassing of the earth took place during the relatively short orogenetic and post-orogenetic periods of successive orogenetic cycles.

Again assuming that this variation in production rate with time was strong enough to be expressed in the balances of atmospheric carbon dioxide, the results will be as indicated in Fig.126. The atmospheric level of carbon dioxide will fall during the geosynclinal periods and will show a sharp rise during the orogenetic and post-orogenetic periods of successive orogenetic cycles.

REFERENCES

BERKNER, L. V. and MARSHALL, L. C., 1965. On the origin and rise of oxygen concentration in the earth's atmosphere. *J. Atmospheric Sci.*, 22:225–261.
BERKNER, L. V. and MARSHALL, L. C., 1966. Limitation of oxygen concentration in a primitive planetary atmosphere. *J. Atmospheric. Sci.*, 23:133–143.
ERICSSON, E., 1963. Possible fluctuations in atmospheric carbon dioxide due to changes in the properties of the sea. *J. Geophys. Res.*, 13:3871–3876.
POSTMA, H., 1964. The exchange of oxygen and carbon dioxide between the ocean and the atmosphere. *Neth. J. Sea Res.*, 2: 258–283.
RUBEY, W. W., 1955. Development of the hydrosphere and atmosphere, with special reference to probable composition of the early atmosphere. *Geol. Soc. Am., Spec. Papers*, 62:631–650.
RUTTEN, M. G., 1966. Geologic data on atmospheric history. *Palaeogeography, Palaeoclimatol., Palaeoecol.*, 2:47–57.
SAGAN, L., 1967. On the origin of mitosing cells. *J. Theoret. Biol.*, 14:225–274.
WHITE, D. E. and WARING, G. A., 1963. Volcanic emanations. Data of geochemistry, 6th ed. *U. S., Geol. Surv., Profess. Papers*, 440-K: 29 pp.

Chapter 16 | The History of Atmospheric Oxygen and Carbon Dioxide

1. TENTATIVE OUTLINE OF ATMOSPHERIC HISTORY FOR OXYGEN AND FOR CARBON DIOXIDE

Based upon the facts reviewed in the preceding chapters a tentative outline of atmospheric history for oxygen and carbon dioxide will be presented now. The reason why only oxygen and carbon dioxide have been considered is that these two gases are the most important for the history of life. The amount of water vapour in the atmosphere will have been relatively constant, once liquid water was present on our planet, as has been discussed in sections 5 and 6 of the preceding chapter, whereas there is a lack of data for the abundance of the other gases throughout the geologic history. The framework for this outline, as presented in Fig.127, is given by the time involved and by the abundance of these gases. The time is given in billions of years before the present, the abundance in fractions of the present atmospheric level (PAL), as discussed in section 5 of the preceding chapter.

As we have assumed that the major orogenetic cycles have played an important part in the history of life, both in providing suitable environments for its development, and in their direct influence on the production and the consumption of oxygen and carbon dioxide, the major orogenies, as shown in Fig.48, have also been indicated. As we saw at that point, there is still considerable uncertainty in absolute dating of orogenetic periods, and the outline taken from that figure must, perhaps, be altered in the future. The important thing is, however, not when these orogenies did exactly occur, but that there has been a rhythmic alternation of orogenetic and geosynclinal periods of successive orogenetic cycles.

The "age of the earth", and the "age of the oldest rocks", both not measured directly but inferred from the isochrons of radiogenic isotopes, are also indicated, as is the first age dated directly by radiometric analysis, that of the oldest crustal rocks (see sections 14 and 16 of Chapter 3). Moreover the major stratigraphic boundary in the history of the earth, that between the Precambrian and the

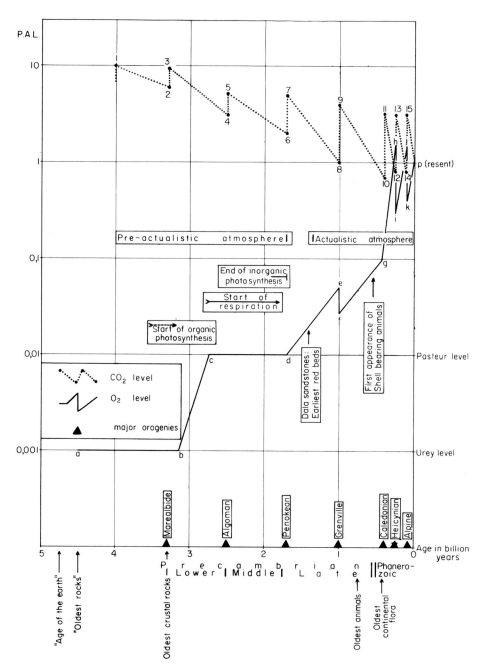

Fig.127. Tentative outline of the history of atmospheric oxygen and carbon dioxide. (For explanation see text.)

Phanerozoic is given, as is the age of the oldest continental flora of higher plants (see section 10 of Chapter 15).

2. HISTORY OF ATMOSPHERIC OXYGEN

We will take the history of atmospheric oxygen first, because we have already a set of data, however vague and unreliable, about the main aspects of this history. We find an earliest level, the so-called Urey level (see section 6 of Chapter 15), at 0.35 cm NTP, that is, below 0.001 PAL, which limits the amount of free oxygen produced through the inorganic photodissociation of water. We may safely assume that it had already been reached at the time the oldest rocks were formed, 4.5 billion years ago, as indicated by a in Fig.127.

At some time in the early history of the earth the Urey level was broken when organic photosynthesis was able to produce enough free oxygen to overcome oxygen losses from contemporaneous oxidation of surface materials. In the preceding chapter we have pointed out that this type of oxygen loss has been heavier during orogenetic periods than during geosynclinal periods. So the Urey level may have been broken either before or after the Marealbide orogenetic period. Since we have no indication when this actually occurred, we have taken a conservative view and placed this event at somewhat over 3 billion years (b).

The actual start of photosynthesis must have occurred earlier than the breaking of the Urey level, because, as we saw, photosynthesis will at first do nothing more than take over a part of the production of oxygen which was earlier produced inorganically through the photodissociation of water. For the absorption of the shorter ultraviolet sunlight in the higher reaches of the atmosphere, above the cold trap in which water vapour is caught, depends only on the total level of oxygen, regardless whether it is produced by abiotic or by biotic processes. We have, however, no data which tell us when it did actually start, nor how long the build-up of biogenic oxygen must have taken, so the point indicated in Fig.127 is arbitrary.

Once the Urey level was broken, it is probable that the production of excess oxygen was fairly rapid, because it was not yet offset by respiration. Moreover, all things remaining equal, the expansion of early life would be exponential and, with expanding photosynthesis, the amount of atmospheric oxygen would also increase exponentially. This has been indicated by the straight line on Fig.127. The slope of this line, which gives the actual rate of expansion, is arbitrary, as it is defined by the position of the points b and c, which are themselves not yet exactly dated.

The next significant level of atmospheric oxygen was reached at about 0.01 PAL, when the Pasteur mechanism started to operate and facultative respirators, able to change their metabolism from fermentation to respiration, could develop (see section 5 of Chapter 8 and section 7 of Chapter 15). As the earliest witness for this level we have taken the Soudan Iron Formation, dated as "over 2.7 billion years old" (see Table XX). Chemical fossils have been interpreted as remnants of former chlorophyll-like molecules, attesting the existence of life capable of photosynthesis. Moreover, the formation of the banded iron formations indicates the existence of a contemporary atmosphere with "a little oxygen" (see sections 12 and 13 of Chapter 13). But although we have thus set the attainment of the level of 0.01 PAL of free oxygen at 2.75 billion years, only slightly earlier than the minimum age given for the Soudan Iron Formation, we must point out that this again is a conservative value. For not only is it possible that actually the Soudan Iron Formation itself is considerably older, but, as we have seen in Chapter 13, there are other banded iron formations which are dated as "older than 3 billion years". As long as the existence of life at such early dates has not yet been proved, I do however, hesitate to extrapolate the existence of the Pasteur level so far back into the history of the earth. The only change in Fig.127 this would entail is that the line b–c, and consequently also the start of organic photosynthesis, must be moved backward in time. It does, however, not alter the general ideas on how atmospheric oxygen developed.

The remarkable coexistence of non-oxidized pyrite sands—formed during post-orogenetic periods—with partly oxidized banded iron formations—formed during geosynclinal periods—in successive orogenetic cycles of the Early and Middle Precambrian points to a persistent low level of atmospheric oxygen and to the existence of an anoxygenic atmosphere all through this part of the earth's history. This is taken to mean that the Pasteur level has not been broken over all of this period, as indicated by the line c–d in Fig.127.

The Pasteur level has eventually been broken. We have assumed that this has happened between the formation of the Blind River gold-uranium ores on the one hand and that of the Dala Sandstones, interpreted as the earliest red beds, on the other (see sections 10 and 17 of Chapter 13). Taking an age of 1.8 billion years for the Blind River deposits and an age of 1.45 billion years for the Dala Sandstones, this rather narrowly limits the time available for the breaking of the Pasteur level. It has been indicated to have occurred after the Penokean orogeny (d in Fig.127).

As we have discussed earlier, the breaking of the Pasteur level may be taken as the end of the primeval anoxygenic atmosphere, which we have provisionally defined as an atmosphere with a maximum of 0.01 PAL of free oxygen. At this point inorganic photosynthesis of "organic" molecules has been terminated. At this point consequently also ends the co-existence of pre-life with early life, which held sway from the start of early life over points b and c, up till point d in Fig.127, a period which has lasted for some two billion years.

The rise and fall of the level of atmospheric oxygen after the breaking of the Pasteur level is indicated in Fig.127 by the line *d–e–f–g–h*. It is assumed that the net production of oxygen was lower than in the earlier period of life's history, before the Pasteur level had been reached, because the production of oxygen is now constantly offset by respiration. It has, moreover, been assumed that the level of atmospheric oxygen has fallen temporarily during the Grenville orogenetic period, because at that time a larger volume of non-oxidized rocks was brought to the surface of the earth, entailing extra oxygen loss through oxidation. These assumptions are, however, not based on fact, but merely indicate one possible model of what might presumably have happened. They should not be taken as historical.

Somewhere along this line the level of oxygen became high enough for animals to develop. We know that this had already taken place before the Cambrian (see section 4 of Chapter 12) but we have no exact dating for the earliest known animal fossils. It seems probable that animals developed only during the Late Precambrian, and were not viable under the anoxygenic atmosphere. But whether this took place before or after the Grenville orogeny, we do not, as yet, know.

One does without difficulty arrive at the next significant level of free oxygen at 0.1 PAL, when the shorter ultra-violet sunlight is absorbed by the atmosphere and "life can conquer the land" (see section 10 of Chapter 15). This can be dated by the oldest major continental flora, which is found in the Silurian and lived 0.44 billion years ago (AXELROD, 1959; STEWART, 1960; CHALONER, 1964).

It is assumed that the expansion of life to the land entailed a significantly higher net production of free oxygen and hence the slope of the line *g–h* is drawn steeper than that of the lines *d–e* and *f–g*. In this way the amount of free oxygen will overshoot its Present Atmospheric Level, which is consistent with the assumption, based on paleontological grounds such as over-sized insects, that the atmosphere was richer in oxygen during the Upper Carboniferous, around 0.3 billion years ago, than it is now. After that the level of free atmospheric oxygen is thought to have oscillated around the present one, under the influence of the Hercynian and the Alpine orogenetic cycles.

3. HISTORY OF ATMOSPHERIC CARBON DIOXIDE

Turning now to the history of atmospheric carbon dioxide we arrive at a model which is even more hypothetical than that for atmospheric oxygen. For one thing, we do not find any regulating mechanisms, such as those which govern the Urey and the Pasteur levels for free oxygen, and concomitantly, we have no indication about the former absolute levels of carbon dioxide. The only thing we can do is to draw a generalized curve, in which accretion and depletion of carbon dioxide varies in relation with, but contrary to, that of free oxygen (see Fig.126).

It seems evident that the level of carbon dioxide, both that of the atmosphere and that of the oceans, will rise during orogenetic periods and will fall during geosynclinal periods.

We have no indication for the absolute values of these variations and the starting point has been arbitrarily set at 10 PAL (Fig.127). According to L. G. Sillèn (personal communication, 1967; see section 7 of Chapter 14) this is probably too high for geochemical reasons, because in that case other minerals would have formed on the sea floors. So it must be remembered as an arbitrary figure only.

It is further assumed that the fall of carbon dioxide during each successive geosynclinal period (the lines *1–2*, *3–4*, etc. of Fig.127) will have increased with time because life and the consumption of carbon dioxide through organic photosynthesis expanded continuously. The slope of each successive line is therefore drawn somewhat steeper in Fig.127. The addition of carbon dioxide during successive orogenetic periods has, on the other hand, been postulated as about the same in each period. Owing to the logarithmic scale of the ordinate in Fig.127, this is expressed by an apparently larger accretion during successive orogenies. We have, as yet, no means of estimating even in a qualitative way, the absolute strength of volcanic activity, nor the amount of CO_2 produced, during any of these cycles. So the accretion has been arbitrarily set at a volume corresponding to three times the Present Atmospheric Level.

The main point in the history of atmospheric carbon dioxide is that its level was probably at its lowest at the end of each geosynclinal period. It is here assumed that somewhere during the long geosynclinal period between the Grenville and the Caledonian orogenies the level of carbon dioxide dipped below the present one. This could have facilitated the secretion of phosphatic and calcitic shells by marine organisms which defines the base of the Cambrian and as such that of the Paleozoic and the Phanerozoic.

4. THE "EXPLOSION OF LIFE AT THE BASE OF THE CAMBRIAN"

The tentative history of atmospheric carbon dioxide, as presented in the last section, leads to a remark on the so-called "explosion of life", apparently found at the base of the Cambrian. BERKNER and MARSHALL (1965) were greatly impressed by this "explosion", by which they dated the attainment of the atmospheric oxygen level of 0.01 PAL, when "life could conquer the seas".

The idea of an "explosion of life at the base of the Cambrian" rests, in fact, on a misconception. The base of the Cambrian, and thus of the Paleozoic and the Phanerozoic, does not coincide with an explosion of life, or even with an explosion of animal life. We have seen in Chapter 12 that rich and diversified Late Precambrian faunae existed. We find their remains, *wherever they have a chance to be fossilized*. The so-called "explosion of life at the base of the Cambrian" is in fact no more than an *explosion of fossils*. This does not rest on a greater overall development of life but merely on the fact that around this time several groups of animals acquired the faculty of secreting hard shells, which fossilize much more readily than the softer parts of animals. Phosphatic at first, these shells later developed into the still harder calcareous shells of mollusks and brachiopods.

The "explosion of life at the base of the Cambrian" seems to be evident from every general textbook on paleontology or historical geology, which, being concerned with visible facts, only relate the history of fossils, not that of life. But, as the Dutch paleontologist I. M. van der Vlerk was wont to say, the sudden abundance of fossils from the Cambrian onwards does not arise from an explosion of life, but only from a new fashion arising in the animal world: that of wearing shells.

The more or less simultaneous acquisition of the faculty of secreting shells in several non-related groups during the Early Cambrian points towards a general, environmental, influence. If, as postulated in the preceding section, this was the depletion of carbon dioxide below its present level, the temporarily more alkaline waters of oceans and lakes would have facilitated the biochemical secretion of phosphates and carbonates. Once this faculty had been acquired, it could easily have been retained by life and defended against a more acid environment due to the rise in the level of carbon dioxide during the next orogenetic period. Biochemical secretion, once developed, can overcome adverse circumstances. As a case in point our present-day fresh water mollusks may be cited. These not only are able to secrete calcareous shells in lake waters severely undersaturated in regard to calcium, but also to defend them against subsequent attack through the development of a thick horny layer or periostracum.

To conclude, the so-called "explosion of life at the base of the Cambrian" is not at all an explosion of life, but an explosion of fossils. It has nothing to do with the history of atmospheric oxygen, but it might be dependent upon a temporary lower level of atmospheric carbon dioxide occurring during the long quiet period between the Grenville and Caledonian orogenies.

A quite different viewpoint has, however, recently been expressed by TOWE (1970), who stressed the importance of collagen synthesis by early animals. Collagen is the main element of the dense, fibrous connective tissue, which forms the support of muscles and organs in every metazoan stock. It is also the base of shells and skeletons. The synthesis of collagen, a biochemically unique protein formed by a triple helix mainly built up by the amino acids glycine, proline and hydroxyproline, requires the presence of molecular oxygen.

In an atmosphere with little oxygen, oxygenation reactions—or oxygenases, which are energetically expensive for the cell, would take place with a higher priority in the physiologically necessary energy-generating reactions, than in the formation of collagen. The evolutionary utilization of collagen in an atmosphere with little oxygen would hence tend to develop only the most immediate priorities, such as the development of support for the coelom, the musculature, and of segmentation. Only when more oxygen became available, would shells, cuticles and carapaces, which, notwithstanding their functional significance, are not a mandatory structural prerequisite of animals, but a sort of low-priority physiological luxury, tend to develop.

Accordingly, in the views of Towe, the explosion of shells at the base of the Cambrian is not related to a temporary low level of carbon dioxide, but to the general rise of the level of oxygen. It is, of course, entirely within the possibilities that both processes have played their part.

REFERENCES

AXELROD, D. I., 1959. Evolution of the psilophyte paleoflora. *Evolution*, 13:264–275.
BERKNER, L. V. and MARSHALL, L. C., 1965. On the origin and rise of oxygen concentration in the earth's atmosphere. *J. Atmospheric Sci.*, 22:225–261.
CHALONER, W. G., 1964. An outline of Pre-Cambrian and Pre-Devonian microfossil records: Evidence of early land plants from microfossils (abstract). *Intern. Botan. Congr.*, *10th, Edinburgh, 1964*, pp.16–17.
STEWART, W. N., 1960. More about the origin of vascular plants. *Plant Sci. Bull.*, 6:1–5.
TOWE, K. M., 1970. Oxygen-collagen priority and the early Metazoan record. *Proc. Natl. Acad. Sci., Wash.*, 65:781–788.

Chapter 17 | Extra-terrestrial Life?

1. INTRODUCTION

Before coming to the final chapter, with our summing up, we must now make a short and inconclusive, but highly interesting detour and study the possible remnants of life, as found in meteorites. The present interest in this subject is of recent date and was started in 1961 by B. Nagy and G. Claus (cf. NAGY et al., 1961; CLAUS and NAGY, 1961). Nagy, an organic geochemist, extracted carbonaceous compounds from a certain meteorite, the Orgueil stone, and found that these were, at least in part, similar to organic compounds found in present-day life. To check whether this could have been the result of terrestrial contamination during the century that samples of this meteorite had lain on museum shelves, Claus, a microbiologist, investigated the sample for microbial remains. He found but a very few remnants of present-day life, but instead numerous, small, morphologically distinct "organized elements", which, although in their outer form resembling certain fossil algae, still were different from all known present-day organisms.

On this basis the conclusion that the carbonaceous compounds and the "organized elements" of the Orgueil and related meteorites represent fossil evidence of extra-terrestrial life seemed warranted. On the other hand, even the most cautious of these proposals (see, for instance, NAGY et al., 1963a,b; MEINSCHEIN et al., 1963; CLAUS et al., 1963) was immediately met by a flood of refutation (see, for instance, FITCH and ANDERS, 1963; ANDERS, 1963). Amongst the counter-arguments the most frequently used was, curiously enough, that of terrestrial contamination, although the possibility of such a contamination had not only from the beginning been in the mind of the scientists investigating the meteorites, and had, in fact, led to the discovery of the "organized elements". This particular objection has been refuted by now. But, as we will see in this chapter, no agreement has been reached yet on the exact nature of these compounds, i.e. whether they are abiogenic or biogenic.

2. METEORITES

Before proceeding further with the study of these particular aspects of a small number of meteorites, we must place the Orgueil and related stones within the general framework of the meteorites. Meteorites (MASON, 1962, 1967) are commonly classified according to two properties, i.e. their composition and the fact whether they possess small—up to about 1 mm in diameter—spherical aggregates, the *chondrules*. According to their composition a division is used into *iron meteorites, stony-iron* and *stony meteorites*. Whereas meteorites containing chondrules, which are only found amongst the stony meteorites, of which they form the majority, are called *chondrites*, in contrast with the *achondrites*. The latter classification is probably genetic in the sense that chondrites represent an earlier form of meteorites, from which achondrites derived through a secondary process of heating, melting and recrystallization-upon-cooling.

By far the larger part of the meteorites has a composition that differs strongly from that of the rocks of the crust of the earth. The dominant minerals of the iron meteorites are formed by nickel and iron, those of the stony meteorites by magnesium-iron silicates, in contrast to the quartz-alumino silicates, which form the dominant material of the earth's crust. The minerals of the normal meteorites are anhydrous and formed in a highly reducing environment, whereas the minerals of the earth's crust are commonly hydrated and/or oxidized.

There is, however, a small number of stones, called the *carbonaceous meteorites*, which in several aspects differ fundamentally from the majority of meteorites. They have normally been called the "carbonaceous chondrites", because they are classed with the stony meteorites, of which the majority is chondritic. But several of the carbonaceous meteorites, and the most typical at that, do not contain chondrules. Although they seem to have a rather complicated history, it is probable that these originally formed in an area in which chondrules —"fiery drops of a rain of molten silicates" as they have been called rather lyrically—could not develop. Thus it seems better to speak of carbonaceous meteorites. These rocks contain a noticeable, and, in some cases, even a large amount of water and also an appreciable amount of carbon. They consist mainly of flaky silicates reminiscent of the clay minerals and with a general formula similar to the common terrestrial mineral chlorite.

The history of one of the best known carbonaceous meteorites, the Orgueil stone, has been studied by BOSTRÖM and FREDRIKSSON (1966), who conclude that three main periods of mineral formation can be recognized. An early stage contains minerals like troilite, a special form of FeS, which are stable at several hundred degrees Celsius. Then follows a second stage with minerals like chlorite and limonite, which formed below 170°C, and a late stage with carbonates and sulphates formed below 50°C. The latter stage developed in a watery environment (see Table XXVI), whereas water vapour was probably present during the second stage also. During the third stage the material of the Orgueil carbonaceous meteorite has been broken up and recemented repeatedly.

TABLE XXVI

THE SUCCESSION OF ORGUEIL MINERALS[1]
(From BOSTRÖM and FREDRIKSSON, 1966)

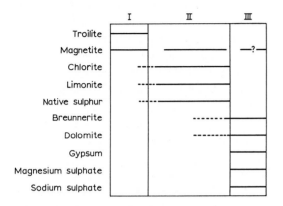

[1] The succession of the minerals is indicated by full lines. Dashed lines indicate possible extensions. During the first period of mineral formation the temperature must have been several hundred degrees Celsius. Period II had moderately high temperatures, probably below 170 °C, whereas during period III vein mineralization occurred in an aqueous environment, probably below 40 °C.

The variation in composition of the meteorites has already in 1866 led G. A. Daubrée (1814–1896) to a theory, later expanded by DALY (1943), that they probably represent fragments of a disrupted former planet, belonging to our own solar system and similar to the earth. For reasons of analogy he then suggested that the earth itself consists of a core formed by nickel-iron minerals comparable to the iron meteorites and a mantle formed by magnesium-iron silicates like the stony meteorites (see Fig.9). It might be remarked parenthetically that the only real "proof" that the earth really has a nickel-iron core to this date rests upon a supposed analogy with the nickel-iron meteorites. In this picture the carbonaceous meteorites would represent samples from the crust of the original planet.

3. THE CARBONACEOUS METEORITES

The number of the carbonaceous meteorites is, as stated above, small. Dependent upon the classification used, as to what is to be exactly considered as a carbonaceous meteorite, a classification which shows some variations with different scientists, their number lies around 20, as against a total number of around 950 stony meteorites. Their overall chemical composition is related to the "high iron" subgroup of the stony meteorites, from which they differ mainly in the argillaceous nature of their minerals, a property dependent upon their high water content. This does, nevertheless, make them quite distinct from the normal stone meteorites, when studied petrographically.

We may cite here BERZELIUS (1834) who received "a very small sample" of the Alais carbonaceous meteorite, and thought it was part of the soil in which the stone had fallen. He only proceeded with his chemical analysis when he was satisfied that it really was a part of the original meteorite, and that all fragments found looked alike. Upon the conclusion of his studies he then stated: "... There is no doubt that the examined stone, notwithstanding all outer differences, is a meteorite, which, in all probability, originated from the normal home of the meteorites." Berzelius, incidentally, was the first to speculate whether the carbonaceous matter found in these meteorites could perhaps indicate the existence of extra-terrestrial life. A possibility which he, without giving further arguments, in the end rejected.

In another way too, the account of Berzelius is illuminating, i.e. in the crudeness of the analytical procedures available to him. The fact that the carbon present was not originally a coal, had, for instance, at that time to be proved by the fact that "the powdered rocks show a greenish brown colour, which, upon dry distillation turns to coalblack", which is a colour scheme different from that found in coals. I will not repeat the rest of the analytical reasoning, but only stress the enormous difference with the sophisticated analytical procedures now at the disposal of organic chemists studying samples of carbonaceous meteorites.

Notwithstanding their small number, carbonaceous meteorites show a large variation and have consequently been grouped into three subclasses by WIIK (1956, see Fig.128). Type I carbonaceous meteorites are the most exceptional, showing the largest amount of water and carbon. They contain 3–7% of carbon; up to 22% of combined water and consist largely of hydrated magnesium-iron silicate, magnetite and magesium sulfate. They contain no chondrule and their density is about 2.2. Type III carbonaceous meteorites show 0.5–2% of carbon and 2% of combined water and consist largely of olivine. Their density is about 3.4. Type II stones are intermediate between types I and III. There are five known meteorites of type I, which are the most important for our study. They are, in chronological order: Alais (1806), Orgueil (1864), Tonk (1911), Ivuna (1938) and Revelstoke (1965).

The latest fall of a carbonaceous meteorite, near Allende in Mexico, has received much publicity, because it consisted of a shower of at least several dozens of stones. The combined weight is over 1,000 kg. It is, however, a class III carbonaceous meteorite, and as such not so important for our story.

The Orgueil meteorite, which exploded in the atmosphere and produced

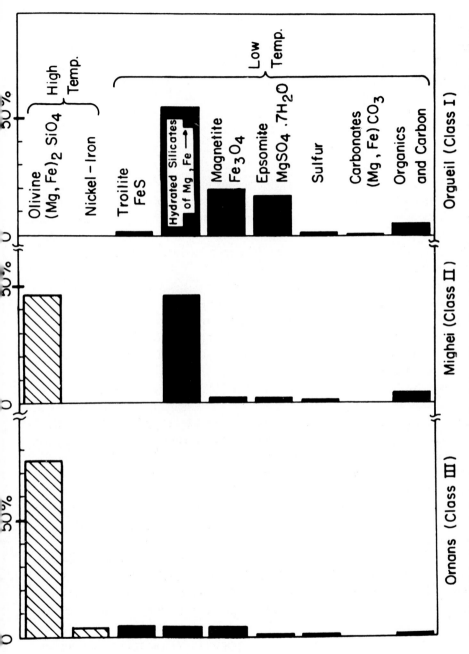

Fig.128. Mineralogy of the carbonaceous chondrites. The ratio of low-temperature minerals (hatched) increases from class III to class I. "Low-temperature" in this presentation includes all three temperature ranges of Table XXVI. (From ANDERS, 1964.)

a meteorite shower witnessed by many persons over a wide district, has produced a large number of meteoritic stones. It is on this material, more abundantly available than the other stones, that most of the recent work has been carried out. Parallel studies have, however, from the onset been done on other stones and have led to a corroboration of the results gained from Orgueil (see, for instance, KAPLAN et al., 1963; TIMOFEJEV, 1963).

4. THE CARBONACEOUS MATERIAL

An important feature of the carbonaceous meteorites is that they all exhibit a thin glassy crust which probably is due to the melting of the meteorite's outer skin by the heat generated during its fall through the earth's atmosphere. This crust, comparable to the heat shield of space vehicles, has protected the inner part of the meteorites from heat. This is, in fact, quite unaffected, as is shown by the presence of minerals such as gypsum.

The study of the carbonaceous material has developed along two lines. These are, first, research into the chemical nature of the extracted organic—or "organic"—material, and secondly the study of the morphology and the nature of the "organized elements". Both routes are beset by their special problems. As to the first line of investigation, it should be noted that only a small part of the carbonaceous material is soluble and can be extracted for further study. The major part is formed by an insoluble residue, comparable to kerogen (see section 3 of Chapter 8) or tar. Although it can be broken up by ozonolysis (BITZ and NAGY, 1966), it has up to now defied attempts at complete analysis. The "organized elements", on the other hand, are so small that they cannot be isolated singly for chemical analysis. So, one never is sure what overall chemical property corresponds to what part of the meteorite; whether it belongs to the ground mass or to the "organized elements". And, furthermore, when two investigators speak about the properties of "organized elements", even of the same meteorite, one is never certain whether they speak of the same subjects, as the discussion between MUELLER (1965) and CLAUS and SUBA-C (1965) has, for instance, shown.

Looking now at the "organic" compounds reported from carbonaceous meteorites, we find the major types listed in Table XXVII. For a full list of the compounds reported from the Orgueil meteorite, complete with structural formulae and bibliographic references, the reader may turn to NAGY (1966), whereas HAYES and BIEMANN (1968) supplied a list of the compounds found in the Murray and Holbrook meteorites. Many of these compounds, such as the entire groups of the saturated hydrocarbons, the carboxylic acids, which include the straight-chain fatty acids, and the nucleotide bases, are also known as common constituents of terrestrial sediments and of living and fossil matter. Other groups, such as those of the aromatic hydrocarbons and the porphyrins also form common constituents of petroleum and of sedimentary rocks, whereas other compounds, notably the group of the triazines, has no known terrestrial biological significance.

TABLE XXVII

MAJOR "ORGANIC" COMPOUNDS REPORTED FROM CARBONACEOUS METEORITES
(From NAGY, 1968)

Hydrocarbons	Non-hydrocarbons
Saturated hydrocarbons	*carboxylic acids*
n-Alkanes	straight-chain fatty acids
Branched chain alkanes	benzene carboxylic acids
Isoprenoids	hydroxybenzoic acids
Cycloalkanes	
Olefinic hydrocarbons	*nitrogen compounds*
	pyrimidines
Aromatic hydrocarbons	purines
Alkyl benzenes	guanylurea
Naphthalene	triazines
Acenaphthenes	porphyrins
Acenaphthylenes	
Phenanthrenes, anthracenes	
Pyrenes	

On the face of it, this imposing array of compounds known also from terrestrial life and from petroleum, of which the biogenic origin seems pretty well proved, presents a strong case in favour of the assumption of extra-terrestrial life, fossilized on a former planet, or on other bodies belonging to the solar system. The fact that one has found some compounds which have no apparent counterparts in terrestrial life does not present a serious obstacle. These could well have been formed during the later history of the carbonaceous matter on the parent body of the meteorites, just as one finds in petroleum compounds, such as pyridines, carbazoles, etc., which have developed from, but have no direct counterpart in, terrestrial life (see, for instance, HENDERSON et al., 1968; MEINSCHEIN et al., 1968).

But there is one serious objection to the assumption of extra-terrestrial life, which lies in the fact that each and everyone of the compounds cited in Table XXVII has been synthesized in experiments using an anoxygenic environment, such as described in Chapter 6. Although the very variedness of the "organic" compounds eeported from carbonaceous meteorites seems to point firmly in favoui of a biological origin of this material, and thus of extra-terrestrial life, it does not offer definite proof of its existence. It remains possible that all of these compounds have børn formed inorganically in an anoxygenic planetary atmosphere, or on the

surface of asteroids too small to retain an important atmosphere. Or even, according to a theory proposed by Professor E. Anders and his collaborators, in the solar
nebulae (STUDIER et al., 1968; HAYATSU et al., 1968).

So notwithstanding the fantastical development of analytical techniques over
the last few years, and although we know at present so very much more about the
nature of these carbonaceous compounds than only a short while ago, we have
not yet found the final answer to their origin. In this respect we have not yet
advanced since 1834, when BERZELIUS, relying only on his rudimentary techniques,
was not able to give a really decisive answer to the question whether these carbonaceous compounds were remnants of extra-terrestrial life or not.

What we consequently need is an even more sophisticated approach. One
possibility lies perhaps in the optical activity of the carbonaceous material, although
even here a more cautious attitude seems warranted (see section 2 of Chapter 14).
Anyhow, the existence of optically active carbonaceous material has as yet not
been proved. MEINSCHEIN et al. (1966) found that terrestrial contamination might
have played a part in earlier analyses. And although NAGY et al. (1964) have
claimed to have definitely found optical activity in uncontaminated material from
Orgueil, other investigators failed to find such activity. This discrepancy may
have been caused either by the fact that these other investigators used analytical
techniques slightly different from those used by Nagy. But, and this is of course
much worse, it might also be due to some artifact arising in Nagy's own laboratory.
So, as long as no other evidence is available, we must, however, regrettably, regard
these investigations also as inconclusive.

Another line of investigation which eventually might supply an answer to
our question lies in the analysis of the non-solvent material. In 1964 DUCHESNE
et al. have compared this coal type material with terrestrial coal and have happily
concluded that: "All these correlations point towards the fact that the carbonaceous matter of the meteorite is, at least in part, the result of biological activity."
This rather pertinent statement is, however, inconclusive too. For, as NAGY (1966)
remarked: "It would be interesting to see the results of similar analyses" (of which
there are none) "of tars obtained from abiological synthesis."

Summarizing, we find that the insights gained from the chemical study of
the carbonaceous material do not, as yet, supply us with definite proof for the
existence of extra-terrestrial life. On the other hand, it seems certain that if this
material has not been formed by life, it must be the product of some sort of pre-life
similar to that which we have assumed has been active in the early history of the
earth.

Although it is thus impossible, at the moment, to decide whether the carbonaceous material of these meteorites is the result of some sort of pre-life or of life,
we can be sure that it represents extra-terrestrial and not terrestrial material. This
has been shown by CLAYTON (1963) who analyzed the ratios of the stable isotopes

$^{18}O/^{16}O$ and $^{13}C/^{12}C$ extracted from carbonate minerals—not carbonaceous matter—of the Orgueil meteorite. These minerals, dolomite $CaMg(CO_3)_2$ and breunnerite $(Fe, Mg)CO_3$, together make up only 0.3% by weight of the meteorite, but even so a successful analysis could be run. The ratio of the stable isotopes of oxygen is not different from terrestrial ratios for oxygen, but that for carbon proved to be quite different. The $^{13}C/^{12}C$ ratio was found to be 6% greater than for any known terrestrial carbonate. Although we have seen that there is a rather large variation for this ratio found in terrestrial rocks (see section 9 of Chapter 14), the ratio for the Orgueil meteorite is so high that it excludes a terrestrial origin of its carbonate minerals.

Parenthetically remarked, this result confirms earlier indications that the interior parts of the carbonaceous meteorites used for these studies have not become contaminated by terrestrial material. The possibility of terrestrial contamination has again caught the attention when it was learned that the crust of the Pueblito de Allende (Mexico) meteorite had become seriously contaminated within one month of its fall (HAN et al., 1969), but this evidently does not apply to those samples obtained by core drilling from the other meteorites.

It is not known why this ratio for carbon in the Orgueil meteorite is so high. Moreover, in other types of meteorites, both iron and stony, it has been found that the ratio of the stable isotopes of sulfur is equal to the terrestrial one. Isotope fractionation by a process uncommon on earth, or incomplete homogeneization in the very early days of nucleosynthesis of the material of the carbonaceous meteorites might have been responsible for this discrepancy. But at present such explanations are no more than guess work.

However, since we are not primarily interested in the history of meteorites, it is enough to retain that the material of the carbonaceous meteorites is of extra-terrestrial origin.

5. "ORGANIZED ELEMENTS"

The morphology of the "organized elements" follows from the pictures in Fig.129–131 and Plates XII and XIII. A full catalogue of the forms known up to 1962 has been presented by MAMIKUNIAN and BRIGGS (1963). They are small, of the order of 5–50 μ, the most common size ranging between 6 and 15 μ, and always "unicellular". They often show a definite organization, such as double walls, and/or of pores and spines in and on their walls. As such they do not belong to the most primitive structures possible. It is therefore not surprising that the major group of scientists who without any doubt accepted them as remnants of extra-terrestrial life were the microbiologists (cf. PAPP, 1963). On the one hand they recognized in these structures a degree of organization which, in their

PLATE XII

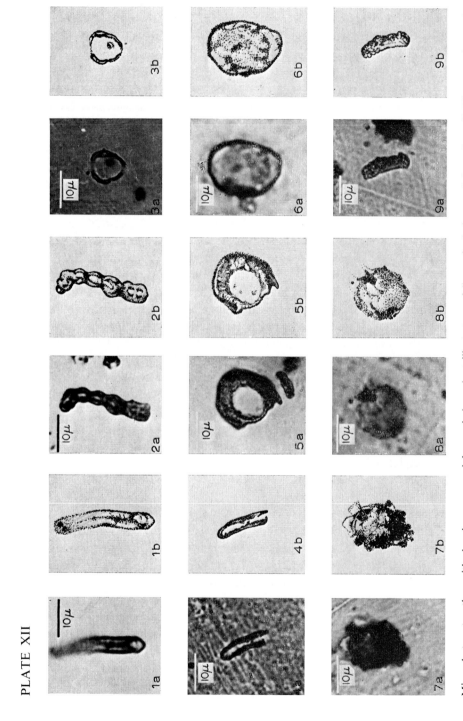

Microphotos, together with drawings prepared by an independent illustrator, of "organized elements" from Orgueil. Electron microprobe analysis has shown that the particles 1, 2, 3, 4, 5 and 9 are partially mineralized with iron and some chlorine. Particle 6 contains iron and nickel, whilst particles 7 and 8 are formed by magnesium silicates. (From NAGY et al., 1963.)

PLATE XIII

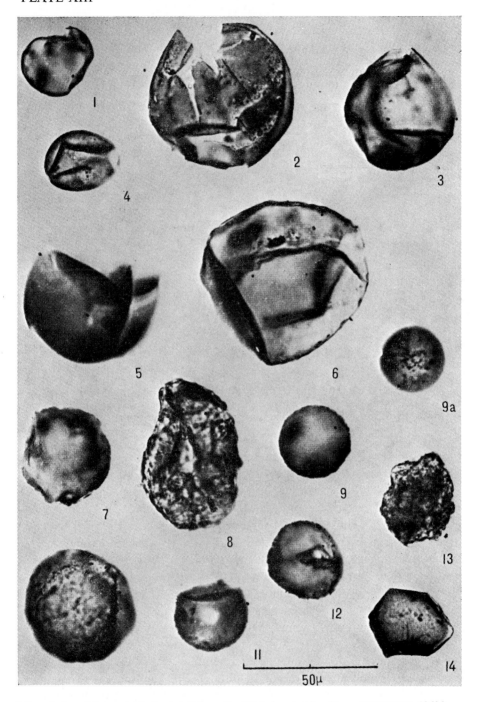

Microphotos of "organized elements" from the Mighei meteorite. (From TIMOFEJEW, 1963.)

experience, is always representative of living matter, and hence must represent fossilized life, while on the other the actual forms were not known from terrestrial life, either living or fossil, and hence the life which had formed them had to be extra-terrestrial.

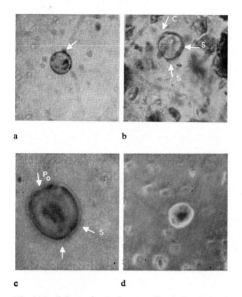

Fig.129. "Organized elements" of the Orgueil meteorite. a–c: "organized elements" separated from a powdered sample. d: "organized element" in thin-section seen under phase contrast. C = collar; Po = pore; S = spine. (From G. Claus, personal communication, 1964.)

One point to be mentioned in regard to the "organized elements" is their extreme abundance. It is not as if "organized elements" are a lucky find, when studying carbonaceous meteorites. Estimates of 1,800 per milligram have been made. In fact Professor A. Papp of Vienna remarked to me in 1964 that the only reason why the "organized elements" perhaps could not be described as fossilized former life forms, was their frequency. He had some trouble in visualizing life having been so much more abundant on the parent body of the carbonaceous meteorites, than on earth. But, as we saw in the discussion of the Precambrian Gunflint flora in section 13 of Chapter 12, terrestrial fossils may locally be preserved just as abundantly, when fossilization permits. Although early life has been very primitive, at least to our standards, there is no reason why it could not have been prolific.

The main objection against the interpretation of the "organized elements" as fossilized remnants of extra-terrestrial life has been that of terrestrial contamination. It has even been stated that 90% of the "organized elements" found would be terrestrial contaminants. Terrestrial pollen, notably that of ragweed which

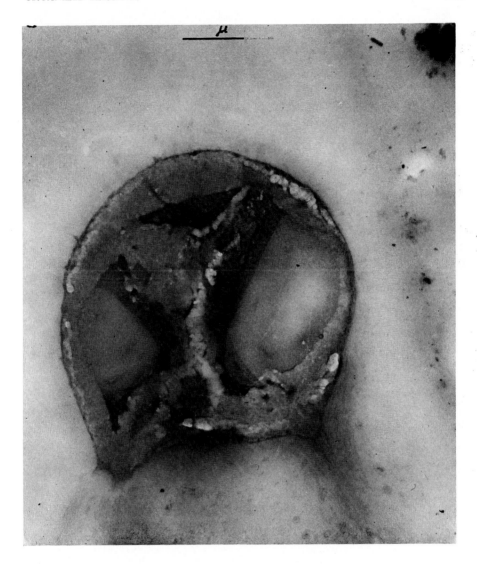

Fig.130. Electron microphoto of a microstructure in an ultrathin section of Orgueil. Before sectioning the sample has been boiled in 6N HCl for one hour. In contrast to the microphotos taken with the light microscope, such as shown in Fig.129 and Plate XII, identification of subjects photographed with the electron microscope is much less positive. This photo may represent an "organized element", or some acid-resistant terrestrial biological contamination, a mineral concretion, or it might even be an artifact formed during the preparation for the electron microscope, as pointed out by NAGY (1966).

10 μ	Type Organized Element	Taxonomical Nomenclature	Occurrence	Size in Microns
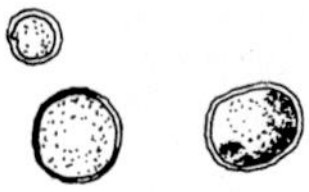	1.	Apolinarisphaera meteoricola (1) Clausisphaera fissa (2)	most abundant	4–12 μ
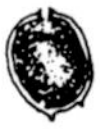	2.	Stemmatopila uniporata (1)	abundant found only (1) Orgueil (2) Ivuna	width: 13–15 μ length: 16–18 μ
	2.	Disacerra sulcata (1) Subreticulate pollen (2)	abundant	diameter: 12–18 μ
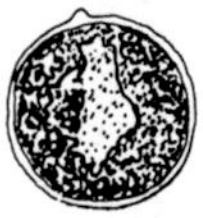	2.	Apophoreta aethrodescensa (1) Incertae sedis type B (2)	not too common	diameter: 16–22 μ
	3.	Ancilicula vetusta (1)	common	width: 10–16 μ length: 14–18 μ
	4.	Dactyliotheca daedala (1) Incertae sedis type A (2)	common	width: 8–14 μ length: 10–20 μ
	5.	Daidaphore brezelii (1) Caelestites sexangulatus (2)	Extremely rare	total diameter: 25–30 μ, solitary body d = 8–16 μ tubular protrusion length: 0.8–1.8 μ width: 0.6–1.2 μ
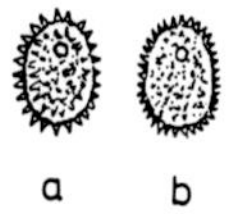 a b	6.	Siderolappa lapillata (1)	Rare	width: 10–21 μ length: 14–16 μ
	7.	Oscenoscaeva proavita (1)	Rare	diameter: 4–6 μ thickness: 2–3 μ
Staplin 1962	8.	Tissue type B (2)	Rare	30–40 μ
Staplin 1962	9.	Protoleio-sphaeriduim A (2)	Rare	25–30 μ

Nomenclature   (1) Claus and Nagy (1962).
(2) Staplin (1962).

Fig.131. The main "organized elements", as known from the work of CLAUS and NAGY (1961) and STAPLIN (1962). (From BRIGGS and MAMIKUNIAN, 1963.)

thrives in the steppes surrounding urbanized areas, can, of course, always contaminate any microscopical preparation. But the contention that by far the larger part of the "organized elements" represents terrestrial contamination is humiliating to the microbiologists concerned. For every microbiologist is confronted with the possibility of contamination from the very first hour of his studies. Whilst in this case they had taken pains to compare these structures with the micro-organisms known to them from present-day life and found no resemblance. The whole study had even been initiated, as we have seen, to look for a possible contamination by present-day life during the stay in the museums of these samples. Postulating that the "organized elements", or at least the majority of them, are ragweed pollen is therefore of the same level as telling a physicist that he cannot read his meters or a chemist that he cannot preserve the purity of his samples.

It must be conceded that microphotos of "organized elements" and of ragweed pollen, such as produced by FITCH and ANDERS, 1963 (cf. Fig.132), show a certain resemblance. But this is a superficial resemblance only, and microphotos of this magnification are notoriously vague. Anyone who will take the trouble of comparing the actual preparations under the microscope will agree that this resemblance is no more than an apparent one and that the "organized elements" have nothing to do with ragweed pollen.

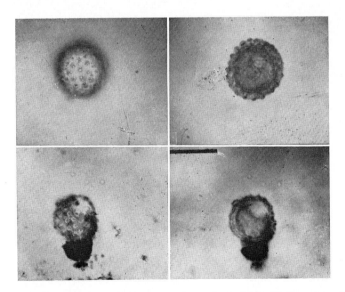

Fig.132. Comparison between an "organized element" from Orgueil (above) and a ragweed pollen; × 720. The left- and righthand pictures are both taken from the same object, but with the microscope focussed slightly different.

The apparent resemblance between the two objects is due to the vagueness of even the best of microphotos at this magnification, which obliterates much detail. When seen directly under the microscope, the objects are distinctly different. (From FITCH and ANDERS, 1963.)

"Organized elements" have also been compared with other forms of terrestrial life and one example is illustrated in Fig.133. Here an "organized element" of Orgueil is compared to a preparation of a fossilized terrestrial dinoflagellate. But in this case, too, the resemblance is a superficial one only, because the magnification of the dinoflagellate is much less than that of the "organized element". I believe one is justified in stating that every time a similarity between "organized elements" and terrestrial life forms has been cited, more detailed studies have shown it to be superficial only.

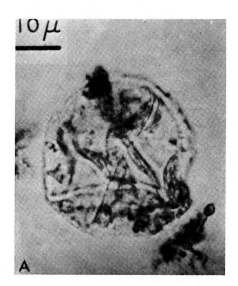 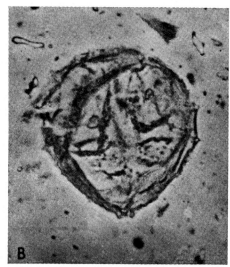

Fig.133. An "organized element" from Orgueil (A), compared with the fossil of a terrestrial dinoflagellate (B). There is a certain superficial resemblance, but the magnification of the dino-flagellate is only about one fourth of that of the "organized element". Moreover, the ultramicro-spectrum of the "organized element" does not resemble the spectra of present terrestrial biological material. Which shows, that, notwithstanding the superficial similarity, there is no identity. The "organized element" is indigenous to the meteorite and not a terrestrial contaminant. (After NAGY, 1966.)

Moreover, at least part of the "organized elements" are mineralized and thus cannot be the result of contamination by terrestrial life forms, which could not become mineralized during the time the samples have lain on museum shelves. Another question is, however, the nature of this mineralization. MUELLER (cf. 1965) has maintained that all "organized elements" consist of olivine, a high temperature mineral. Mineralization by olivine posterior to the formation of "organized elements" by pre-life or by life would have destroyed the original structures, and what we see at present could be no more than artifacts. This contention has, however, been countered by CLAUS and SUBA-C (1965) and it seems established by now that these authors have looked at different sets of structures. Moreover Nagy and his collaborators (NAGY et al., 1963a,b) have analyzed "organized elements" by electron probe microanalysis and have found a mineralization with iron, chlorine and some other elements, which indicated an aqueous, low-temperature environment of the "organized elements" during the time of their formation.

Another objection is that small, organized objects, even with double walls, can form inorganically, as has been shown by Fox (see section 6 of Chapter 6). Personally I am, however, convinced that the intricacy of the wall structures of the "organized elements" strongly indicate their biological origin. So I am still prepared to follow microbiologists and micropaleontologists who, from their studies, have always come up convinced by the reality of these structures. This does, however, depend upon a kind of feeling, developed over the years in working with biological systems. As such it can hardly be expressed in formulas, and so it fails to convince investigators working in the so-called exact sciences.

To conclude we must therefore admit, that notwithstanding the strong indications for a biological origin supplied by the "organized elements", the information at hand is still insufficient to decide whether they represent fossilized remnants of extra-terrestrial pre-life or life.

6. DISCUSSION

As has been remarked by almost everyone of the recent investigators of carbonaceous meteorites, the discussion as to the origin of the carbonaceous material would be simplified if one only knew where these meteorites come from. Unfortunately, the only concensus about this problem is that they belong to our own solar system. A difficulty in this story lies in the fact that many meteorites, the carbonaceous meteorites included, show signs of a complex history in which the development at one time does not need to show a relation to their earlier or later history.

The theory of a disrupted former planet, once situated between Mars and Jupiter, so authoritatively developed by DALY (1943; see section 2 of this chapter) is at present no longer followed, mainly owing to the large variations in the chemical composition and the structure of the meteorites. It is generally conceded that this variation is so large that it would be impossible for a single planet to be so diversified. Moreover the pressures at which the nickel minerals in the iron meteorites must have formed are lower than would be compatible with the core of a large planet (ANDERS, 1964). So another theory has been put forward, which postulates not a single parent body but a smaller or larger number of discreet parent bodies. These are thought to have existed in the same general area, i.e. in the asteroidal belt between Mars and Jupiter (cf. ANDERS, 1963). In this view it is thought that meteorites originate from the scattering of colliding asteroids. The chemical and structural variations found in the meteorites can then be correlated with those supposedly existing amongst the asteroids. A variation which would, in turn, depend upon the place of formation of the asteroid within the primeval stellar gas cloud. The high-temperature asteroids, from which the iron meteorites are thought to be derived, would have been formed near the center; whereas low-temperature asteroids, including the parent bodies of the carbonaceous meteorites, would have been formed near the flanks of the stellar gas clouds.

I will not try to further evaluate these theories on the origin of meteorites, but return to the facts gathered from the study of the carbonaceous material. We there did find that this material, as far as its extractable part is concerned, contains many kinds of "organic" molecules which also belong to present-day terrestrial life. But, as far as is known now, all these compounds may form inorganically too, under the proper circumstances in a reducing environment. They thus represent fossilized remains of some sort of either pre-life or life. Other investigations, notably on the ratio of the stable isotopes of carbon, have moreover shown that, whether pre-life or life, it was extra-terrestrial.

The "organized elements" are shown, at least in part, to consist of carbon, whilst they are also, at least in part, mineralized. They are indigenous to the meteorites and no terrestrial contaminants. Their morphology is such that here again a definite answer to the question "pre-life or life?" cannot be given although microbiologists generally seem to prefer the second answer. However, just as in the chemical analysis it can be maintained that their origin is, without doubt, extra-terrestrial. This time the conclusion is based on the morphology of the "organized elements", which differs from the morphology of biological structures produced by terrestrial life.

We may therefore summarize the evidence in stating that the most plausible explanation for the origin of the carbonaceous material in these meteorites is that it was formed extra-terrestrially, perhaps by some sort of pre-life, but more probably by some sort of early life. This does not mean that the former existence of life on a meteorite parent body has been proved. However, at present it offers the best explanation of the facts described in this chapter, whereas, as the many unsuccessful tries over the last years have proved, it is difficult to disprove.

If the "organized elements" would really represent fossil forms of extra-terrestrial life, the question may crop up why their chemistry is so similar, their morphology so different from terrestrial life. We touch here upon the differences between chemical compounds and morphological structures, already touched upon in section 9 of Chapter 4 in the discussion of the Pirie drawing. The simple molecules under consideration are governed by a rigid set of physico-chemical equations, and under similar circumstances only a limited number of compounds is viable. As we saw already, the parent body of the carbonaceous meteorites must have had an aqueous, low-temperature environment, similar to the terrestrial one. It is, of course, possible, although this seems improbable at present (BITZ and NAGY, 1966), that the more complicated molecules of the non-extractable carbonaceous material show different structures, but for the simple compounds analyzed so far, similarity of structure is nothing to be wondered at. In regard to morphology, things are, however, quite different. Even the most simple morphological structures such as the pores and the spines shown in Fig.129, can develop in many places, in many ways and in many forms in parallel lines of evolution. There are no simple, outspoken reactions, which favour one or the other mode of development. The development of such structures depends on the interaction between the mutations available within a population at a given time and the selection pressure executed by its surroundings. The number of independent variables in such reactions seems to preclude the possibility of a repetition. So, when such a structure develops in one evolutionary line, there is hardly a chance that it will be exactly duplicated in another line.

The probability that extra-terrestrial life has been recorded in the carbonaceous meteorites undermines one of the most cherished notions held for a long time by man, that of the uniqueness of terrestrial life. It might even be argued that the involvedness of the outcry against the first interpretations by Nagy and Claus has been conditioned by this background. But if we look further than our own solar system, the chances are that life has developed elsewhere in the universe. As Mamikunian (in MAMIKUNIAN and BRIGGS, 1966, p.vii) has summed up our knowledge: "There is a large statistical probability that many at present undetectable planets exist throughout the universe which have environments capable of sustaining life. Perhaps one star in a million has a planet that meets all the necessary conditions".

SHAPLEY (1953) listed these conditions as follows:

(1) Water, the practical solvent for living processes, must be available in liquid form. The kind of life we are talking about and thinking of does not occur in uncondensing steam or unmelting ice. The basic requirement therefore is that the living planet must be at the proper distance from its star—in the liquid-water belt—, not as close as Mercury is to the sun, nor as remote as Jupiter.

(2) The planet must have a suitable rotation period, so that nights do not overcool, nor days overheat.

(3) The orbital eccentricity must be low to avoid excessive differences in insolation as the planet moves from perihelion to aphelion and back (most cometary orbits would be lethal for organisms).

(4) The chemical content of air, ocean and land surface must be propituous, and not perilously polluted with substances inimical to biological operations.

(5) The controlling star must not be variable by more than 4 or 5%. It must

not be a double star, and of course not be subject to catastrophic explosions like those of the novae.

And, again following Mamikunian's calculations: "Since there are 10^{11} stars in our galaxy, this implies that there must be 100,000 planets in the Milky Way capable of supporting higher organisms. ... For the entire universe, consisting of perhaps 10^{22} this gives a total estimate of 10^{16} habitable planets." So there is enough probability to warrant some speculation on the existence of extra-terrestrial life !

7. LIFE ON THE MOON?

The possibility of the existence of life on the moon, either at present, or in the past, has been much discussed. It has been flatly answered in the negative by the study of the samples brought back by the astronauts of the Apollo 11 mission.

The detailed descriptions of these studies can be found, along with the reports on studies in other fields, in number 3918 of *Science* of 30 January 1970. Not only were no indications of life, either present or fossil, found, but moreover there have not been found any remnants of a possible lunar pre-life. From the reports in *Science* it will be seen how a large number of distinguished organic geochemists studied the samples, using various very refined techniques. No pre-biological compounds, such as purines, pyrimidines, amino acids or porphyrines were detected, although the analytical procedures were capable of registering such materials even if they occurred in quantities of less than ten parts per billion. Parenthetically remarked, it has been possible to register the contamination by the exhaust gases of the retrorockets of the lunar landing module, but no such compounds which were indigenous to the moon have been found.

It follows that the moon has known neither life nor pre-life, which probably is the result of the total absence of water on the moon.

REFERENCES

ANDERS, E., 1963. On the origin of carbonaceous chondrites. *Ann. N. Y. Acad. Sci.*, 108:514–533.
ANDERS, E., 1964. Origin, age and composition of meteorites. *Space Sci. Rev.*, 3:583–714.
BERZELIUS, J. J., 1834. Ueber Meteorsteine. *Ann. Phys. Chem.*, 33:113–148.
BITZ, M. C. and NAGY, B., 1966. Ozonolysis of "polymer-type" material in coal, kerogen and in the Orgueil meteorite. A preliminary report. *Proc. Natl. Acad. Sci. U.S.*, 56:1383–1390.
BOSTRÖM, K. and FREDRIKSSON, K., 1966. Surface conditions of the Orgueil meteorite parent body as indicated by mineral associations. *Smithsonian Inst. Miscell. Collections*, 151:1–39.
BRIGGS, M. H. and MAMIKUNIAN, G., 1963. Organic constituents of the carbonaceous chondrites. *Space Sci. Rev.*, 1:647–682.
CLAUS, G. and NAGY, B., 1961. A microbiological examination of some carbonaceous chondrites. *Nature*, 192:594–596.

CLAUS, G. and SUBA-C, E. A., 1965. Interpretation of micro-structures in carbonaceous meteorites. *Nature*, 205:1201.

CLAUS, G., NAGY, B. and EUROPA, D. L., 1963. Further observations on the properties of the "organized elements" in carbonaceous chondrites. *Ann. N. Y. Acad. Sci.*, 108:580–605.

CLAYTON, R. N., 1963. Carbon isotope abundance in meteorite carbonates. *Science*, 140:192–193.

DALY, R. A., 1943. Meteorites and the earth-model. *Bull. Geol. Soc. Am.*, 54:401–456.

DUCHESNE, J., DEPIREUX, J. and LITT, C., 1964. Sur la nature des radicaux libres de la météorite Cold Bokkeveld. *Compt. Rend.*, 259:1891.

FITCH, F. W. and ANDERS, E., 1963. Observations on the nature of the "organized elements" in carbonaceous chondrites. *Ann. N. Y. Acad. Sci.*, 108:495–513.

HAN, J., SIMONEIT, B. R., BURLINGAME, R. L. and CALVIN, M., 1969. Organic analysis of the Pueblito de Allende meteorite. *Nature*, 222:364–365.

HAYATSU, R., STUDIER, M., ODA, A., FUSE, K. and ANDERS, E., 1968. Origin of organic matter in early solar system. II. Nitrogen compounds. *Geochim. Cosmochim. Acta*, 32:175–190.

HAYES, J. M. and BIEMANN, K., 1968. High resolution mass spectrometric investigations of the organic constituents of the Murray and Holbrook chondrites. *Geochim. Cosmochim. Acta*, 32:239–268.

HENDERSON, W., EGLINTON, G., SIMMONDS, P. and LOVELOCK, J. E., 1968. Thermal alteration as a contributory process to the genesis of petroleum. *Nature*, 219:1012–1016.

KAPLAN, I. R., DEGENS, E. T. and REUTER, J. H., 1963. Organic compounds in stony meteorites. *Geochim. Cosmochim. Acta*, 27:805–834.

MAMIKUNIAN, G. and BRIGGS, M. H., 1963. A catalog of microstructures observed in carbonaceous chondrites. *Jet. Prop. Lab., Pasadena, NASA Tech. Rept.* 32-398, 75 pp.

MAMIKUNIAN, G. and BRIGGS, M. H. (Editors), 1966. *Current Aspects of Exobiology*. Pergamon, London, 420 pp.

MASON, B., 1962. *Meteorites*. Wiley, New York, N.Y., 247 pp.

MASON, B., 1963. The carbonaceous chondrites. *Space Sci. Rev.*, 1:621–646.

MASON, B., 1967. Meteorites. *Am. Sci.*, 55: 429–455.

MEINSCHEIN, W. G., NAGY, B. and HENESSY, D. J., 1963. Evidence in meteorites of former life: The organic compounds in carbonaceous chondrites are similar to those found in marine sediments. *Ann. N. Y. Acad. Sci.*, 108:553–579.

MEINSCHEIN, W. G., FRONDEL, C., LAURA, P. and MISLOW, K., 1966. Meteorites: Optical activity in organic matter. *Science*, 154:377–380.

MEINSCHEIN, W. G., STERNBERG, Y. A. and KLUSMAN, R. W., 1968. Origins of natural gas and petroleum. *Nature*, 220:1185–1189.

MUELLER, G., 1965. Interpretation of microstructures in carbonaceous meteorites. *Nature*, 205: 1200–1201.

NAGY, B., 1966. Investigations of the Orgueil carbonaceous meteorite. *Geol. Fören., Stockholm, Förhandl.*, 88:235–272.

NAGY, B., 1967. The possibility of extraterrestrial life: ultramicroscopical analyses and electron-microscopic studies of microstructures in carbonaceous meteorites. *Rev. Paleobotan. Palynol.*, 3:237–242.

NAGY, B., 1968. Carbonaceous meteorites. *Endeavour*, 27:81–86.

NAGY, B., MEINSCHEIN, W. G. and HENNESSY, D. J., 1961. Mass spectrometric analysis of the Orgueil meteorite. Evidence for biogenic hydrocarbons. *Ann. N. Y. Acad. Sci.*, 93:25–35.

NAGY, B., MEINSCHEIN, W. G. and HENNESSY, D. J., 1963a. Aqueous, low temperature environment of the Orgueil meteorite parent body. *Ann. N. Y. Acad. Sci.*, 108:534–552.

NAGY, B., FREDRIKSSON, K., UREY, H. C., CLAUS, G., ANDERSEN, C. A. and PERCY, J., 1963b. Electron probe microanalysis of organized elements in the Orgueil meteorite. *Nature*, 198:121–125.

NAGY, B., MURPHY, M. T., MODZELESKI, V. E., RAUSER, G., CLAUS, G., HENESSY, D. J., COLOMBO, U. and GAZZARRINI, F., 1964. Optical activity in saponified organic matter isolated from the interior of the Orgueil meteorite. *Nature*, 202:228–233.

ORÓ, J., NOONER, D. W. and ZLATKIS, A., 1966. Paraffinic hydrocarbons in the Orgueil, Murray, Mokoia and other meteorites. *Life Science, Space Science, Res., Spartan Books, Washington*, 4:63–100.

PAPP, A., 1963. The identity of "organized elements". *Ann. N. Y. Acad. Sci.*, 108:613.

SHAPLEY, H., 1953. On climate and life. In: H. SHAPLEY (Editor), *Climatic Change*. Harvard Univ. Press, Cambridge, Mass., pp.1–12.

STAPLIN, F. L., 1962. Microfossils from the Orgueil meteorite. *Micropaleontology*, 8:343–347.

STAPLIN, F. L., 1964. Organic remains in meteorites. In: G. MAMIKUNIAN and M. H. BRIGGS (Editors), *Current Aspects of Exobiology*. Pergamon, London, pp.77–92.

STUDIER, M., HAYATSU, E. and ANDERS, E., 1968. Origin of organic matter in early solar system. I. Hydrocarbons. *Geochim. Cosmochim. Acta*, 32:151–174.

TIMOFEJEV, B. W., 1963. Lebensspuren in Meteoriten. Resultate einer microphytologischen Analyse. *Grana Palynologica*, 4:92–99.

UREY, H. C., 1965. Meteorites and the moon. *Science*, 147:1262–1265.

UREY, H. C., 1966. Biological material in meteorites. A review. *Science*, 151:157–166.

UREY, H. C., MEINSCHEIN, W. G. and NAGY, B., 1968. Comments on meteoritic hydrocarbons. *Geochim. Cosmochim. Acta*, 32:665.

WIIK, H. B., 1956. The chemical composition of some stony meteorites. *Geochim. Cosmochim. Acta*, 9:279–289.

Chapter 18 | Conclusions

1. WHAT WE KNOW OR PRESUME TO KNOW

The time has come to summarize our considerations as set forth in the preceding chapters and to state the result of our quest into a possible origin of life through natural causes. In this summary we will have to distinguish between facts undoubtedly known and those processes and events which we may reasonably infer to have taken place. We have been guided throughout by the Principle of Actualism, which, as we saw in Chapter 2, allows us to elucidate the geologic history of the earth by comparing ancient rocks with the results of the processes we find now in operation.

We then saw in Chapter 3 how we are now able to measure the age of certain types of rocks. Within the limits of the method applied we found that the oldest reliably dated crustal rocks are between 3.3 and 3.4 billion years old, a figure which might possibly be extended by some tenth's of billions. Still older ages have been in fact reported, but these have been measured by an indirect method, using the so-called isochrons, and stand in need of confirmation. In a similar way the oldest rocks of the mantle, which underlies the crust of the earth, have been dated at 4.5 billion years old, whereas for the age of the earth figures of 4.5 and 4.75 billion years have been published. And though these figures may be found to stand in need of further correction—and correction in absolute dating commonly is towards older dates—they still supply us with a framework within which the history of life on earth must be contained.

In Chapter 4 we realized that, according to modern biological viewpoints, early life could only develop through natural causes after "organic" material had been synthesized by inorganic processes. Such inorganic formation of "organic" material is at present impossible on any practical scale, since the energy for the reactions required is not now generally available. Moreover, even if, through some chance combination, such material was exceptionally formed, it would be attacked immediately by oxidation, and hence stand no practical chance of escaping destruction.

This does, however, only apply to our present oxygenic environment. If the earth has possessed, sometime during its earlier history, a primeval atmosphere which was anoxygenic, such inorganic reactions synthesizing "organic" material would have been possible. The required energy would then have been available in the shorter ultraviolet sunlight, which would be able to penetrate such an anoxygenic atmosphere, and reach the surface of the earth. Moreover, although the primeval "anoxygenic" atmosphere was never completely devoid of free oxygen, the destruction of these "organic" compounds by oxidation would have been so much slower than in an oxygenic environment, that for practical purposes we may consider it as non-existent. Accordingly, the quest for a possible origin of life through natural causes has centered upon the probability of a primeval anoxygenic terrestrial atmosphere and upon the environmental conditions which must have prevailed under such an anoxygenic atmosphere.

Astronomical considerations, as related in Chapter 5, have taught us that both interstellar matter and primary planetary atmospheres are of a reducing character, due to the dominant abundance of free hydrogen. And although the terrestrial atmosphere probably is of a secondary nature, not directly derived from interstellar or solar matter, but from the later outgassing of the earth's interior, similar considerations apply to this secondary atmosphere, which in turn is the "primeval" atmosphere of geologic history. The earth is at present quite exceptional amongst the planets in its oxygenic atmosphere and the obvious explanation is that the free oxygen of this atmosphere is biogenic, produced by organic photosynthesis, through dissociation of CO_2.

As set forth in Chapter 13, the geology of the Precambrian provides evidence that the earth has indeed possessed in its early history an atmosphere low in free oxygen. We arrive at this conclusion by the study of certain types of old sediments, notably the pyrite sands of the gold-uranium ores and the banded iron formations of the Lake Superior type. From the ages of these sediments it follows that the anoxygenic primeval atmosphere has persisted up to the end of the Middle Precambrian, 1,800 million years ago. Oxidized sediments of the Red Bed type, occurring from 1,450 million years ago, on the other hand attest to the presence of the modern oxygenic atmosphere. The transition from the primeval anoxygenic atmosphere to the modern oxygenic atmosphere has consequently taken place sometime between 1,800 million years and 1,450 million years ago. Even then, the oxygen content of the atmosphere remained considerably lower than its present value, reaching one tenth of this value only with the beginning of the Silurian, some 440 million years ago.

These considerations have led to a large number of extremely varied laboratory experiments, in which, starting from simulated primeval atmospheric environments, numerous "organic" compounds have been synthesized inorganically by the application of various forms of energy. As related in Chapter 6, these compounds, not only the smaller "organic" molecules, but even such complicated structures as nucleotides and proteinoids, form so readily that one must conclude that their synthesis has been both common and abundant under an anoxygenic atmosphere.

The actual development of early life out of this inorganically formed pre-life is a link in the story which is at present not yet well known. The experimental difficulties in simulating this transition are much greater than those encountered in the in-vitro synthesis of "organic" compounds. Many possible processes, which would have been able to bridge this gap between pre-life and life have been proposed, but the transition has, as yet, not been duplicated. One salient point stands out, however, that is that early life will not have been nearly as complex in its metabolism, as present-day life is. Even the crudest of "rate-enhancing" processes of chemical reactions will already have given early life a decisive survival value over pre-life, in accordance with the proverb that "in the country of the blind the one-eyed man is king".

We arrive again at more solid ground when we ask where to look for the remains of early life. The origin of life is situated so far back in the geologic history that our research is limited to the Precambrian old shields. And even within the old shields we have to look for areas in which sediments formed during the very early history of the earth have remained relatively non-disturbed since. The oldest of such "cratonic" areas known today are found in South Africa. Sediments of the Onverwacht and Fig Tree Series of the Swaziland System, dated as older than 3.2 billion years, are at present the oldest non-metamorphosed sediments known. It is doubtful, whether still older sediments have been preserved on any of the other old shields.

Among the earliest vestiges of life found in these old sediments, we may distinguish between biogenic deposits, either macroscopic, such as algal limestones, or molecular, the so-called molecular fossils, and real fossils. From the oldest sediments mentioned above, the Onverwacht and Fig Tree Series, neither their "organized elements"—globular structures of carbonaceous material—nor this material itself, can as yet be accepted as proof of indubitable biological origin. Although they probably are, and although further research may well prove them to be, of biogenic origin, an inorganic genesis of both the "organized elements" and the carbonaceous material cannot, at present, be excluded. The oldest indubitable biogenic deposits, both the algal limestones of Southern Rhodesia, and the molecular fossils found in the Soudan Iron Formation of the Canadian Shield, are older than 2.7 billion years. So, even if further research in the Transvaal Series would prove its deposits to be of biological origin, this would hardly alter the general picture. All of these formations belong to the earliest period of geologic history, i.e. to the Early Precambrian, so an even older date for the "earliest recorded life on earth" will not basically alter our present views.

There is, however, another reason to assume that life, and life capable of photosynthesis at that, has developed already much earlier than the minimum age of 2.7 billion years, which is attested to the Dolomite Series of Southern Rhodesia and the Soudan Iron Formation of the Canadian Shield. Banded Iron Formations of the Lake Superior type, which, as we saw in Chapter 13, are supposed to have been laid down under an atmosphere containing free oxygen at about one percent of the present atmosphere, date back to well over 3 billion years in age. Since, as we saw in Chapter 15, free oxygen can only be produced up to one thousandth of its present atmospheric level by the inorganic photodissociation of water, all free oxygen above that level must have been produced biologically, through organic photosynthesis and photosynthetic life must have been existent before the deposition of the oldest banded iron formation.

The oldest real fossils known at present belong to the microscopic, but already well diversified, Gunflint flora of Ontario, which is rather loosely dated at about 2 billion years old. From then onwards, an albeit fragmentary, but continuous number of fossil flora localities is known from the various old shields, whereas animals seem only to have developed during the later part of the Late Precambrian. With the base of the Cambrian several groups of early animals developed the capacity of secreting shells, a feature possibly related to a temporary low level of atmospheric CO_2, as discussed in Chapter 16. It is only by way of these animal shells, hard enough to become fossilized regularly, that fossil remnants of early life have become abundant in the geologic history. It is at this point that the base of the Phanerozoic is generally placed. The development of continental floras of higher, vascular plants then took place only at the base of the Silurian.

The Phanerozoic, with its relative abundance of fossils, is the normal, "documented" period of geologic history, such as it is mostly treated in the text-books. It seems well at this point to oppose the relatively short period of the Phanerozoic, which is not more than the history of the further evolution of life (Fig.134), to the far more distant events relating to the origin of life (Fig.135).

2. CO-EXISTENCE OF PRE-LIFE AND EARLY LIFE

One of the important aspects in the development of early life from pre-life has only recently become fully appreciated. It is the long time span during which life, as attested by fossils, has existed under the primeval anoxygenic atmosphere, as attested by the ancient types of sedimentary ores described in Chapter 13. From the postulate that the inorganic reactions leading to the synthesis of the "organic" materials of pre-life took place throughout the time of the anoxygenic atmosphere follows the long time during which pre-life and early life must have been co-existent.

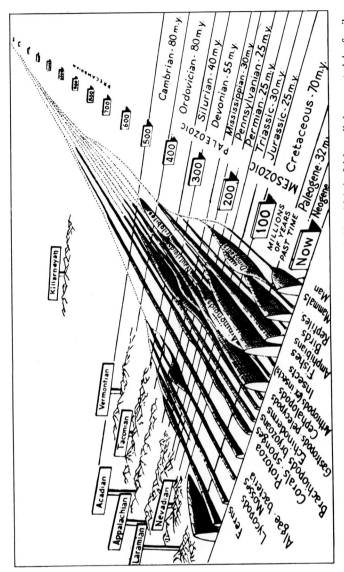

Fig.134. Schematic representation of the evolution of that part of life which is fairly well documented by fossils, during the Phanerozoic (including Paleo-, Meso- and Neozoic). (From MOORE, 1958.)

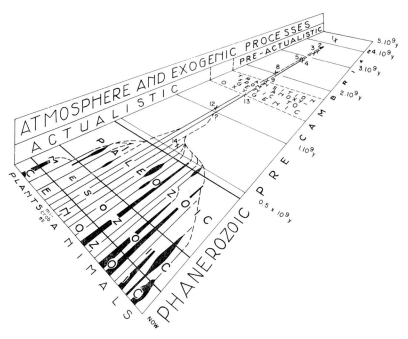

Fig.135. Schematic representation of the history of life in which the whole story is given. It is apparent how much longer early life has taken than higher life, which only developed during the last billion years of the earth's history.

Numbers refer to the world map of Fig.53, and to Table XIV. The origin of animals, about which very little is known, is indicated by a question mark. (From RUTTEN, 1962.)

In the literature one often finds a simple picture of the development of life from pre-life. Basically the story runs as follows. In the beginning the "organic" compounds of pre-life were formed inorganically. Then life developed, and, being heterotrophic, fed on this pre-formed "organic" material. Once this was consumed life had to develop autotrophic feeding, that is organic photosynthesis, to be able to sustain itself on inorganic, mineral substances. This type of reasoning is, for instance, expressed in the famous hourglass picture of the development of life by Pirie (see Fig.13). This is, of course, a far too simple picture of such a complicated event as the origin of life. From the basis of what we know from the further evolution of life, I have argued earlier (RUTTEN, 1962, see Fig.136) that this transition must have been much more complicated and gradual. I supposed that in pre-life, too, several more or less parallel developments have occurred, of which only a certain number, or perhaps even only one, eventually led to the transition to early life.

At that date this was no more than a theoretical contention. But at present, through our better understanding of the qualities of sedimentary ores (RUTTEN,

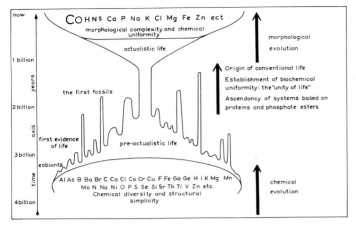

Fig.136. Schematic representation of the origin and further evolution of life. The "hourglass" picture of Pirie re-interpreted. (From RUTTEN, 1962.)

1969) and our more exact knowledge of the absorption of ultraviolet sunlight by atmospheres of various composition (BERKNER and MARSHALL, 1965, 1966), we have a much better factual basis to assess the relation between pre-life and life during the period of their co-existence. As related in Chapter 13, we know that sedimentary ores formed under an atmosphere containing "only a small amount of oxygen" were laid down all through the Early and Middle Precambrian, up to around 1.8 billion years ago. Moreover we saw in Chapter 15 that the influence of the Pasteur mechanism regulated the amount of free oxygen of this anoxygenic atmosphere at about one percent of its present level. And finally, we have seen that at this level the shorter ultraviolet sunlight, active in the synthesis of the "organic" material of pre-life, still reaches the surface of the earth.

It follows that all through the existence of the primeval anoxygenic atmosphere the products of pre-life were continuously synthesized inorganically at the surface of the earth. Life, which also existed at that time, as attested by fossils, could not yet exist at the surface of the earth, but only under a cover of soil or water. Thus pre-life and early life, although co-existent during all this time, did not occupy the same habitat. But even when life had already developed auto-trophic organisms, capable of organic photosynthesis, this does not mean that the "organic" materials had been depleted. Nor does it mean that primitive heterotrophic organisms had become superseded. It seems even probable that primitive, heterotrophic organisms, both chemo- and photo"organo"trophic, survived during all of this time, feeding on those products of pre-life which were swept into early life's habitat.

Moreover, to return to the opening paragraph of this section, it is probable that biopoesis, the development of life from pre-life went on, either continuously or intermittently, during the entire period of co-existence of pre-life and early life. This has been schematically illustrated in Fig.137. During this time there will not only have been competition between various modes of pre-life, such as indicated on Fig.136, but also between forms of life which arose at an earlier and at a later date from pre-life. And, just as in the later evolution of life there have existed entire groups of organisms, both plants and animals, which have since died out complete-

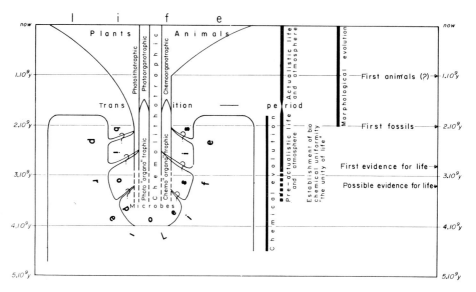

Fig.137. Schematic representation of the development of life from pre-life.

The first biopoesis, the formation of life from pre-life, must already have taken place well over 3 billion years ago. However, inorganic synthesis of "organic" molecules, i.e. the formation of pre-life, has proceeded simultaneously with the development of life up to the end of the Middle Precambrian, about 1.8 billion years ago. It is quite possible, and perhaps even probable, that during this time biopoesis still occurred, either continuously or intermittently. This possibility is indicated by the flame-like tongues connecting pre-life with life.

If such a later transition by chance took place in an environment which differed markedly from the more normal ones, the newly formed life could well be quite distinct from that already existing. We have, as yet, no data to base this contention on, but we have, for instance, indicated the possibility that eucaryotic organisms (see section 8 of Chapter 8) arose during such a newer biopoesis. In that case it might be due to their more productive metabolism that enough oxygen was produced to break the Pasteur Level (see section 7 of Chapter 15), and the reign of the anoxygenic atmosphere finally came to an end.

The duration of the period of transition between the oxygenic and the anoxygenic atmosphere is a matter of definition. If one takes the transition from anoxygenic to oxygenic strictly at the point where atmospheric oxygen rises above 1% of its present value (point d in Fig.127), there is no period of transition at all. If one would define the period of transition as the time when the level of atmospheric oxygen rose from 1 to 10% of its present value, the period of transition would only end with the beginning of the Silurian, 0.44 billion years ago. I have taken an intermediate position here, accepting the formation of red beds as indicating the existence of an oxygenic atmosphere. In this way the period of transition ends with the formation of the Dala Sandstones (see sections 16 and 17 of Chapter 13), 1.4 billion years ago. (From RUTTEN, 1970.)

ly, it is probable that evolutionary lines of early life which had developed at an earlier date from pre-life, have since been superseded by other life forms which only developed at a later date.

The period during which pre-life and early life were co-existent has stretched from the earliest development of life—somewhere between 4 billion and 3 billion years ago—unto the end of the anoxygenic atmosphere. That is to about 1.8 billion years ago (see point d in Fig.127), a period of the order of two billion years. If the reader will find the deductions summarized in Fig.137 hypothetical, I do fully agree. But if we take into consideration the Phanerozoic, where during a period of hardly more than half a billion years group after group of organisms deployed themselves only to die out some tens or hundreds of million years later, we must take such processes into consideration for the development of early life also.

3. WHAT WE HOPE TO LEARN

During the decade just ended both our knowledge of, and our insight into the origin of life have made enormous strides. And, in view of the very active research going on in many fields, we will again learn a great deal more of it during the next decade. Although some of this will be quite unsuspected at present, it still seems appropriate to give a short overview of those areas in which further results are likely.

To take absolute dating first, the ages of the oldest rocks will certainly become more exactly and more reliably known. The ages of the sedimentary series of the Precambrian, which are the most important for our topic, will become more narrowly bracketed when their relation to well-dated igneous rocks becomes better understood. And, may be, better ways of direct dating of sedimentary rocks might become available. In the same way we may hope for a better knowledge of the age of the oldest mantle rocks, of the earth itself, and of the constitution of the new-born earth which governed the outgassing and the formation of the primeval atmosphere. First attempts in this direction, such as that by RUNCORN (cf. 1962), are still too schematic to be of much value.

In the field of laboratory experiments duplicating the synthesis of compounds formed during pre-life, much more data will certainly become available in the coming years, in view of the many organic chemists who are active in this field. With the basic chemistry of the "organic" molecules relatively well understood, attention will turn more and more to their combination and polymerization and to the development of life itself. It is hoped that we will also get more data on the viability of such compounds, when once formed, i.e. on their half life.

Progress can also be expected in regard to the molecular fossils. Two fields seem to be wide open here, one the study of other organic materials than the alkanes, such as lipids and pigments, the other a better understanding of what compounds, or what combination of compounds, can be accepted as really biogenic, and what might also have been synthesized inorganically, and thus belong to pre-life. With a word which has become fashionable recently, the latter field can be described as the study of biogenecity. A major point of attack in the study of molecular fossils lies in the analysis of kerogen, in which only the first steps have been taken (see, for instance, BURLINGAME and SIMONET, 1969).

It follows that we may expect to arrive at a more qualified understanding of what must be considered truly abiogenic, what is really biogenic, and perhaps also about the transitional steps. Not so long ago it was generally thought that all carbon compounds found in sediments were exclusively biogenic. The organic chemists in the mean time have made us aware of the possibility of abiogenic synthesis of "organic" material, whilst on the other hand they have pushed backwards farther and farther the age of the sediments from which biogenic materials have been reported. This has resulted in the blurring of the borderline between non-living and living, and perhaps even, just as in biology, in a sort of fading of this line into a not too well defined zone.

With the realization that not every carbon compound found in ancient sediments is of biological origin, other criteria have been sought. For instance, if one had to concede that amino acids could be synthesized inorganically under an anoxygenic atmosphere, then perhaps the occurrence of proteins or of nucleic acids could be accepted as proof for early life? As we have seen, not a single one of such discrete criteria seems at present really acceptable, whilst even such venerable properties as the ratio of stable isotopes of carbon, or optical activity, have become suspect.

Personally I am convinced that we have not to look so much for a single criterion to decide whether a carbon compound found in an ancient sediment is a molecular fossil, the remain of early life, or not, and thus the remain of pre-life. Instead, we will have to decide on a broader spectrum of properties.

The history of the appreciation of the isoprenoid alkanes, short as it may be, offers a case in point. Originally, as we saw in section 9 and 10 of Chapter 12, the statistical relationship of the amount of compounds of various length around C_{17} was taken as proof that they represented degradation products of biological materials, like chlorophyll. Then it was learned that these compounds also formed in industrial processes. And although these take place under conditions which are admittedly far from a simulated primeval atmosphere, and equally far removed from conditions under which sediments normally form, still this fact has shaken the faith. Now, in his latest book, Professor CALVIN (1969) stresses the heterogeneity of the alkane spectra from ancient sediments, which is thought, if not to offer proof, at least to give a strong indication of the biological nature of these materials. A continuous spectrum of carbon compounds, showing a Gaussian distribution, would, on the other hand, indicate the presence of abiogenic materials. Such thoughts, which have, of course, been expressed earlier in various forms, seem to point the way in which the border between non-living and living in geology will eventually be described.

It must, of course, always be kept in mind that these sediments have an extremely long post-depositional history. During the degradation of the original material, be it "organic" or organic, separation and heterogeneization might easily have occurred, which as a sort of artifact, may mask the original composition.

As to the real fossils, we may also expect many new results in the years to come. Here again, we may distinguish two main directions. First there is the —admittedly slender—hope that very old non-metamorphosed sediments, older than known by now, will be encountered on one or more of the old shields. And second we might expect better preserved remains from the fossiliferous areas known by now. After all, the results obtained so far are only known for several years, whereas the lore of fossil hunting in the Phanerozoic tells us how chance finds of important fossil localities occur again and again in areas which had already been exploited for centuries. There are a number of specially important questions to be answered here. First comes the uncertainty whether the primitive "organized elements", such as those found in the Onverwacht and Fig Tree Series of South Africa (see section 17 of Chapter 12) can be proved to be biogenic, or if they have been formed by pre-life. If the first assumption would be true, this would push the proved occurrence of life on earth backward for about half a billion years. But perhaps more important than this type of record hunting would be to have more information about the qualities and the composition of life. We are, for instance interested in the time of appearance of the oldest eucaryotic organisms (see section 8 of Chapter 8 and section 10 of Chapter 14), because their development will have given an entirely new aspect to early life, whilst another important step in the evolution of life is, of course, the development of the earliest animals, another feature whose data we are badly in want of.

Both in relation to the molecular fossils, as in the understanding of the "organized elements", the further study of the carbonaceous meteorites might be expected to yield important results for our insight in the development of life on earth. This study is so important, because it may, as it has done in the past, cross-fertilize related studies on tellurian material. It has, for instance, been maintained that the paper by NAGY et al. (1961) on "biogenic material" of the Orgueil meteorite has been the starting point of the whole flood of interest in terrestrial molecular fossils.

Finally there are two fields of study, which although not of an organic nature, nevertheless may give new insights in the origin of life. These are the geochemistry of ancient sediments and the study of atmospheric physics and chemistry. It is to be hoped that a better and specially a more qualitative understanding of the relationship between ancient sediments and the contemporary atmosphere might be worked out. Whereas for atmospheric physics we may expect that new data from space research will deepen the basic studies by BERKNER and MARSHALL related in Chapter 15. In this vein also, to return again to life and its functions, we would warmly welcome a better understanding of the Pasteur mechanism (see section 7 of Chapter 15) and the level at which it is operative.

So even without those developments which are at present still unsuspected, but which will certainly crop up in a field so strongly interdisciplinary as the study of the origin of life, we may conclude this section in the certainty that the very active research will yield many new results over the coming decade. I am myself convinced that a large part of the story as told in this book will be outdated pretty soon.

4. WHAT WE PRESUMABLY WILL NEVER KNOW

Although this is perhaps superfluous, I think it well to state once again in this summary how, all the possibilities for further research mentioned in the preceding section notwithstanding, the history of the origin of life will eventually be known in a schematic and incomplete way only. This is the inevitable result of the deficiency of the geological record, a deficiency which, just as in human history, grows stronger the further one goes back in time. There are less and less sediments preserved, the farther one goes back into the geological record. Whereas, the farther one goes back into the origin of life, the more primitive the forms of early life become, and thus the greater the difficulty to recognize them as such and the less they reveal.

As to the first aspect mentioned, it has been argued that it is improbable that we will find sufficiently well-preserved sediments older than those of the South African Transvaal System and dated at more than 3.2 billion years old. This pessimistic prognosis implies that no record at all has been preserved of the earliest period of life on earth, because this must date back far earlier in time. As to the second aspect, we have noted the difficulty in distinguishing between inorganically synthesized globules of "organic" material and fossilized remnants of the primitive forms of early life, which have presumably formed similar globules. The term "organized elements" had even to be coined expressly to exclude from exact evaluation all forms morphologically transitional between pre-life and early life.

A final deficiency of the geological record is that if and when fossilization preserves the morphological expression of life, it still gives next to no clues about its metabolism. This deficiency, which is also encountered in the study of life during the Phanerozoic, becomes the more pressing for the Precambrian, not only because the data are so much scarcer, but also because the metabolism of microbial life is potentially so much more varied than that of higher organisms such as plants and animals (see section 4 of Chapter 8).

5. OPTIMISTIC OUTLOOK

However, let us not become dismayed by the deficiencies encountered in every study of the history of the earth. Similar deficiencies are, for instance, also encountered in the study of life during the Phanerozoic and still it must be conceded that we have come to know a great deal about organic evolution over the last half billion years.

If we compare the tentative hypotheses put forward only a few decades ago by pioneers such as Oparin, Haldane and Bernal, with, for instance, the state of the art in 1962, when I first tried to present an overview, and again with what we know at present, we see that from a set of rather thinly documented hypotheses both our knowledge and our theorizing have progressed enormously. Not only do we have a wealth of new data, but we have also learned where the first theories went wrong. That is, we have in many instances learned what to retain and what to reject, a process which has considerably narrowed down the possibilities between which the origin of life must have taken place.

So, even without the new insights into the origin of life which will surely become available in the future, we may already now claim a high level of probability for the theory of a development of life through natural causes out of a sort of pre-life consisting of inorganically formed "organic" compounds. This theory is in agreement with a large number of well-attested scientific facts and observations, drawn from many different disciplines of the natural sciences. It thus compares favourably with earlier theories that involve many inconsistencies and improbabilities, even if one is willing to accept the postulate of extra-terrestrial, or even supernatural, influence. As far as the origin of life is concerned, these postulates have become not merely unsatisfactory, but superfluous; which is a sound scientific reason for discarding them. Life on earth could, and by all available evidence did, arise through natural causes.

REFERENCES

BERKNER, L. V. and MARSHALL, L. C., 1965. On the origin and rise of oxygen concentration in the earth's atmosphere. *J. Atmospheric Sci.*, 22:225–261.
BERKNER, L. V. and MARSHALL, L. C., 1966. Limitation on oxygen concentration in a primitive planetary atmosphere. *J. Atmospheric Sci.*, 23:133–144.
BURLINGAME, A. L. and SIMONET, B. R., 1969. High resolution mass spectrometry of Green River Formation kerogen. *Nature*, 222:741–747.

CALVIN, M., 1969. *Chemical Evolution.* Clarendon, Oxford, 278 pp.

MOORE, R. C., 1958. *Introduction to Historical Geology,* 2nd ed. McGraw-Hill, New York, N. Y., 270 pp.

NAGY, B., MEINSCHEIN, W. G. and HENESSY, D. J., 1961. Mass spectrometric analysis of the Orgueil meteorite: Evidence for biogenic hydrocarbons. *Ann. N. Y. Acad. Sci.,* 93:25–35.

RUNCORN, S. K., 1962. Convection currents in the earth's mantle. *Nature,* 195:1248–1249.

RUTTEN, M. G., 1962. *The Geological Aspects of the Origin of Life on Earth.* Elsevier, Amsterdam, 146 pp.

RUTTEN, M. G., 1969. Sedimentary ores of the Early and Middle Precambrian and the history of atmospheric oxygen. *Proc. Inter.-Univ. Congr., 15th. 1967, Leicester,* pp.187–195.

RUTTEN, M. G., 1970. The history of atmospheric oxygen. In press.

Index